Buildings Energy Efficiency and Innovative Energy Systems

Buildings Energy Efficiency and Innovative Energy Systems

Editor

Vítor Leal

MDPI • Basel • Beijing • Wuhan • Barcelona • Belgrade • Manchester • Tokyo • Cluj • Tianjin

Editor
Vítor Leal
Department of Mechanical
Engineering
University of Porto
Porto
Portugal

Editorial Office
MDPI
St. Alban-Anlage 66
4052 Basel, Switzerland

This is a reprint of articles from the Special Issue published online in the open access journal *Energies* (ISSN 1996-1073) (available at: www.mdpi.com/journal/energies/special_issues/Building_Energy_Efficiency_Energy_Systems).

For citation purposes, cite each article independently as indicated on the article page online and as indicated below:

LastName, A.A.; LastName, B.B.; LastName, C.C. Article Title. *Journal Name* **Year**, *Volume Number*, Page Range.

ISBN 978-3-0365-2068-1 (Hbk)
ISBN 978-3-0365-2067-4 (PDF)

Contents

About the Editor . . . . **vii**

Preface to "Buildings Energy Efficiency and Innovative Energy Systems" . . . . **ix**

Fátima Lima, Paula Ferreira and Vítor Leal
A Review of the Relation between Household Indoor Temperature and Health Outcomes
Reprinted from: *Energies* **2020**, *13*, 2881, doi:10.3390/en13112881 . . . . **1**

Dania Ortiz and Vítor Leal
Energy Policy Concerns, Objectives and Indicators: A Review towards a Framework for Effectiveness Assessment
Reprinted from: *Energies* **2020**, *13*, 6533, doi:10.3390/en13246533 . . . . **25**

Ali Mohammed AL-Dossary and Daeung Danny Kim
A Study of Design Variables in Daylight and Energy Performance in Residential Buildings under Hot Climates
Reprinted from: *Energies* **2020**, *13*, 5836, doi:10.3390/en13215836 . . . . **51**

Taesub Lim, Woong Seog Yim and Daeung Danny Kim
Evaluation of Daylight and Cooling Performance of Shading Devices in Residential Buildings in South Korea
Reprinted from: *Energies* **2020**, *13*, 4749, doi:10.3390/en13184749 . . . . **67**

Tiago Souto, Margarida Almeida, Vítor Leal, João Machado and Adélio Mendes
Total Solar Reflectance Optimization of the External Paint Coat in Residential Buildings Located in Mediterranean Climates
Reprinted from: *Energies* **2020**, *13*, 2729, doi:10.3390/en13112729 . . . . **81**

Vasco Granadeiro, Margarida Almeida, Tiago Souto, Vítor Leal, João Machado and Adélio Mendes
Thermochromic Paints on External Surfaces: Impact Assessment for a Residential Building through Thermal and Energy Simulation
Reprinted from: *Energies* **2020**, *13*, 1912, doi:10.3390/en13081912 . . . . **99**

Wongeun Lee, Taesub Lim and Daeung Danny Kim
Thermal and Energy Performance Assessment of the Prefab Electric Ondol System for Floor Heating in a Residential Building
Reprinted from: *Energies* **2020**, *13*, 5723, doi:10.3390/en13215723 . . . . **115**

Taesub Lim, Yong-Kyu Baik and Daeung Danny Kim
Heating Performance Analysis of an Air-to-Water Heat Pump Using Underground Air for Greenhouse Farming
Reprinted from: *Energies* **2020**, *13*, 3863, doi:10.3390/en13153863 . . . . **127**

Vítor Leal and Raul Teixeira
PoDIT: Portable Device for Indoor Temperature Stabilization: Concept and Theoretical Performance Assessment
Reprinted from: *Energies* **2020**, *13*, 5982, doi:10.3390/en13225982 . . . . **137**

Dania Ortiz, Vítor Leal and Isabel Azevedo
A Review of the Measures and Instruments to Promote Efficiency and Renewable Energy in Domestic Water Heating
Reprinted from: *Energies* **2020**, *13*, 5370, doi:10.3390/en13205370 . . . . **153**

Isabel Azevedo and Vítor Leal
Decomposition Analysis of the Evolution of the Local Energy System as a Tool to Assess the Effect of Local Actions: Methodology and Example of Malmö, Sweden
Reprinted from: *Energies* **2021**, *14*, 461, doi:10.3390/en14020461 . . . . **175**

Abdullahi Ahmed, Monica Mateo-Garcia, Andrew Arewa and Kassim Caratella
Integrated Performance Optimization of Higher Education Buildings Using Low-Energy Renovation Process and User Engagement
Reprinted from: *Energies* **2021**, *14*, 1475, doi:10.3390/en14051475 . . . . **197**

About the Editor

Vítor Leal

Vítor Leal is an Assistant Professor at the Department of Mechanical Engineering of the Faculty of Engineering of the University of Porto (FEUP), where he got his Ph.D. in Engineering Sciences. He lectures courses in Thermodynamics, Energy Management, Energy in Buildings, Energy Planning, and Research Methods to both undergrad and post-grad students. He is the director of the Advanced Studies program in Sustainable Energy Systems at FEUP.

He participated and co-coordinated many national and international research projects in the areas of Sustainable Buildings and Energy Planning, such as TIP-VENT, SOLVENT, CA –EPBD, PROT-OVT and PROT-RN, THINK Smartcities, OBSERVA, Climaespaço DHC, HEALTHVENT, Energy Matrixes and Sustainability plans for AMP-N municipalities, and ProgressHeat.

He has supervised over 30 Ph.D. and M.Sc. theses and is co-author of over 50 papers published in research journals with peer review, with over 1000 independent citations in the scientific literature.

Preface to "Buildings Energy Efficiency and Innovative Energy Systems"

It is well known that energy use in buildings can be greatly reduced by adopting strict energy efficiency measures at the levels of envelope design, thermal insulation, heat recovery, and ventilation and by adopting renewable-based heating and cooling. While building codes and practices have steadily been moving forward, there is still room for more innovation, to achieve both a greater effect and better cost-efficiency.

Yet, the energy in buildings cannot be decoupled from broader aspects such as cold homes/fuel poverty, impacts on health/indoor temperature requirements, integrating buildings into cities, or even how policies for buildings are designed and evaluated. This justifies a comprehensive approach to the theme of Energy Efficiency and Innovative Energy Systems in Buildings. This approach was adopted for the Special Issue of the journal *Energies*, which is adopting a book format.

Vítor Leal
Editor

energies

MDPI

Review

A Review of the Relation between Household Indoor Temperature and Health Outcomes

Fátima Lima [1], Paula Ferreira [2,*] and Vítor Leal [1]

1 Faculty of Engineering (FEUP), University of Porto, 4200-465 Porto, Portugal; up201608016@fe.up.pt (F.L.); vleal@fe.up.pt (V.L.)
2 ALGORITMI Research Centre, School of Engineering, University of Minho, Campus de Azurém, 4800-058 Guimarães, Portugal
* Correspondence: paulaf@dps.uminho.pt

Received: 18 April 2020; Accepted: 2 June 2020; Published: 4 June 2020

check for updates

Abstract: This paper provides a review of research that addresses the relationship between indoor temperatures and health outcomes, taking into consideration studies that focus heat or cold exposure within the household context. It aims to extend previous research by considering both indoor temperatures from existing housing, and empirical studies that focus on energy efficiency measures and subsequent health impacts. To achieve this aim, a literature review was undertaken, combining engineering and health databases. The review established that, overall, inadequate indoor temperatures are associated with poor health status, whereas energy efficiency measures have been associated to improved indoor temperatures and occupant's health namely regarding cardiovascular, respiratory and mental health disorders. These health conditions are among the most prevalent non-communicable diseases (NCD). The review also highlighted the need for more empirical studies with an extended timeframe to deal with climate change challenges. It underlined the potential advantages of the convergence between health and energy efficiency studies, for better modelling and planning.

Keywords: indoor environment; indoor temperature; health; buildings; elderly; fuel poverty

1. Introduction

Health has been increasingly recognised as one of the areas that could be most adversely affected by climate change. It is currently estimated that the impacts of human activity could lead to a loss or reversal of health benefits conquered over the last five decades [1]. As many if not most of the world's households do not have fully functional HVAC systems, the changes in outdoor temperatures are expected to influence also the indoor temperatures. Among the most susceptible to these variations in temperature are the elderly. It has been noted that this segment of the population tends to spend a greater amount of time indoors [2], which emphasises the relevance of studying the health outcomes in the context of the urban built environment and more specifically within the household.

In Europe, concerns regarding the potential health impact of heatwaves date back to the 2003 heatwave with a reported excess of 30,000 deaths [3] and have drawn attention to the expected increase in extreme events, which leveraged studies on the potential adverse effects on health. When addressing mortality and morbidity associated to environmental factors triggered by climate change, Liu et al. [4] have noted the relevance of indirect health impacts, related to respiratory and cardiovascular causes, in contrast to direct health impacts, related to hypothermia or hyperthermia [4].

A recent review of temperature-related mortality and morbidity effects, performed by Bunker et al. [5] reported that a likely rise in ambient temperature, induced by climate change, would result in an increased risk of cardiovascular, cerebrovascular and respiratory outcomes among the ageing population [5].

The upsurge of heatwaves, however, does not mean that winter mortality should be neglected and neither health impacts linked to non-extreme ambient temperature [6]. In fact, a cross-country comparison to quantify mortality has attributed the greatest death burden to outdoor cold, mostly associated with moderate rather than extreme temperature events [7].

Therefore, although studies have reinforced the existing association between temperature and health outcomes, the previously mentioned background points towards the pertinence of considering the association between indoor temperature and health outcomes within the building context.

The built environment (with a focus on housing conditions) has been considered a key area of concern in establishing indicators to assess the population's health for different stakeholders in Europe [8]. Furthermore, too high or too low indoor temperatures in the households have been recently considered by Howden-Chapman, Roebbel, and Chisholm [9] as two key focal areas for establishing healthy housing guidelines (HHGL). In addition, based on World Health Organisation (WHO) guidance on housing, energy and thermal comfort, Ormandy and Ezratty [10] have concluded that health protection from high or low indoor temperatures should be taken into consideration for the development of policies in an energy context, such as energy efficiency, fuel poverty and climate change. Recently, Graff and Carley [11] have also emphasised the need for authorities to recognise and deal with the prevalence of energy insecurity at the household level. The inability of households to keep comfortable indoor temperatures might result in adverse health outcomes, particularly for vulnerable groups, such as children and the elderly population [11].

Although a few review studies already addressed the interlinkage between health, energy efficiency and outdoor temperature, the relation between energy efficiency, indoor environment (i.e., temperature), and health outcomes still require further research. Therefore, we conducted a review to answer the two research questions focused on these interlinkages:

- How do indoor temperatures affect health outcomes?
- How does housing energy efficiency improve health outcomes?

Taking into consideration this background, the present work aims to critically review literature which addresses the link between indoor temperatures and health. The contribution is twofold: (1) To address the association between indoor temperature and health-related outcomes; (2) to ascertain how indoor temperature improvements resulting from household energy efficiency measures impacts health.

The remainder of this paper is organised in the following sections: Section 2 provides a review of the links between indoor temperature and health outcomes in health and energy efficiency studies.; Section 3 outlines the main research methodological steps of the current study A critical analysis based on key indicators and dimensions is described in subsequent Section 4. Main conclusions and further research needs are presented in Section 5.

2. On the Relevance of Indoor Temperature and Housing Energy Efficiency for Health Outcomes

2.1. Relevance of Indoor Temperature for Health Outcomes

Focus on epidemiological studies has gained relevance in the context of climate change, given that temperature-related events have also been considered, as one of the main reasons of future concern by the Intergovernmental Panel on Climate Change (IPCC) [12].

In this section, the importance of indoor temperature for health outcomes is briefly contextualised. Several aspects of diversified background have been emphasised as being influencing factors in the association between indoor temperature and a given health outcome.

In the case of heat, the human response to temperature exposure is influenced by external and internal factors. External factors are related to environmental aspects, such as air temperature and humidity level, while internal factors are more linked to physiological and behavioural aspects, such as gender, ageing, adaptation, fitness, hydration and chronic diseases—diabetes; hypertension and obesity, among others [13].

Heat and cold exposure, have both been known to increase the risk for adverse health effects in elderly particularly those suffering from cardiorespiratory conditions, such as chronic obstructive pulmonary disease (COPD) [14]. Medical conditions such as cardiovascular (e.g., heart attack and stroke), respiratory diseases (e.g., COPD and asthma) and diabetes are a few examples of long-term chronic disabilities that have been categorised as non-communicable diseases (NCDs).

According to the World Health Organisation [15], NCD's have accounted for 70% of premature deaths for age ranges between 30 and 69 years old, at a worldwide level. In view of their increasing relevance in developed and developing countries, they have been considered a priority field of action, influenced by environmental and behavioural risk factors with interlinks to housing and urban expansion. Among driving forces for NCD are aspects related to sedimentary lifestyles, such as tobacco use, unhealthy diet and lack of physical activity [15,16].

These driving forces may adversely influence relevant biomarkers for cardiovascular conditions among for elderly, such as raised blood pressure. However, the relevance of other influencing factors, such as the built environment, has been increasingly recognised in terms of physiological responses to temperature [17–20].

2.2. Relevance of Housing Energy Efficiency for Health Outcomes

Recently, higher excess winter mortality has been associated to temperate regions with fewer energy-efficient houses, such as the Mediterranean countries, emphasising the links to housing, indoor temperatures and socioeconomic background [21,22]. According to Miguel-bellod, et. al, high rates of excess winter mortality in Southern European countries are determined, to a great extent, by the high incidence of poverty rates, energy inefficient building stocks and high energy prices [23].

These aspects give a glimpse of the complexity and multiple causalities that can be associated with the assessment of health outcomes in the context of variable temperatures, in keeping with the multifaceted concept of healthy housing. The World Health Organisation has established that healthy housing is associated not only to physical but also to mental and social wellbeing [24].

In this context, emerging research has looked to identify the explanatory pathways that connect the urban housing environment features to health outcomes. The need to simultaneously reduce energy consumption while ensuring comfortable and healthy indoors has led to consider a new integrated goal (wellbeing), resulting from the convergence of strictu-sensu health and of thermal comfort [25].

Associated to the evaluation of housing energy efficiency programs, Willand et al. [26] have identified benefits in health outcomes as being related to three main explanatory factors: indoor warmth; affordability of fuel; and, psycho-social focal areas. These explanatory factors illustrate the different theoretical ways, or pathways, in which energy efficiency could potentially impact health.

The improvement of indoor temperatures, stipulated within the warmth pathway, also contemplated effects from warmth on mortality and morbidity (respiratory, cardiovascular and general health outcomes). Meanwhile, mental health symptoms, as well as energy consumption, have been comprised within the affordability pathway. Whereas the perception of the householder regarding its home has been focused on the psychosocial pathway.

More recently, Armstrong et al. [27] have developed an empirical study, also considering the main pathways. The author's aim was to determine which sets of energy efficiency measures affect energy and health outcomes, i.e., how changes in housing energy efficiency measures impact different categories of health outcomes. This study found that an upscale of energy efficiency measures by the building stock is required, in order to reach health full potential benefits and climate change reaching climate change mitigation targets. The building envelope (walls, roofs, floors and windows) insulating properties; ventilation control, energy efficiency of the heating, lighting and other appliances, as well as energy sources have been considered the main four energy efficiency categories. While the main health outcomes have been associated with cardiorespiratory conditions, winter mortality and morbidity, thermal comfort, psychosocial wellbeing and nutrition outcomes. Additionally, to be associated to different efficiency pathways, these outcomes have also been classified according to their time horizon, as short term /immediate or long-term impacts (timeframe greater than 10 years) [27].

Figure 1 represents an approach which emphasises the need not only to assess health outcomes associated to already established energy efficiency pathways, as proposed by Willand et al. [26] but also to study this association throughout time, as suggested by Armstrong et al. [27]. Such a complementary approach would enable to further understand interactions between pathways through time while accounting for short term and long-term impacts for all health stages, being compliant with "healthy household" definition.

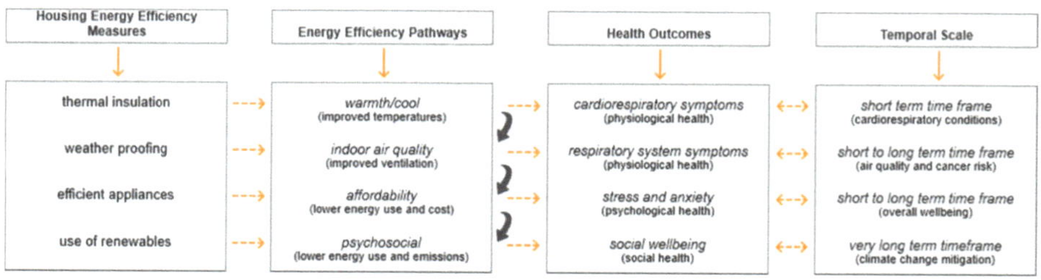

Figure 1. Housing energy efficiency pathways and related health outcomes (based on References [26,27]).

This may imply that for instance, the improvement of the building envelope may be linked to the improvement of indoor temperatures through housing energy features, as illustrated in Figure 1. The interaction between pathways is represented by the grey arrows: it shows that theoretically, a warmer household could positively influence affordability and psychosocial pathways. Based on the hypothesis posited by Willand [26] and Armstrong [27], improvements at building envelope level should lead to improvements in indoor temperature that may contribute to reducing energy consumption and related costs, *ergo* promoting energy affordability, particularly for low income households.

A recent review by Kolokotsa and Santamouris [28] has considered improvements to the building envelope as one of most cost-effective and efficient technologies to deal with indoor environmental quality and energy consumption issues, for low income households in Europe.

Improvements to indoor temperature may also affect the psychosocial pathway, by improving householder perception of his home, reinforcing social interaction with family and friends, based on the increasing use of his house. For instance, Poortinga et al. [29] have reported evidence that improvements to indoor temperatures were associated with increased use of more rooms in the house, enabling frequent visits from relatives and friends.

If achieved through efficiency, this could be possible by simultaneously requiring less energy use, expenditures and emissions, contributing to improving health outcomes at different time frames. For instance, better cardiorespiratory conditions on a short time frame, as well as less stress, anxiety and overall social health in the longer term. Benefits from improved air quality may imply an almost immediate reduction in air pollutants, but with long term implications (≥10 years), in terms of cancer risk [27].

3. Method

The steps that have led to the identification of key indicators and dimensions to be considered in the research are illustrated in Figure 2. This figure includes two lines for the literature review (1 and 2) and another two for critical analysis and results (3 and 4). Level 1 shows the three databases used in the research, level 2 shows the keywords considered in the search, which have contributed to establishing the criteria for the analysis, defined in level 3, as key indicators and key dimensions.

Key indicators are defined as general topic areas that are common to studies identified in previous steps (1 and 2), such as geographic location, focal area and health benefits. Key dimensions more specific, such as temperature exposure; health outcome, energy efficiency measures etc. These key features are at the basis of the critical analysis, identified in Figure 2 as level 4.

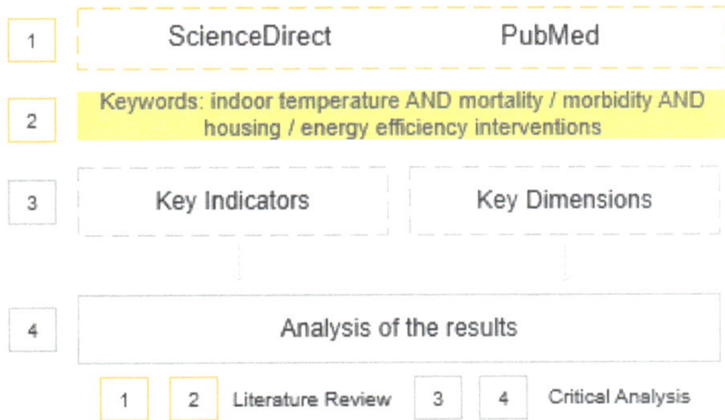

Figure 2. Search and review outline.

The present review has looked to promote a search that encompasses both engineering and health databases. In addition to keywords in Figure 2, the following exclusion criteria from the critical analysis was applied in order to select studies from different databases:

1. Studies for which full paper is not available on the database;
2. Studies addressing health outcomes not related to human and other health outcomes (e.g., vector-borne diseases);
3. Studies that are duplicated;
4. Studies that are not written in the English language;
5. Studies that are not empirical;
6. Studies that do not monitor indoor the temperature in loco, with the exception of studies that use empirical indoor temperatures from national surveys;
7. Studies that do not assess concurrently indoor temperatures and health outcomes;
8. Studies for which the focal point is to assess indoor air quality-related health outcomes;
9. Time restrictions are not applied.

Primary literature research aimed to identify peer-reviewed papers (n = 292), according to the selected keywords. The aforementioned exclusion criteria were then applied, and an approach based on Willand et al. [26] was subsequently used. This approach consisted of screening the references or the 'cited-by sections' of the documents found on primary research which enabled to locate new 'low-profile' studies that were added to the original selection. From this process resulted the final set of selected studies (n = 15).

The scarcity of energy efficiency studies comparatively to the topic of indoor temperature and human health should be seen under the light of the exclusion criteria undertaken and justifies the pertinence of this study. This is not an uncommon problem, for example, Mauree et al. faced also a similar issue of shortage of studies when reviewing the future implications of climate change on urban design and thermal comfort [18]. Although relevant from an energy and thermal comfort perspective, a more detailed approach to studies that monitor indoor temperature and energy consumption, but disregard its health implications, goes beyond the scope of this review, since the association to a health outcome would be lacking.

The use of such systematised guidelines for conducting the current report synthesis has been considered helpful to deal with validity concerns. Validity concerns have been gaining increasing importance associated with the growth of systematic reviews in the literature [30]. Following the study from Zhou et al. [31], the main threats to validity were considered and mitigation actions included to minimise their effect, which is reflected in the specified research exclusion criteria.

Literature from trustworthy intergovernmental Organisations, such as the World Health Organisation (WHO), have also been considered, within this context.

Prior reviews have highlighted relevant issues on the intersection between energy and health. For instance, Ormandy and Ezratty have emphasised that while thermal comfort and household energy efficiency are directed towards protecting the health householders, particularly the vulnerable ones, many challenges lie ahead in integrating strategies for key concerns such as energy efficiency, fuel poverty and climate change [10]. More recently, reviews by Kolokotsa and Santamouris have emphasised several energy efficiency alternatives to address the energy and indoor environmental issues for low-income households in Europe [28].

Meanwhile, Willand et al. have emphasised the complexity of energy efficiency interventions, and that the interactions between energy efficiency measures, householders and health-related outcomes is still misunderstood [32]. Liddell and Morris further claim that health impact assessment needs to be extended to incorporate mental besides physical wellbeing [33].

In this sense, Ortiz et al. have highlighted the concept of wellbeing should be interpreted as the overlap of health and comfort, which is linked to both environmental and behavioural aspects [25]. Hoof et al. considers that financial constraints may lead older people subject to inadequate indoor temperatures with adverse health outcomes, go currently undetected [34]. The evidence of the impacts of cold indoor temperature thresholds on human health has been recently reviewed by Jevons et al. that showed that despite scarce available evidence, minimum indoor temperature threshold should be at least 18 °C for the whole population [35].

Compared to existing reviews, namely References [24,26,36], the current study extends prior research by considering both indoor temperatures from existing housing, and studies that focus on energy efficiency measures and subsequent health impacts. By integrating these elements we intend to demonstrate the beneficial association between energy efficiency measures and health conditions. This should contribute to show the relevance of following a multi-disciplinary research agenda with approaches that tie health and engineering to support policymaking.

For this study, the housing and energy efficiency pathways of Figure 3 were then taken into consideration to establish the links between indoor temperature and health outcomes illustrated Figure 3, which served as the basis for the subsequent critical analysis.

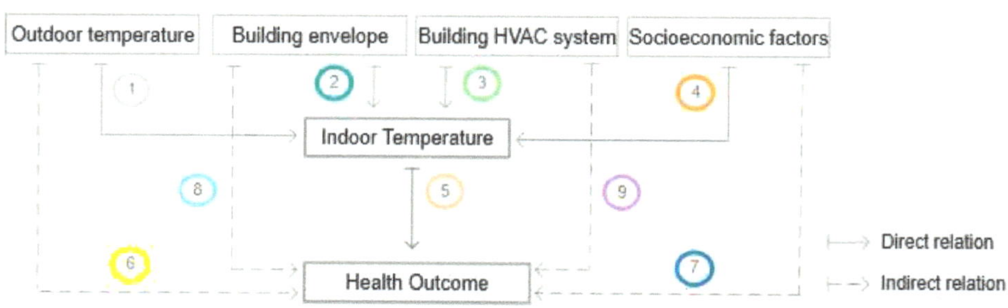

Figure 3. Links between indoor temperatures and health outcomes.

Indoor temperature is considered to be influenced by environmental, technological and socio-economic aspects (direct relation: links 1 to 4) which then impacts health directly (link 5) or indirectly (links 6 to 9). The main focus of the literature is on empirical studies that monitor the indoor temperature and its direct relation to health outcomes. However, four indirect links have also been identified (links 6 to 9).

For content analysis, particular focus was given to non- communicable diseases (NCD) health outcomes, namely from the cardiorespiratory system, given their relevance under the Sustainable

Development Goals (SDG's) scope ([37,38]) and the increasingly focused contribution of energy efficiency measures to improve these health outcomes ([27,29]).

4. Results and Critical Analysis

The literature review allowed us to identify relevant links that were used to draw a distinction of studies based on the type of relationship between indoor temperature and health outcomes.

4.1. Analysis of Direct and Indirect Links between Indoor Temperature and Health Outcomes

Table 1 presents the studies addressing each one of the links described in Figure 3 and summarises the identified energy efficiency-indoor temperature-health outcomes.

Table 1. Distribution of studies on indoor temperature by link to health outcomes.

Links		Studies on Indoor Temperature
1	outdoor- indoor temperature	[27,29,39–50]
2	building envelope- indoor temperature	[27,29,40,44,46–48]
3	HVAC system- indoor temperature	[22,27,29,44,46,48]
4	socioeconomic factor- indoor temperature	[27,29,39–50]
5	indoor temperature- health outcome	[27,29,39,40,47–50]
6	outdoor temperature- health outcome	[27]
7	building envelope- health outcome	[22,27,45,46]
8	HVAC system- health outcome	[22,27,45,46]
9	socioeconomic factor- health outcome	-
Cause-Effect assessed		
[22]	energy efficiency upgrade—low indoor temperature—emergency hospital admissions cardiovascular and respiratory conditions and injuries	
[27]	energy efficiency upgrade—low indoor temperature—cardiovascular and respiratory health related conditions	
	outdoor temperature—cardiovascular and respiratory-related mortality and morbidity	
[29]	high indoor temperature—emergency hospital admissions cardiovascular and respiratory conditions; psychosocial impacts	
[39]	low indoor temperature—emergency care for cardiovascular and respiratory conditions	
[40]	low indoor temperature—cardiovascular and mental health related biomarkers	
[41]	low indoor temperature—cardiovascular and mental health related biomarkers	
[42]	low indoor temperature—cardiovascular and mental health related biomarkers	
[43]	low indoor temperature—cardiovascular related biomarkers	
[44]	high indoor temperature—heat-related health symptoms	
[45]	low indoor temperature—long term health disabilities	
[46]	energy efficiency upgrades—low indoor temperature cardiovascular and respiratory-related mortality	
[47]	energy efficiency upgrade—low indoor temperature general health; respiratory-related conditions and general practitioner's hospital admissions	
[48]	energy efficiency upgrade—low indoor temperature—respiratory related conditions, general health status hospital admissions	
[49]	low indoor temperature—cardiovascular related biomarkers	
[50]	low indoor temperature—cardiovascular related biomarkers	

The most prevalent steps that establish the direct relation between indoor temperature and health outcomes are links 1 to 5. This means that most studies take into consideration the influence and interconnection with outdoor temperatures or climate (link 1). The higher relevance assigned to socioeconomic factors-indoor temperature and indoor temperature-health outcome (links 4 and 5) compared to other direct relations, namely building envelope-indoor temperature and HVAC system-indoor temperature (links 2 and 3) is clear.

The influence of outdoor temperatures on the indoor environment has been assessed in studies such as Uejio et al. [39] and Osman et al. [40]. Several studies such as [41–44] have stressed that indoor temperatures were more significantly related to a given health indicator than the outdoor temperature.

Amongst the reviewed studies, only one study [27] measured the association between temperature and health outcome indirectly based on outdoor temperature (link 6), without directly monitoring indoor temperature. However, changes in indoor temperature and health outcomes from energy efficiency studies have also been considered when resorting to existing empirical datasets. Energy Follow Up Survey or the English Housing Survey are examples of databases that integrate indoor temperature and energy efficiency used by studies associated to different links (e.g., References [27,45,46] from the outdoor temperature-health outcome; building envelope-health outcome and HVAC system-health outcome—links 6, 7 and 8). The studies in References [46] and [47] are connected on the other hand, as they are based on the same empirical fieldwork data. As for, Rodgers et al. [22] their study adopts a retrospective approach to investigate if empirical improvements to housing standards could lead to better health in householders.

Conversely, as expected given the empirical nature of the studies, most of the cause-effect relationships assessed are based on link 5 (indoor temperature-health outcome association). Within this scope, fewer studies have considered building envelope aspects (link 2), though when considered, this link is often associated with energy efficiency upgrades. From these, only one study [44] has accounted for the existence of air conditioning (link 3). Though currently not considered a widespread feature, some studies forecast the increase of its relevance in the context of climate change [51]. A large majority of studies in HVAC system-indoor temperature (link 3) (e.g., References [22,27,29,46,48]), is therefore associated with heating in contrast to cooling systems. Many studies address both building envelope/HVAC system-indoor temperature (link 2 and link 3) because studies often consider simultaneously multiple energy efficiency improvements, such as thermal insulation and upgrades to heating systems. Although the focal point of the present work is indoor temperature, building HVAC systems (link 3) also contemplates ventilation issues. It should be mentioned that of the abovementioned studies, only Armstrong et al. [27] take into consideration how insulation improvements may change ventilation and indoor air quality along with changes in winter indoor temperatures.

The number of studies in socioeconomic factor-indoor temperature (link 4) is indicative that socioeconomic influence seems to be more accounted for than building envelope aspects (link 2). The difference between these two links could be attributed to the fact that studies with health as a focal area tend to account for sociodemographic while largely disregarding building envelope aspects. Whereas studies focusing on energy efficiency take into consideration building variables from the upgrades or alterations to building envelopes, such as wall insulation or double glazing. This issue is consistent with existing literature, which has emphasised the relevance of household characteristics for health outcomes and energy efficiency contexts and how it has often been overlooked and considered a drawback from a health perspective ([34–36]).

Nevertheless, a significant amount of studies from link 2 are also featured in link 4, being associated to social housing or the lower income segment of the population, where the probability of occurrence of worst housing quality is higher [22,29,47,48]. Yet, the highlight goes to one study [29] that assessed relevant issues for both energy and health, such as fuel poverty status, financial difficulties and stress, food security, social interaction, thermal satisfaction and self-reported housing conditions.

These results show the segmentation between these focal study areas and denote the need to consider both socioeconomic and building aspects, in order to promote a better understanding of the association between indoor temperatures and health outcomes to contribute towards identifying potential energy efficiency measures to improve the characterisation of indoor temperature-health outcome (link 5), building envelope–indoor temperature (link 2) and socioeconomic–indoor temperature (link 4). The consideration of both latter links and fields of knowledge is desirable and could contribute to shifting health sector's perception regarding the need for energy efficiency measures for healthcare reasons. In this sense, Jonathan Wilson et al. [52] claims that a shift towards a more opened and receptive attitude from health sector would require more empirical evidence.

This insight is also in keeping with a key challenge pointed out by Haines et al. [53], which consists on the need for the public health sector to establish partnerships with other relevant areas (e.g., city planners) and various stakeholders (research institutions, governmental and non-governmental bodies, public and private), in order to provide decision makers with well-grounded research evidence.

Meanwhile, studies that circumvent in loco monitoring of indoor temperatures have been categorised in HVC system–health outcomes and socioeconomic–health outcomes (links 7 and 8), as favouring less direct relationships to health outcomes.

No identified studies featured the interconnection between building HVAC system and health outcomes (link 9). The low number and/or absence of studies from HVAC system-indoor temperature (links 3) and HVAC system-health outcome (link 9) might be related to the fact that in Europe, in contrast to the United States, there is not a widespread adoption of air conditioning in the residential building stock [51].

However, a recent review by Willand et al. has also cautioned that the householder's response may also undermine the outcome of residential energy efficiency interventions, among which limited technical knowledge to deal with energy efficiency measures is highlighted [32]. The relevance of the impact of technical aspects such as filtration on residential energy use has been further explored by Alavy et al. [54]. While the integration of human dynamics in the control of HVAC systems has been studied by Jung and Jazizadeh [55].

Yet the use of air conditioning is also closely linked to socioeconomic status, namely to household income. The Howden-Chapman [56] study on energy poverty and health emphasised the increased vulnerability of low-income households with elderly. This segment of the population spends a high share of their income on energy. They are also more likely to be hospitalised for respiratory and cardiovascular conditions [56]. Furthermore, Xu and Chen concluded that low income households have fewer energy efficiency appliances and less access to energy efficiency programs and require tailored policy measures to make energy more affordable and accessible [57].

The strong association between income and household energy may also play a relevant role in a household's adaptation to climate change and the choice between air conditioning and thermal insulation choices. De Cian et al. [58] found that the future adoption of thermal insulation might be more difficult given that the adoption of air conditioning is promoted by income, urbanisation and demographic trends.

These studies anticipate the relevance of the interconnection between socioeconomic variables (link 4) to building envelope/HVAC system/indoor temperature and health (links 2, 3 and 9) with climate change. They also reinforce the need for conceptual frameworks to consider a time scale in the assessment of the relationship between household energy efficiency, indoor temperature, and health outcomes.

4.2. Analysis of the Relevance of Climate Change Timeframe for Health and Energy Efficiency

Moreover, though cardiovascular and respiratory health outcomes seem to prevail in the cause-effect column, different conditions and biomarkers are specific to each study. A more detailed perspective of these aspects is provided below.

The assessment of cause-effect relations on Table 1 has enabled us to emphasise the high number of studies that focus health outcomes from exposure to low indoor temperatures (12 out of 15 studies) in contrast to Uejio et al. [39] and Loenhout et al. [44] that feature outcomes related to exposure to high indoor temperatures (2 out of 15 studies). Only one study reviewed (Armstrong et al. [27]) considers both cold indoor and overheating exposures. The geographical distribution of these studies and indoor temperature, focus on overheating (in orange) versus cold indoors (in blue) or both (in yellow), is illustrated in Figure 4.

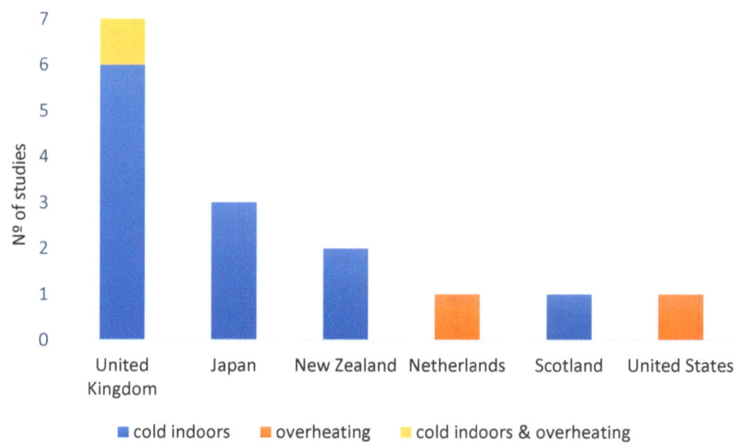

Figure 4. Geographic distribution and indoor temperature focus of the studies reviewed.

Armstrong et al. [27] present a longer timeframe than all others, contemplating winter and summer seasons, that is aligned with the nature of the health indicators used. This timeframe (of 10 years) is consistent with climate change concerns, and this is the only study that has looked to assess the perception of householders regarding home energy efficiency and climate change.

In addition, it is noticeable that studies focused mainly on developed countries. Noteworthy is the considerable amount of research developed in the United Kingdom (UK), which might result from the available national datasets with empirical indoor temperature measurements, as emphasised by Huebner et al. [45].

Even within developed countries, results show that there is a considerable lack of empirical research in countries that have been greatly affected by cold homes and excess winter mortality such as Portugal, Malta, Spain or Greece. A recent body of research has emphasised that these countries are at the top rate of excess winter mortality out of a total of 30 European countries [59]. These and other mild climate counties have been targeted as experiencing unacceptably low indoor temperatures [60,61].

Another observation is the scarcity of studies featuring indoor temperature monitoring during summertime, as emphasised in Figure 4. This field of research has been considered scarce, despite the growing interest and concern within academic and local communities, as well as overall society. This argument has been supported by recent studies developed within either the health or more energy efficiency and indoor temperature oriented scopes [27,29,39].

Departing from previously identified cause-effects in Table 1, it is also possible to establish that a greater number of studies features morbidity outcomes comparatively to mortality. Figure 5 illustrates the relation between the number of studies by health outcomes and study timeframe (in years). It is possible to see that there is a large diversity in study period considered for morbidity outcomes, that constitute 87% the of total number of studies. Yet, mortality outcomes (13% of total studies) are only associated with studies with longer timeframes (≥5 years). Once more, this result is in accordance with prior studies [27,46].

Given that longer timeframes in the research of indoor temperatures are also compatible with climate change research, results might suggest that mortality and morbidity outcomes could be considered in the context of climate change pathway if studies consider very long-term implications. However, in the current review, only one study, Armstrong et al. [27], has a time frame beyond 10 years.

From a householder perspective, health has been considered a more relevant issue for the implementation of energy efficiency measures than climate change [27]. Despite this, there is not much research to understand the impacts on health on longer timeframes, compatible with climate change issues, as illustrated by Figure 5. Therefore, it is possible to imply that currently, householders might be misinformed and unaware of the real impact of the exposure to inadequate indoor temperatures on health and the relevance of energy efficiency in the context of climate change.

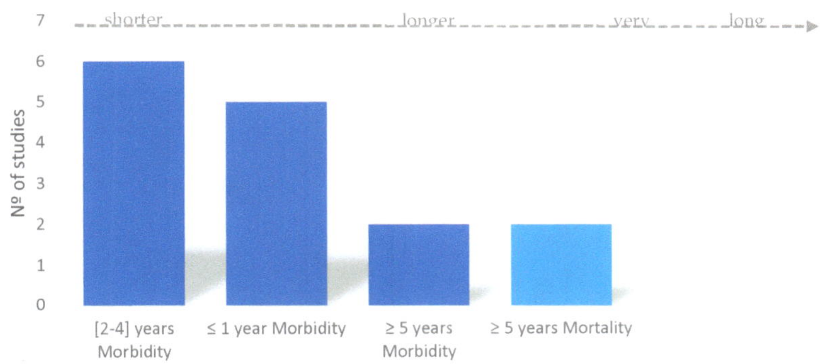

Figure 5. Mortality and morbidity health outcomes by study timeframe.

Consequently, there is an opportunity to leverage on the interest of people for health and promote studies with longer timeframes, that may lead to a better understanding of the impact of heatwaves on indoor temperature and health outcomes. Studies with very long timeframes would also contribute to understanding which energy efficiency measures could help improve health outcomes while mitigating climate change.

This result is aligned with Rodgers et al. findings, that emphasised that some of the co-benefits of energy efficiency may not be immediately perceived and that currently there is a scarcity of existing research in these terms, of the lack of long-term period studies [22].

Armstrong et al. also claim that small sample sizes of empirical indoor temperature monitoring and lack of pre and post-intervention monitoring, make it difficult to determine accurately the impact of energy efficiency on indoor temperature and upon health outcomes [27]. This may have implications in appropriately conveying health co-benefits from energy efficiency to decision and policymakers or even local communities. The consideration of longer timeframes could be crucial to address these issues and provide the scientific community with a more accurate and reliable empirical database to study the impacts of climate change.

Thus, very long-term studies could contribute to increased public awareness about climate change and its impacts and inform, based on empirical data, policymakers towards best available energy efficient solutions to mitigate them.

These results are in line with Willand et al. that have emphasised the need to integrate health goals into low carbon energy transition as a crucial aspect to develop an effective strategy for the housing sector [62].

4.3. Analysis of Health Outcomes by Study

Four main categories of health indicators have been identified as being related to mental health disorders, cardiovascular or respiratory conditions, or other health outcomes. A more detailed listing of health indicators, specific for each study, is provided in Table 2.

Table 2. Summary of health indicators by study.

Health Indicators	Mentioned	Direct Assessment (Measured)
Mental health disorders	[42]	[22,29,39,45]
Altered mental health status	-	[39]
Depression	[42]	-
Long term mental disability (LTD)	-	[45]
Common mental health disorders	-	[22,29]
Cardiovascular	[27,42,43,48,50]	[22,29,39,45,46]
Cardiac condition	-	[39]
Cardiac arrest	-	[39]
Cardiovascular condition	-	[22,29,39,46]
Myocardial infarction	[41,43]	-
Coronary heart disease	[27,42,43]	-
Heart attacks	[43,50,51]	-
Stroke/Cerebrovascular disease	[27,43,48,50]	-
Long term heart disability (LTD)	-	[45]
Respiratory	[44,46,47,50]	[22,29,39,40,45–49]
Difficulty breathing	[39]	[39,45]
Asthma	[46,47]	[22,39]
Respiratory conditions	-	[22,29,39,45–49]
Pneumonia	[39]	[39]
Insufficient cardiac blood flow	-	[39]
Lung disease	-	[39]
Emphysema/Chronic Obstructive Pulmonary Disease (COPD)	-	[22,39,40,48]
Lung function	-	[49]
long term breathing disability (LTD)	-	[45]
Other health indicators	[27,39,41–44,48]	[22,27,29,40–45,47–50]
Hypertension	[39,42–44]	-
Diabetes	[39,41–43]	-
Unconscious	[39]	-

Table 2. *Cont.*

Health Indicators	Mentioned	Direct Assessment (Measured)
Epileptic status	[39]	
Annoyance	-	[44]
Thirst	-	[44]
Sleep disturbance	-	[44]
Excessive sweating	-	[44]
Dehydration	[44]	-
Hyperthermia	[44]	-
Malaise	[44]	-
Hyponatremia	[44]	-
Renal colic and renal failure	[44]	-
Dry mouth	[44]	-
Impaired endurance	[44]	-
Fatigue	[44]	-
Sleep onset latency (SOL)	-	[41]
All-cause mortality	[27,44,47]	-
Excess winter mortality	-	[27]
Nocturia	-	[43]
Reduced quality of life	[43]	[29]
Falls and fractures/injuries	[43]	[22]
Blood pressure	-	[50,52,63]
Mean arterial pressure	-	[50]
Handgrip	-	[50]
Blood low-density lipoprotein level	-	[50]
Vitamin D level	-	[50]
Blood insulin-like growth factor	-	[50]
Blood haemoglobin level	-	[50]
White Blood cell count	-	[50]
Increased Blood viscosity	[52]	-
Platelet count (PTL)	-	[44]
General health status	-	[40,51,52]
Long term vision disability (LTD)	-	[46]
Long term hearing disability (LTD)	-	[46]
Long term mobility disability (LTD)	-	[46]
Long term learning disability (LTD)	-	[46]

The relationship between the identified health outcomes, energy efficiency pathways (from Figure 1) and indoor temperature is summarised by the Sankey diagram in Figure 6.

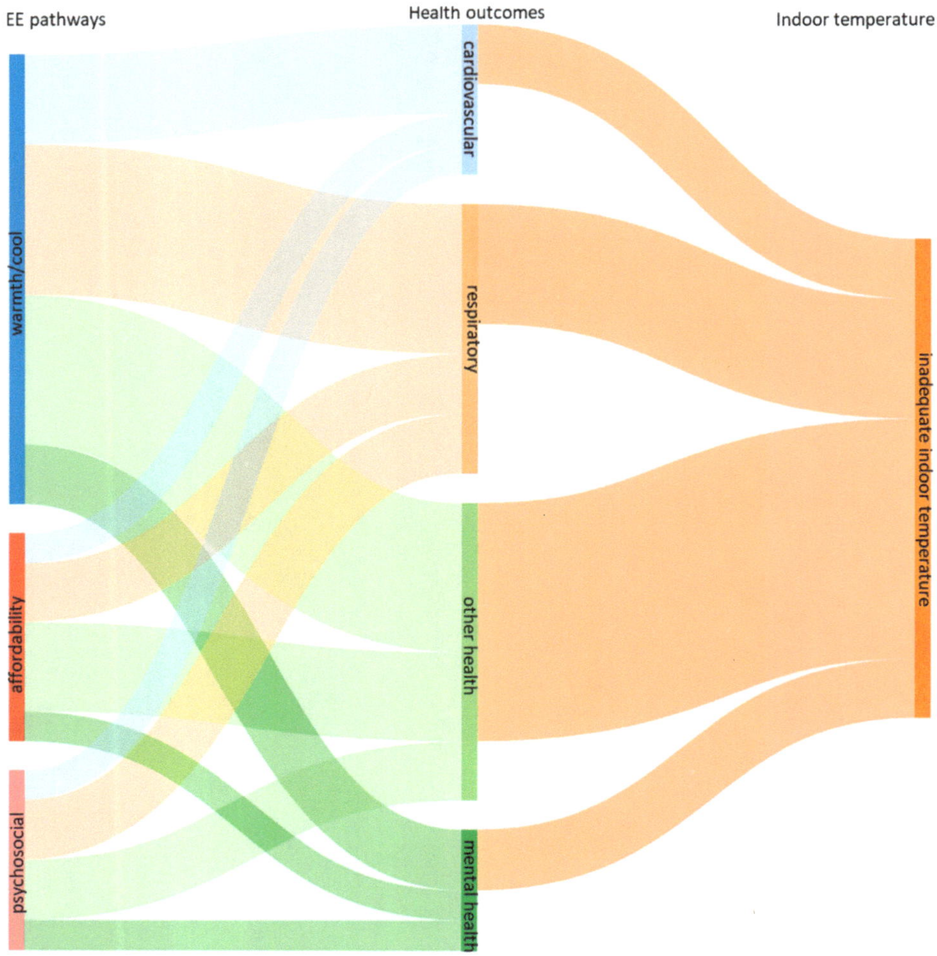

Figure 6. Sankey diagram between energy efficiency pathways, health outcomes and indoor temperature (elaboration using SankeyMATIC [64]).

The Sankey diagram represents fluxes between nodes. These fluxes show the transition from one node to another, here displaying possible energy efficiency pathways, health outcomes and indoor temperature identified in the reviewed studies. It is possible to see the extreme nodes (EE pathways and indoor temperature) are intermediated by a set of common health outcome nodes. The fluxes in a Sankey diagram are representative of the relevance of each health outcome as well as of the associated energy efficiency pathway. This relevance is estimated by the number of studies addressing each of the nodes. For instance, amongst the health outcomes, the least focused category is that of mental health disorders, with a thinner flux. However, its increasing relevance is recognised, as it is mentioned in studies that address diverse health outcomes, such as nocturia health indicator directly assessed by Saeki et al. [42] or sleep onset latency (SOL) or difficulty in falling asleep assessed by Saeki et al. [41].

In contrast to mental health outcomes, respiratory and cardiovascular conditions were the second and third most assessed health indicators, with wider fluxes compared to mental health.

Within cardiovascular and respiratory nodes relevant chronic conditions have been reported. According to the WHO [15], the top positions of deadliest NCD at the worldwide level are occupied by cardiovascular disorders (e.g., heart attack and stroke) in the first place followed by cancer in second place and respiratory conditions (e.g., asthma and COPD) and diabetes in the third and fourth places. These four major NCD disease groups have currently affected approximately 17.9 million and 3.9 million people annually, with cardiovascular and respiratory conditions, respectfully [15].

Yet the node for "other health" indicators category seems to be most representative, from indoor temperature to health outcome and from health outcomes to efficiency pathways, as illustrated in Figure 6. This result is indicative of both the diversity and complexity of direct assessment of health impacts. A few examples of the complex interconnection between different health outcome nodes are given below and are detailed in Table 2.

Most of the indicators categorised as "other health indicators" seem to be interconnected to other categories, namely for cardiovascular health conditions. For example, hypertension, mean arterial or blood pressure, and platelet count have been either mentioned or used as different health biomarkers in studies that aim to associate low indoor temperatures to health biomarkers for cardiovascular conditions (e.g., References [41,43,49,50]). Whereas respiratory cases have been in terms of relevance the second most assessed health condition with the least amount links to "other health indicator" categories. Its relevance comes from being directly assessed in empirical studies (e.g., References [33,45,51]).

Since exposure to inadequate indoor temperatures (too cold or too hot) may imply adverse health impacts, some studies have suggested that through the improvement of indoor temperatures, health gains for householders could be achieved (e.g., References [41–43]). Therefore in Figure 6, health outcomes departing from inadequate indoor temperature are related to warmth/cool EE pathway. However, the fluxes are not directly connected, given that the studies tend to focus on each of the extreme nodes. This lack of connection between studies reinforces, once more, the need for a more holistic approach to address the relationship between energy efficiency-health-indoor temperature that better supports policy-making for efficient and healthy households.

The improvement of indoor temperatures on existing building stock is often, as previously shown in References [22,27,46–48], linked to building envelope aspects (link 2 and link 3), associated to the adoption of energy efficiency measures.

4.4. *Analysis of Housing Energy Efficiency and Health*

In this section, the relationship between indoor temperature and health outcomes linked to the building envelope is assessed.

A total of 6 out of the 15 studies in Table 2 are household energy efficiency related. A more detailed examination of energy efficiency measures adopted or mentioned is provided in Table 3.

Based on Table 3, it is possible to establish that thermal insulation measures were featured in all studied energy efficiency interventions, with wall insulation being the most adopted one. Improvements in household appliances namely heating systems and to windows and doors were also widely implemented. Other energy efficiency measures are residual comparatively to the previous categories for the studies considered in this review.

Wall insulation and heating systems contributed to increase the indoor temperature but not always to decrease relative humidity (RH), as shown in Table 3.

Table 3. Summary of energy efficiency measures and their effects by study.

EE Measures	Studies	Observation Summary
Insulation wall floor ceilling/roof loft	[22,27,29,48] [46–48] [46,47] [22,27,48]	[22] 65.8% of installed wall insulation met housing quality standard; changes in indoor temperature not specified; [21] 31.8% of installed loft insulation met housing quality standard, changes in indoor temperature not specified [27] loft insulation associated to lower increases in temperature than increases associated to cavity and wall insulation; overall modest change in standard indoor temperature, averaging *0.09 °C [29] external wall insulation was the EE alternative that most contributed to increase indoor temperature (1.12 °C, 95% CI); overall indoor temperature increased on average by *0.84 °C; did not change indoor RH levels (−0.60% RH, 95% CI) [46]; [47] insulation retrofits increased on average bedroom temperatures by 0.5 °C, reduced time exposed to temperatures below 10 °C by 1.7 h per day and decreased relative humidity by 2.3% [a] [48] no adoption of EE alternatives implied lower baseline rating [b] than EE adoption (4.8 vs. 5.6); fewer hours of baseline hours of warmth above 21 °C in living rooms (48 h vs. 69 h) *
Windows & doors	[22,27,29,47]	[22] 52.4% of new windows and doors met housing quality standard; changes in indoor temperature not specified [27] double glazing was adopted and not associated with appreciable energy savings; overall modest change in standard indoor temperature, averaging *0.09 °C [29] new windows and doors did not increase indoor air temperatures significantly on average by −0.02 °C, 95% CI); new windows and doors increased indoor relative humidity (RH) on average by (5.15% RH, 95% CI) [47] EE adoption included draught stopping around windows and doors; changes in indoor temperature not specified by this measure
Appliances heating systems	[22,29,46,48]	[22] 77.4% of new heating systems met housing quality standard; changes in indoor temperature not specified [29] new boiler or heating system did not increase indoor air temperatures significantly on average by (−0.19 °C, 95% CI); did not change indoor RH levels (−1.59% RH, 95% CI) [46] heating retrofits with baseline underfloor and ceiling insulation recorded an increased average living room temperature by1.1°C [a] [48] EE adoption implied improvements in EE rating and fuel costs
Others fuel switching kitchens bathrooms electrical systems [c] garden paths	[29] [22] [22] [22] [22]	[29] gas network connection significantly increased indoor temperature on average by (0.69 °C, 95% CI); increased indoor RH levels by (3.86% RH, 95% CI) [22] 81.1% of kitchen improvements met housing quality standard [22] 81.9% of bathroom improvements met housing quality standards [22] 91.6% of electric system adoption met housing quality standard [22] 30.2% of garden path improvements met standard housing quality

* average value for all interventions; [a] empirical data from large scale intervention Warm Up New Zealand: Heat Smart (WUNZ: HS); [b] National Home Energy Rating (NHER); [c] electrical system upgrades include adding power sockets, and extractor fans in kitchens and bathrooms; RH- Relative Humidity; CI- Confidence Interval; EE- Energy Efficiency.

Although each energy efficiency alternatives contributed differently for the increase in indoor temperatures, overall increases—for all intervention (e.g., References [27,29])—have been small, on average below 1 °C. However, even this slight increase has contributed in certain studies, such as for Poortinga et al. [29], to reduce the number of hours exposed to very low indoor temperatures (<18 °C or <16 °C). Osman et al. [52] also claim the adoption of EE alternatives has contributed to have fewer hours with unwanted temperatures (<21 °C) in the living room of chronically ill patients. It is also noteworthy that the largest improvements in indoor temperatures have been reached in critical living spaces in the household, such as the living room and the bedroom, where people spend their daytime and night time.

Besides the indoor environment, adopted measures have contributed to improving housing quality and efficiency standards [22,48], namely by reducing energy costs and increasing affordability. Furthermore, combinations of multiple energy efficiency alternatives, such as cavity wall and loft insulation with condensing boiler for the heating system, have reached considerable reductions (11.2%) in gas demand [27]. Taking into consideration information from Tables 2 and 3, regarding health indicators, it is also possible to say that to a large extent, a large share of studies featuring energy efficiency measures tend to focus on NCD health outcomes, compared to studies without energy efficiency measures.

Table 4 displays the most adopted energy efficiency measures vs. the most assessed health outcomes and impact on indoor temperature. The impact scale definition was inspired by Rodgers et al. [22] and adapted for this case. The impact level represents, for each energy efficiency alternative undertaken, if its adoption contributed to improve or worsen a given health outcome. It can range from low to high impact level, with low level (–) being indicative of an undesirable change like an increase in hospital admissions. Conversely, a high impact level (++) is associated with a desirable change in the health outcome, such as a decrease in hospital admissions or medical appointment. Both these conditions are associated with a level of significance, to the *p*-value for each study. No change (nil impact level) in the health outcome implies no association with a specific EE alternative with non-significant p-value. Most health outcomes reported in Table 4 have been assessed individually for each EE alternative, with emphasis for insulation and heating systems. However, some results have been reported aggregately, either as a combination of all health outcomes (e.g., Reference [27]) or as resulting from all intervention (e.g., References [29,48]).

It should also be noted that some of these studies have resorted to proxies in order to establish health status. For instance, Rodgers et al. [22] and Armstrong et al. [27] have resorted to hospital health statistics to assess the impact on health services such as emergency hospital admissions for COPD, asthma and mental disorders as a proxy for health outcomes. Besides hospital admissions Viggers et al. [47], also considers self-reported health status and days off school and work, as well as visits to a general practitioner's office.

From all studies reviewed, only one reported a negative association while the other eleven associations reported improvements in health, though with different levels of impact. Yet, the results also show that a significant number of conducted studies where the association between health indicators, energy efficiency and indoor temperatures was inconclusive. This is particularly highlighted for mental health and for cardiovascular health indicators, given the lower number of studies comparatively to respiratory conditions, as illustrated in Table 4.

Among energy efficiency alternatives, insulation measures have contributed most for the improvement of respiratory and cardiovascular conditions, followed by alterations to windows and doors.

This beneficial association between energy efficiency measures and identified NCD's could denote an effective course of action or opportunity to tackle some of the challenges in the health sector. A more detailed account of key findings for each study is provided in Table 5.

Table 4. Changes in health outcomes in studies with energy efficiency measures.

EE Alternatives / References	Health Outcome: Cardiovascular: Insulation	Cardiovascular: Windows & Doors	Cardiovascular: Heating Systems	Respiratory: Insulation	Respiratory: Windows & Doors	Respiratory: Heating Systems	Mental Health: Insulation	Mental Health: Windows & Doors	Mental Health: Heating Systems	All Health *: Insulation	All Health *: Windows & Doors	All Health *: Heating Systems	Indoor Temperature
[22]	++ a 0 b	0	0	+ a 0 b	++	0		0 d		+ a 0 b	++	∩	↑
[27]											+ d		↑
[29]		–			0 d			0 d					↑
[46]	++ c		0	+ c		0							↑
[47]				++ c						++ c			↑
[48]				++ d		++ d							↑

Impact level:

--	-	0	+	++
low	medium low	nill	medium high	high

worsen ←——→ improvement

* All health = combined health outcomes; EE alternatives: insulation: [a] wall insulation; [b] loft insulation; [c] floor and ceiling insulation; [d] all intervention (combined EE alternatives).

Table 5. Summary of key findings for EE alternatives upon health outcomes.

Studies	Summary of Key Findings
	Cardiovascular Condition—Older residents:
[22]	- wall insulation was significantly associated with 27% less emergency admissions for cardiovascular conditions - no changes associated (p-value > 0.01) to upgrades for windows and doors, new kitchens and bathrooms, loft insulation, electric system upgrades or heating upgrades
[29]	- for people aged ≥ 60, all intervention measures were significantly associated ($p < 0.009$) to an increase in emergency admissions for cardiovascular conditions - for people aged ≥ 60 and for all intervention measures no changes associated (p-value > 0.10) in emergency admissions for cardiorespiratory and respiratory conditions

Table 5. *Cont.*

Studies	Summary of Key Findings
	Respiratory Condition—Older residents:
[22]	- upgrades to windows and doors were significantly associated to 39% less emergency hospital admissions for respiratory conditions - wall insulation associated with 24%, electrical system upgrades associated to 57% and garden path improvements to 38% fewer emergency hospital admissions for respiratory conditions - no changes associated (p value > 0.01) in emergency admissions for upgrading kitchens and bathrooms, loft insulation or heating
	Injuries (falls and burns)—Older residents:
[22]	- upgrades to windows and doors were significantly associated to 39% less emergency hospital admissions for respiratory conditions - wall insulation associated with 24%, electrical system upgrades associated to 57% and garden path improvements to 38% fewer emergency hospital admissions for respiratory conditions - no changes associated (p value > 0.01) in emergency admissions for upgrading kitchens and bathrooms, loft insulation or heating
	All health outcomes—All ages:
[22]	- for people of all ages, households with electrical system upgrades had 34%; upgrade to windows and doors had 22%; wall insulation associated with 20% and garden path with 19% less combined emergency admissions than the reference group - no changes associated (p value > 0.01) in emergency admissions from heating upgrades, new kitchen and bathrooms or loft insulation
[27]	- the gains in winter temperatures by 0.09 °C are associated with an estimated annual reduction of ≈280 cold-related deaths in England
	Respiratory Condition—All ages:
[22]	- for all ages, prescribed medication for respiratory conditions, such as asthma or COPD have reduced 8% for households with upgrades for windows and doors; electrical system upgrades were associated with 9% fewer general practice attendance
	Cardiovascular and respiratory Condition—All ages:
[46]	- for all ages and all intervention, no changes (p-value > 0.10) in emergency admissions for cardiovascular, cardiorespiratory and respiratory conditions
[46]	- the insulation group relative to the control group, interpretable as a 32.7% reduction in mortality risk during the period studied - no additional health benefit from heating system (not significant, $p = 0.122$)
[47]	- insulated homes were significantly associated with less fair or poor self-rated health, self-reports of wheezing, fewer days off of school and work and visits to a general health practitioner
	Mental health—All ages:
[22]	- no changes associated (p value > 0.01) for any cointervention for prescribed mental health medications
[29]	- no change in self-reported mental health from energy efficiency improvements
	Wellbeing, thermal satisfaction and social interaction:
[29]	- participants who received intervention reported significant association ($p < 0.004$) for improved subjective wellbeing; significant associations were also registered for higher thermal satisfaction ($p < 0.003$) and increased social interaction ($p < 0.012$)

There are also reports of other health outcomes, though less assessed and mentioned than NCD's. This is the case of injuries and falls, that have had significant improvements, translated into less emergency hospital admissions for elderly householders that received upgrade to windows and doors [22]. These results seem to indicate that more vulnerable groups have greater sensitivity to indoor temperatures and that even small readjustments could lead to significant changes regarding temperature-related diseases. This is true particularly for the elderly population, that tends to spend more time indoors. Other previous reviews have also highlighted the potential health benefits associated with even small increases in temperatures [26]. Less objective outcomes such as "wellbeing" and "improved social interaction" have also been associated with improved indoor temperature from energy efficiency measures and might contribute indirectly towards better psychological health [29].

5. Conclusions

This paper has described the main key findings of studies that address the relation between indoor temperature and health outcomes.

It was found that inadequate temperatures (too low or too high) are associated with poor health status, whereas energy efficiency measures have been associated to improved health biomarkers for several health outcomes, namely cardiovascular, respiratory and mental health disorders. The analysis also highlighted that these health conditions are considered among the most prevalent non-communicable diseases (NCD), further emphasising the relevance of adopting housing energy efficiency measures for improving the occupant's health.

Thermal insulation, heating systems and improvements to windows and doors were widely implemented energy efficiency measures and have contributed to small increases in temperatures leading to fewer hours of exposure to low indoor temperatures. Among energy efficiency alternatives, insulation measures have contributed most for the improvement of respiratory and cardiovascular conditions, followed by alterations to windows and doors.

On the methodological front in assessing the problem, the reviewed studies demonstrate there is the need for an integrated approach, that conciliates medical and energy efficiency knowledge. This could promote a better understanding of areas such as mental health and wellbeing, as well as fostering additional evidence-based research geared towards anticipating impacts of climate change.

Further research should be developed to better understand available energy efficiency alternatives, to avoid the environmental and economic burden of improving indoor temperatures, particularly for low income households. Additionally, a longer timeframe should be considered in empirical research, to better attend to these concerns in a climate change context as this time frame seems to play a major role in the assessment of health impacts and its relation to energy use and efficiency, indoor temperatures and climate change. However, this is far from being fully explored in the literature as most studies seem to focus on immediate to short term health impacts and do not effectively address the challenges that changes on the climatic conditions will pose to health.

Joint research of health and energy fields would contribute also to improve the quality of data on housing and health. A recent assessment of available household data databases performed by Alkire and Samman highlighted that high quality and timely surveys, combining household surveys and lighter interim surveys, could provide in-depth information about core indicators regarding socioeconomic conditions such as poverty and deprivation in the household contexts [63]. A multidisciplinary approach could also be a step forward to improve the issue raised by Wilson et al. [52] of translating energy efficiency improvements into benefits for healthcare utilisation.

Thus, a shift towards a more interdisciplinary approach is suggested as future research considerations for policymaking, ultimately contributing to the development of tailored solutions, from promoting the convergence between housing energy efficiency and potential health outcomes, into the local and national planning process.

Author Contributions: Conceptualisation, F.L., P.F., V.L.; methodology, F.L.; validation, P.F. and V.L.; draft preparation, visualisation F.L.; writing—review and editing, supervision, P.F. and V.L. All authors have read and agreed to the published version of the manuscript.

Funding: The researchers further like to acknowledge the financial support by FCT–Fundação para a Ciência e Tecnologia within MIT SES Portugal Doctoral Program under the scholarship PD/BD/128029/2016. This work has also been supported by FCT–Fundação para a Ciência e Tecnologia within the R&D Units Project Scope: UIDB/00319/2020.

Conflicts of Interest: The authors declare no conflict of interest.

List of Acronyms

EE	energy efficiency
HVAC	heating, ventilation and air-conditioning systems
HHGL	healthy housing guidelines
NCD	non-communicable diseases
WHO	World Health Organisation
COPD	chronic obstructive pulmonary disease

References

1. Watts, N. Review the Lancet Countdown on health and climate change: From 25 years of inaction to a global transformation for public health. *Lancet* **2018**, *391*, 581–630. [CrossRef]
2. Vardoulakis, S. Impact of climate change on the domestic indoor environment and associated health risks in the UK. *Environ. Int.* **2015**, *85*, 299–313. [CrossRef]
3. Vandentorren, S. August 2003 heat wave in France: Risk factors for death of elderly people living at home. *Eur. J. Public Health* **2006**, *16*, 583–591. [CrossRef]
4. Liu, C.; Yavar, Z.; Sun, Q. Cardiovascular Responses to Environmental Stress Cardiovascular response to thermoregulatory challenges. *Physiol. Heart Circ. Physiol.* **2015**, *309*, H1793–H1812. [CrossRef]
5. Bunker, A. Effects of Air Temperature on Climate-Sensitive Mortality and Morbidity Outcomes in the Elderly; a Systematic Review and Meta-analysis of Epidemiological Evidence. *EBioMedicine* **2016**, *6*, 258–268. [CrossRef]
6. Schneider, A.; Breitner, S. Temperature effects on health—Current findings and future implications. *EBioMedicine* **2016**, *6*, 29–30. [CrossRef]
7. Gasparrini, A. Mortality risk attributable to high and low ambient temperature: A multicountry observational study. *Lancet* **2015**, *386*, 369–375. [CrossRef]
8. Freitas, Â.; Santana, P.; Oliveira, M.D.; Almendra, R.; Bana, J.C.; Bana, C.A. Indicators for evaluating European population health: A Delphi selection process. *BMC Public Health* **2018**, *18*, 557. [CrossRef]
9. Howden-chapman, P.; Roebbel, N.; Chisholm, E. Setting Housing Standards to Improve Global Health. *Int. J. Environ. Res. Public Health* **2017**, *14*, 1542. [CrossRef]
10. Ormandy, D.; Ezratty, V. Health and thermal comfort: From WHO guidance to housing strategies. *Energy Policy* **2012**, *49*, 116–121. [CrossRef]
11. Graff, M.; Carley, S. COVID-19 assistance needs to target energy. *Nat. Energy* **2020**, *5*, 352–354. [CrossRef]
12. Intergovernmental Panel for Climate Change (IPCC). In Proceedings of the First Joint Session of Working Groups I and accepted by the 48th Session of the IPCC, Global Warming of 1.5 °C—Summary for Policymakers, Incheon, Korea, 1–5 October 2018.
13. Kenny, G.P. Towards establishing evidence-based guidelines on maximum indoor temperatures during hot weather in temperate continental climates. *Temperature* **2018**, *6*, 11–36. [CrossRef] [PubMed]
14. Hansel, N.N.; Mccormack, M.C.; Kim, V. The Effects of Air Pollution and Temperature on COPD ABSTRACT. *COPD J. Chronic Obstr. Pulm. Dis.* **2016**, *13*, 372–379. [CrossRef] [PubMed]
15. World Health Organisation. Non Communicable Diseases (NDC)-Key Facts. Fact Sheets. 2018. Available online: https://www.who.int/news-room/fact-sheets/detail/noncommunicable-diseases (accessed on 31 December 2018).
16. World Health Organisation. *Noncommunicable Diseases Country Profles 2018*; World Health Organisation: Geneva, Switzerland, 2018.

17. Wang, N.; Phelan, P.E.; Harris, C.; Langevin, J.; Nelson, B.; Sawyer, K. Past visions, current trends, and future context: A review of building energy, carbon, and sustainability. *Renew. Sustain. Energy Rev.* **2018**, *82*, 976–993. [CrossRef]
18. Mauree, D.; Naboni, E.; Coccolo, S.; Perera, A.T.D.; Nik, V.M.; Scartezzini, J.-L. A review of assessment methods for the urban environment and its energy sustainability to guarantee climate adaptation of future cities. *Renew. Sustain. Energy Rev.* **2019**, *112*, 733–746. [CrossRef]
19. Taylor, J. Comparison of built environment adaptations to heat exposure and mortality during hot weather, West Midlands region, UK. *Environ. Int.* **2018**, *111*, 287–294. [CrossRef]
20. Taylor, J. Mapping the effects of urban heat island, housing, and age on excess heat-related mortality in London. *Urban Clim.* **2015**, *14*, 517–528. [CrossRef]
21. Healy, J.D. Excess winter mortality in Europe: A cross country analysis identifying key risk factors. *J. Epidemiol. Community Health* **2003**, *57*, 784–789. [CrossRef]
22. Rodgers, S.E. Health impact, and economic value, of meeting housing quality standards: A retrospective longitudinal data linkage study. *Public Health Res.* **2018**, *6*, 1–83. [CrossRef]
23. Miguel-bellod, J.S.; González-martínez, P.; Sánchez-ostiz, A. The relationship between poverty and indoor temperatures in winter: Determinants of cold homes in social housing contexts from the 40s–80s in Northern Spain. *Energy Build.* **2018**, *173*, 428–442. [CrossRef]
24. World Health Organisation. *Who Housing and Health Guidelines*; World Health Organisation: Geneva, Switzerland, 2018.
25. Ortiz, M.A.; Kurvers, S.R.; Bluyssen, P.M. A review of comfort, health, and energy use: Understanding daily energy use and wellbeing for the development of a new approach to study comfort. *Energy Build.* **2017**, *152*, 323–335. [CrossRef]
26. Willand, N.; Ridley, I.; Maller, C. Towards explaining the health impacts of residential energy efficiency interventions—A realist review. Part 1: Pathways. *Soc. Sci. Med.* **2015**, *133*, 191–201. [CrossRef] [PubMed]
27. Armstrong, B. The impact of home energy efficiency interventions and winter fuel payments on winter- and cold-related mortality and morbidity in England: A natural equipment mixed-methods study. *Public Health Res.* **2018**, *6*, 1–138. [CrossRef] [PubMed]
28. Kolokotsa, D.; Santamouris, M. Review of the indoor environmental quality and energy consumption studies for low income households in Europe. *Sci. Total Environ.* **2015**, *536*, 316–330. [CrossRef]
29. Poortinga, W. The health impacts of energy performance investments in low-income areas: A mixed-methods approach. *Public Health Res.* **2018**, *6*. [CrossRef]
30. Ampatzoglou, A.; Bibi, S.; Avgeriou, P.; Verbeek, M.; Chatzigeorgiou, A. Identifying, categorizing and mitigating threats to validity in software engineering secondary studies. *Inf. Softw. Technol.* **2019**, *106*, 201–230. [CrossRef]
31. Zhou, X.; Jin, Y.; Zhang, H.; Li, S.; Huang, X. A map of threats to validity of systematic literature reviews in software engineering. *Proc. Asia-Pac. Softw. Eng. Conf. APSEC* **2016**, 153–160. [CrossRef]
32. Willand, N.; Maller, C.; Ridley, I. Understanding the contextual influences of the health outcomes of residential energy efficiency interventions: Realist review. *Hous. Stud.* **2020**, *35*, 1–28. [CrossRef]
33. Liddell, C.; Morris, C. Fuel poverty and human health: A review of recent evidence. *Energy Policy* **2010**, *38*, 2987–2997. [CrossRef]
34. van Hoof, J.; Schellen, L.; Soebarto, V.; Wong, J.K.W.; Kazak, J.K. Ten questions concerning thermal comfort and ageing. *Build. Environ.* **2017**, *120*, 123–133. [CrossRef]
35. Jevons, R.; Carmichael, C.; Crossley, A.; Bone, A. Minimum indoor temperature threshold recommendations for English homes in winter—A systematic review. *Public Health* **2016**, *136*, 4–12. [CrossRef] [PubMed]
36. Thomson, H.; Thomas, S.; Sellstrom, E.; Petticrew, M. Housing improvements for health and associated socio-economic outcomes. *Cochrane Database Syst. Rev.* **2013**, 1–159. [CrossRef] [PubMed]
37. United Nations. *The Sustainable Development Goals Report*; United Nations-Department of Economic and Social Affairs: New York, NY, USA, 2018.
38. Nugent, R. Series The Lancet Taskforce on NCDs and economics 1 Investing in non-communicable disease prevention and management to advance the Sustainable Development Goals. *Lancet* **2018**, *391*, 2029–2035. [CrossRef]
39. Uejio, C.K.; Tamerius, J.D.; Vredenburg, J.; Asaeda, G.; Isaacs, D.A.; Braun, J.; Freese, J.P. Summer indoor heat exposure and respiratory and cardiovascular distress calls in New York City, NY, US. *Indoor Air* **2016**, *26*, 594–604. [CrossRef]

40. Osman, L.M.; Ayres, J.G.; Garden, C.; Reglitz, K.; Lyon, J.; Douglas, J.G. Home warmth and health status of COPD patients. *Eur. J. Public Health* **2008**, *18*, 399–405. [CrossRef]
41. Saeki, K.; Obayashi, K.; Tone, N.; Kurumatani, N. A warmer indoor environment in the evening and shorter sleep onset latency in winter: The HEIJO-KYO study. *Physiol. Behav.* **2015**, *149*, 29–34. [CrossRef]
42. Saeki, K.; Obayashi, K.; Kurumatani, N. Indoor cold exposure and nocturia: A cross- sectional analysis of the HEIJO-KYO study. *BJUI-BJU Int.* **2016**, *117*, 829–835. [CrossRef]
43. Saeki, K.; Obayashi, K.; Kurumatani, N. Platelet count and indoor cold exposure among elderly people: A cross-sectional analysis of the HEIJO-KYO study. *J. Epidemiol.* **2017**, *27*, 562–567. [CrossRef]
44. van Loenhout, J.A.F. The effect of high indoor temperatures on self-perceived health of elderly persons. *Environ. Res.* **2016**, *146*, 27–34. [CrossRef]
45. Huebner, G.M.; Hamilton, I.; Chalabi, Z.; Shipworth, D. Comparison of indoor temperatures of homes with recommended temperatures and effects of disability and age: An observational, cross-sectional study. *BMJ Open* **2018**, *8*, e021085. [CrossRef]
46. Preval, N.; Keall, M.; Telfar-barnard, L.; Grimes, A.; Howden-chapman, P. Impact of improved insulation and heating on mortality risk of older cohort members with prior cardiovascular or respiratory hospitalisations. *BMJ Open* **2017**, *7*, e018079. [CrossRef]
47. Viggers, H. Effect of insulating existing houses on health inequality: Cluster randomised study in the community. *BMJ* **2007**, *334*, 460. [CrossRef]
48. Osman, L.M.; Ayres, J.G.; Garden, C.; Reglitz, K.; Lyon, J.; Douglas, J.G. A randomised trial of home energy efficiency improvement in the homes of elderly COPD patients. *Eur. Respir. J.* **2010**, *35*, 303–309. [CrossRef]
49. Shiue, I. Cold homes are associated with poor biomarkers and less blood pressure check-up: English Longitudinal Study of Ageing, 2012–2013. *Environ. Sci. Pollut. Res.* **2016**, *23*, 7055–7059. [CrossRef]
50. Shiue, I.; Shiue, M. Indoor temperature below 18 °C accounts for 9% population attributable risk for high blood pressure in Scotland. *Int. J. Cardiol.* **2014**, *171*, e1–e2. [CrossRef]
51. van Hooff, T.; Blocken, B.; Timmermans, H.J.P.; Hensen, J.L.M. Analysis of the predicted effect of passive climate adaptation measures on energy demand for cooling and heating in a residential building. *Energy* **2016**, *94*, 811–820. [CrossRef]
52. Wilson, N.J.; Jacobs, N.D.; Reddy, N.A.; Tohn, T.E.S.E.; Cohen, D.J.; Jacobsohn, D.E. *Home RX: The Health Benefits of Home Performance—A Review of the Current Evidence*; National Renewable Energy Laboratory for the U.S. Department of Energy (DOE): Columbia, MD, USA, 2016.
53. Haines, A. Health and Climate Change 6 Public health benefi ts of strategies to reduce greenhouse-gas emissions: Overview and implications for policy makers. *Lancet* **2009**, *374*, 2104–2114. [CrossRef]
54. Alavy, M.; Li, T.; Siegel, J.A. Energy use in residential buildings: Analyses of high-efficiency filters and HVAC fans. *Energy Build.* **2020**, *209*, 109697. [CrossRef]
55. Jung, W.; Jazizadeh, F. Human-in-the-loop HVAC operations: A quantitative review on occupancy, comfort, and energy-efficiency dimensions. *Appl. Energy* **2019**, *239*, 1471–1508. [CrossRef]
56. Howden-Chapman, P.; Viggers, H.; Chapman, R.; O'Sullivan, K.; Barnard, L.T.; Lloyd, B. Tackling cold housing and fuel poverty in New Zealand: A review of policies, research, and health impacts. *Energy Policy* **2012**, *49*, 134–142. [CrossRef]
57. Xu, X.; Chen, C.F. Energy efficiency and energy justice for U.S. low-income households: An analysis of multifaceted challenges and potential. *Energy Policy* **2019**, *128*, 763–774. [CrossRef]
58. de Cian, E.; Pavanello, F.; Randazzo, T.; Mistry, M.N.; Davide, M. Households' adaptation in a warming climate. Air conditioning and thermal insulation choices. *Environ. Sci. Policy* **2019**, *100*, 136–157. [CrossRef]
59. Guertler, P.; Smith, P. *Cold Homes and Excess Winter Deaths a Preventable Public Health Epidemic That Can No Longer Be Tolerated*; February 2018; E3G and National Energy Action (NEA): Newcastle upon Tyne, UK, 2018.
60. Daniel, L.; Baker, E.; Williamson, T. Cold housing in mild-climate countries: A study of indoor environmental quality and comfort preferences in homes, Adelaide, Australia. *Build. Environ.* **2019**, *151*, 207–218. [CrossRef]
61. Magalhães, S.M.C.; Leal, V.M.S.; Horta, I.M. Predicting and characterizing indoor temperatures in residential buildings: Results from a monitoring campaign in Northern Portugal. *Energy Build.* **2016**, *119*, 293–308. [CrossRef]
62. Willand, N.; Maller, C.; Ridley, I. Addressing health and equity in residential low carbon transitions—Insights from a pragmatic retrofit evaluation in Australia. *Energy Res. Soc. Sci.* **2019**, *53*, 68–84. [CrossRef]

63. Alkire, S.; Samman, E. *Mobilizing the Household Data Required to Progress toward the SDGS*; University of Oxford: Oxford, UK, 2014.
64. Bogart, S. SankeyMATIC. 2020. Available online: http://sankeymatic.com/ (accessed on 20 May 2020).

energies

Article

Energy Policy Concerns, Objectives and Indicators: A Review towards a Framework for Effectiveness Assessment

Dania Ortiz * and Vítor Leal *

Faculty of Engineering, University of Porto (FEUP), Rua Dr. Roberto Frias, 4200-465 Porto, Portugal
* Correspondence: up201608014@fe.up.pt (D.O.); vleal@fe.up.pt (V.L.)

Received: 11 November 2020; Accepted: 5 December 2020; Published: 10 December 2020

Abstract: This work presents a review that aims to characterize the policy evaluation practices regarding the public policies on energy, with a focus on the metrics: concerns, objectives, and indicators. As key novelty, emphasis was put into finding attributes and metrics that can be used to assess effectiveness, not only efficacy or efficiency. The concerns and objectives were organized into four categories: Institutional, Environmental, Economic, and Social. For every category, detailed and condensed concerns were identified. It was attempted to find indicators for every condensed concern, which resulted in 15 core indicators.

Keywords: energy policy evaluation; energy policy effectiveness; energy policy indicators; energy planning; sustainability

1. Introduction

Differences among the policy objectives and the policy results are often found. This situation can be intensified if the energy policies (EP) are formulated or implemented, without considering the particular characteristics and capabilities of different countries or regions. EP evaluation has previously been proposed as a tool not only to compare the policy achievements with their objectives but also to assess the "goodness" of policy programs. This is becoming increasingly important as countries, regions, and even municipalities are pressed to develop carbon neutrality plans. Regardless of the jurisdiction, authorities are organizing to face the challenge of climate change mitigation. The efforts can be seen from top-level international agreements, as the Paris Agreement [1], and the Sendai Framework [2], to city-level climate-neutral plans (e.g., the plans developed in Berlin, Germany [3], Ann Arbor, USA [4], Bayside, Australia [5] or Freiburg, Germany [6]). Additionally, the United Nations 2030 Agenda [7] has established the Sustainable Development Goals (SDGs) as crosscutting international alignment principles. They represent a set of universal goals that meet the urgent environmental, political, and economic challenges facing the world. To meet these goals, Governments will need better planning tools and to monitor progress towards the targets; more specifically, tools that enable evaluating policies in multidisciplinary ways, allowing decision makers to analyze the policies in an integral way.

As an intended contribution to the planning and management of climate action plans, this work addresses the problem of EP evaluation, with a focus on effectiveness [8]. It starts with the conceptual differences between effectiveness, efficiency, and efficacy, and then moves on to find specific indicators for a comprehensive assessment. Such a set would be useful for ex-post evaluations, but in some cases, it could also be useful for ex-ante evaluation and thus assist the design of better policy programs. Therefore the main motivations of this work have been to create a support document that enables a more structured and comprehensive planning process (and that does not leave important aspects forgotten); planning programs that have a higher chance of succeeding, in practice; and supporting the good use of resources (public and/or private).

With the aim of evaluating EP and provide information that facilitates their adjustment to distinctive characteristics, a set of concerns was intended to be identified through this review. Concerns that were organized under the different proposed categories and subcategories. These concerns were organized under condensed concerns by identifying their final objective.

An indicator identification process was followed. Indicators were selected as the assessment metrics that can provide a comparison between the policies defined objectives and their achievement, therefore selected as an effectiveness measurement tool. The identification of concerns ends objectives has also been done through this report, serving as cornerstones for the identification of indicators.

The objective of this work was to review and systematize with a scientific-based methodology, the available information regarding EP effectiveness and its assessment indicators. The review led to the identification of trends in the subject. The main contribution of the document is the identification of the multidisciplinary attributes to assess energy policy effectiveness. No other review has been found available in the literature in which EP effectiveness is characterized holistically.

The paper is structured into 6 sections. Section 2 addresses the concepts of policy success: efficacy, efficiency, and effectiveness. The different EP effectiveness categories reported through the selected documents are presented in Section 3. Section 4 focuses on the systematization of energy policy concerns and objectives. Section 5 presents energy policy indicators, where the concerns of Section 4 have been matched with corresponding indicators. Section 6 includes a Discussion, and finally, the conclusions are presented in Section 7.

2. Policy Success: Efficacy, Efficiency and Effectiveness

A sensible goal of policy evaluation is to assess whether or not they are "successful". "Success" is defined as "the accomplishment of an aim or purpose" [9]. This is very much in line with the definition of "efficacy", which is defined as "the ability, especially of a medicine or a method of achieving something, to produce the intended result" [10].

A strict or "hard" approach to the evaluation of "success" or "efficacy" could therefore rely on the comparison between what is achieved and what was the target or goal; e.g., a program that has the target of promoting the installation of 80,000 solar collectors would be "successful" or "efficacious" if it achieves 80,000 or more installations, while it would be "unsuccessful" if it achieves equal or less than 79,999 installations.

The example above illustrates as well that such a "hard" approach to success evaluation is probably too strict. Under this approach, failing by a large amount or by a small amount is put under the same label. It could also serve as a stimulus to establish artificial and sometimes counter-productive low targets, to increase the probability of achieving the results.

Therefore, more flexible and comprehensive metrics are needed to assess policy success. This is where the concept of "effectiveness" may come into play. Although, sometimes seen as equivalent to "success", it is formally defined as "the degree to which something is successful in producing a desired result" [11]. The word "degree" opens room for a more gradual evaluation, clearly allowing for differentiation between failing by a large amount vs. failing by a small amount, as well as for succeeding by a small amount vs succeeding by a large amount.

Furthermore, "effective" has already been used, e.g., in the field of medicine as different from "effective" in the sense that it is something that is evaluated from the real-world use or implementation, including informal feedback from doctors and patients, rather than from controlled experiments alone [12].

It may be of interest to bring into consideration also the concept of "efficiency". This is clearly distinct from both efficacy as from effectiveness, assessing what was achieved concerning the means that were put into the program. While of course being efficacious is more important than being efficient, being efficient is also important. It implies that resources (e.g., funding to invest), which typically are limited, are being well used. By being efficient, more effect can be achieved from a given amount of resources.

As a more nuanced concept than efficacy, effectiveness thus has the potential to encompass the components of taking into consideration contextual variables and of assessing "success" in a quite modulated manner. It can even be advocated that, while privileging the evaluation of the achievement of results, a truly effective program should assess those results concerning the contextual variables (geographical, economic, and cultural) rather than by just comparing it to the initial goals of the program. This is so because these often are not declared and, even when they are, their choice is inherently somewhat arbitrary. Effectiveness can thus incorporate some dimension of efficiency into the assessment.

3. Materials and Methods

The academic databases Sciencedirect, Google Scholar, and JSTOR, together with the web browser Google were used for gathering documents. All the sources were useful to identify documents that characterize energy policy effectiveness. The expressions under which the search was done were "Energy policy evaluation" (academic and open-source); "Energy policy effectiveness" (academic source); "Energy policy assessment" (academic source); "Energy policy efficiency" (academic source). The selection of the documents to be reviewed was done by their title, abstract content, and keywords. After doing so, the criteria for the document selection for the analysis was having at least one concern that assesses energy policies. A summary of this process is presented in Table 1.

Table 1. Search terms by source and documents reviewed and adopted by each.

Search Term	Source	Documents Reviewed	Documents Adopted
Energy Policy Evaluation	Sciencedirect	18	9
Energy Policy Evaluation	Google	17	10
Energy Policy Assessment	Sciencedirect	16	10
Energy Policy Effectiveness	Sciencedirect	9	5
Energy Policy Effectiveness	Google Scholar	20 (3 repeated)	15
Energy Policy Effectiveness	JSTOR	5	3
Energy Policy Evaluation	Google Scholar	19 (3 repeated)	13
Energy Policy Evaluation	JSTOR	5 (1 repeated)	1

The detail of each search source is presented in Appendix A, for which reference and order numbers have been provided and that have been used for the organization and analysis of the documents presented in the document.

Analysis: Geographical Scope

When analyzing the scope or level of application of the energy policies evaluations (local, regional, national, international, or a combination among them), it is noted in Table 2 that the highest number of the reviewed documents regards the national level (with 40 documents), with a considerable difference above the other alternatives. The national scope is followed by the international scope (with eleven documents). Some documents regard all the levels of application or a combination of more than one level. Another observation is that only three locally focused documents were found.

Concerning the time frame shown in Table 3, almost all of the retrieved documents state or enabled us to infer the period of the evaluation (summing 53 documents), against 13 documents where that was not possible. The time framing of the process, according to the reviewed documents, can be ex ante, in progress, or ex post. The ex post energy policy evaluation approach was the most often found (34 documents), followed by the ex ante approach (10 documents). It can be noticed that only two of the reviewed documents ("Moving Toward Energy Efficiency: A Results-Driven Analysis of Utility-Based Energy Efficiency Policies", Theel and Westgaard, 2017 [13] and "Evaluating Consistency in Environmental Policy Mixes through Policy, Stakeholder, and Contextual Interactions Lieu et al., 2018 [14] declared a combined approach (with more than one possible timing period).

Table 2. Energy policy evaluation documents found per geographical level of analysis.

Level	Quantity	Document Numbers (from Tables in Appendix A)
National	40	1–4; 6–11; 13; 16–18; 20; 21; 24–26; 28; 30–32; 34; 35; 45; 46; 50; 52; 53–55; 58–64; 66.
Regional (State)	9	14; 15; 19; 27; 36; 38; 39; 47; 56.
International	11	29; 33; 37; 40; 41; 43; 44; 49; 51; 57; 65.
Combined	3	12; 23; 22.
Local	3	5; 42; 48.

Table 3. Energy policy evaluation documents found per time-frame.

Time Framing of the Process	Quantity	Documents Numbers (from Tables in Appendix A)
Ex ante	10	1; 5; 15; 22; 23; 29; 31; 40; 47; 51.
Ex post	34	4; 8–10; 18; 21; 27; 32; 33; 35; 36; 38; 39; 41–63; 65; 66.
In progress	7	19; 28; 30; 34; 46; 48; 64.
Combined	2	14; 17.
Not inferred	13	2; 3; 6; 7; 11–13; 16; 20; 24–26; 37.

Table 4 shows the participation of stakeholders in the EP evaluation. Stakeholder participation is related to the involvement of actors along the evaluation process. Less than one-third of the documents retrieved report some form of participation from stakeholders (19 out of 66 documents).

Table 4. Energy policy evaluation documents found per stakeholder participation.

Stakeholder Participation	Quantity	Documents Numbers (from Tables in Appendix A)
Yes	19	3–6; 13; 14; 17; 20; 23; 29; 30; 37; 40; 53; 54; 58; 59, 64.
No	47	1; 2; 5-12; 16; 18; 19; 21; 22; 24–28; 31–36; 38; 39; 41–52; 55–57; 60–63; 65; 66.

The methodological approaches for evaluation were also reviewed and are presented in Table 5. As expected, very diverse approaches were found from document to document. The qualitative-based methodology was found as the most popular approach (19 documents). Descriptive and qualitative analysis methods have been considered as qualitative-based. The second more frequent approach is statistical (17 documents). Scenario and modeling techniques have also been identified (12 documents). The category others include the approaches that were found only once e.g., cost–benefit analysis (CBA), return on assets (ROA), etc.

Table 5. Energy policy evaluation documents found per methodological approach.

Methodology	Quantity	Documents Numbers (from Tables in Appendix A)
Statistical	17	10; 14; 20; 27; 32; 33; 36; 38; 39; 42–44; 48; 52; 56; 61; 63.
Scenario	12	1; 5; 9; 16; 18; 21; 22; 34; 53; 60; 65; 66.
Qualitative-based approach	19	2–4; 11; 13; 17; 23–25; 28; 31; 37; 40; 41; 45; 46; 57; 58; 64.
Indicators	4	8; 12; 35; 59.
Others	14	6; 7; 15; 19; 26; 29; 30; 47; 49-51; 54; 56; 63.

In Table 6, it can be seen that when analyzing the motivation stated in the reviewed documents, the most common one is to conduct a policy evaluation itself, observed in 34 documents. Four documents clearly defined their interest in supporting policymakers, while 16 documents target the evaluation of policy success. It can be noted the interrelation between the identified motivations.

Table 6. Energy policy evaluation documents found per expressed objectives.

Motivation	Quantity	Documents Numbers (from Tables in Appendix A)
Policy evaluation	34	6–8; 13–16; 18–22; 27; 30; 32–34; 37; 39; 40; 45; 46; 49; 53–55; 57–62; 64-66.
Policy review	7	23; 24; 26; 29; 35; 36; 51.
Support policy makers	4	3; 5; 17; 20.
Evaluate policy success	16	1; 9–12; 38; 41–44; 47; 48; 50; 52; 56; 63.
Creation of a methodology/framework	5	2; 4; 25; 28; 31.

4. Results

4.1. Energy Policy Concerns and Objectives

The identification of the concerns most frequently included in the literature and how they are defined led to forming a list of more than 60 concerns. They have been classified into four broader categories, and some subcategories have been formed.

4.1.1. Environmental Concerns

As shown in Table 7, two subcategories were defined: greenhouse gas (GHG) and other environmental impacts. The burden among the categories could be very thin, as some concerns are related to both, as renewable energy (RE) sources related concerns (e.g., RE standards, RE Deployment and RE Tech Efficiency).

Table 7. Energy Policy Environmental concerns found.

Document	GHG	Other Impacts
Lieu et al., 2018 [14].	-	Environmental stakeholders
Abotah, 2015 [15].	Low carbon economy	Environmental protection
Chapman et al., 2016a [16].	GHG emissions	RE Deployment, RE Tech Efficiency
Chapman et al, 2016b [17].	GHG emissions	-
Chen, 2011 [18].	-	Eco-label system
Alyamani et al., 2016 [19].	GHG emissions	-
Lee & Shih, 2010 [20].	-	Environmental protection
Nikolaev & Konidari, 2017 [21].	GHG emissions	-
Hafeznia, Aslani, Anwar, & Yousefjamali, 2017 [22].	-	Climate information
S. J. Li et al., 2017 [23].	-	RE energy production
Yu et al., 2015 [24].	CO2 emissions	Energy savings
Aized et al., 2018 [25].	GHG emissions	-
Arababadi et al., 2017 [26].	GHG emissions	-
Burke & Stephens, 2017 [27].	-	RE minimum requirements
Chang & Fang, 2017 [28].	-	Sustainability in energy supply
García-Álvarez et al., 2018 [29].	-	General environmental awareness
Mirjat et al., 2017 [30].	-	General environmental awareness
Atalla et al., 2018 [31].	Climate change mitigation	-
Delmas & Montes-sancho, 2011 [32].	GHG emissions	Renewable resources
Menz & Vachon, 2006 [33].	-	Resources capacity and quality
Arbolino et al., 2019 [34].	GHG emissions	Renewable share
Thornley & Cooper, 2008 [35].	-	Resource capacity
Jenner et al. 2013 [36]	-	Resources share
Dong, 2012 [37].	GHG emissions	-
Sancho, 2010 [38].	GHG emissions	-
Bersalli, et al., 2020 [39].	GHG emissions	-
Stagl, 2006 [40].	Global warming and climate change	Renewable energy use; Less-pollutant transportation
Hoffman et al., 1977 [41].	GHG emissions	Environmental impact, energy resources
Carley, 2009 [42].	-	Natural resource endowments, RE share
Kilinc-Ata, 2016 [43].	GHG emissions	-

Greenhouse gas (GHG) emissions reduction was found as the predominant concern for evaluating energy policies under an environmental perspective and gathered under the subcategory Environmental-GHG. The GHG emissions concern consideration has been incentivized by the continuous and reinforced influence produced by the definition of international agendas, national targets, agreements, and other international efforts to reduce GHG emissions and minimize climate change.

Other impacts include the concerns that assess the conservation and improvement of the environment, excepting GHG. RE related concerns (e.g., RE standards, RE Deployment and RE Tech Efficiency) have been included in this subcategory.

4.1.2. Economic Concerns

A large set of concerns was found under the economic category, as shown in Table 8. It should be noticed that almost all the reviewed documents included at least one aspect under this category. The economic dimension regards all the concerns that assess policies in monetary terms, from which two subcategories were formed.

Table 8. Energy Policy Economic concerns found.

Document	Cost	Competitiveness
Chapman et al., 2016a [16].	Energy price performance	Market growth and maturity
Theel & Westgaard, 2017 [13].	Energy price and demand	GDP
Abotah, 2015 [15].	Economic feasibility	-
Qiang et al., 2016 [44].	Cost-Effectiveness	-
Lieu et al., 2018 [14].	Supply chain Stakeholders and Service providers	-
Chapman et al., 2016b [17].	Generation Cost, Electricity Price Impact, Levelized cost of electricity (LCOE)	GDP Impact and Market Impact
Energy and Environmental Economics, 2014 [45].	Net Cost	-
Chen, 2011 [18].	Energy demand, energy price	-
Bukarica & Tomšić, 2017 [46].	-	Market barriers
Alyamani et al., 2016 [19].	Energy demand, cost, and price	Utility profits and revenue
Lee & Shih, 2010 [20].	-	Macroeconomic parameters
Collado & Díaz, 2017 [47].	Energy savings	-
Nikolaev & Konidari, 2017 [21].	Internalized total cost	Attractiveness in RES investment
A.J. Chapman [48].	Generation Cost, Electricity Price	GDP Impact and Market Impact
X. Zhao et al., 2016 [49].	Power demand, energy substitution, technological change	-
S. J. Li et al., 2017 [23].	Economic data	-
S. Ma et al., 2020 [50].	Carrier price, demand	Patents
Shrimali et al., 2017 [51].	Cost-effectiveness potential	Investment incentives
Aized et al., 2018 [25].	Costs, supply, and demand	-
Burke & Stephens, 2017 [27].	Financial measures	-
Chang & Fang, 2017 [28].	Economic context	-
Liu et al., 2018 [52].	Costs and profits	-
García-Álvarez et al., 2018 [29].	Electricity price trends, economic situation, security of supply	-
Mirjat et al., 2017 [30].	Economic efficiency	-
Atalla et al., 2018 [31].	Welfare and Economic impact	-
Lund, 2007 [53].	Cost-effectiveness	Market State
Dong, 2012 [37].	GDP per capita, Oil imports, electricity consumption	-
Agnolucci, 2007 [54].	Financial support	-
Mitchell et al., 2006 [55].	Risk reduction	-
Sancho, 2010 [38].	GDP	-
Qiu et al., 2019 [56].	City Infrastructure	-
Bersalli et al., 2020 [39].	Energy Production, Energy dependence rate	Access to domestic credit, Income per capita
Economics, 2011 [45].	Cost-effectiveness	-
Yuan et al., 2011 [57].	Cost-effectiveness	
Rosenow 2016 [58].	Energy Savings	-
Nicholson-crotty & Carley, 2016 [59].	Energy generation	-
Stagl, 2006 [40].	Energy efficiency; energy costs	-
Zhang et al., 2014 [60].	Energy costs, renewable energy costs (and subsidies), Carbon prices	-
Hoffman et al., 1977 [41].	Cost, generation, dependence on imports	-
Carley, 2009 [42].	Gross state product per capita, electricity use, electricity price	-
Neij & Kerstin, 2006 [61].	Technology characteristics, electricity costs	-
J. Zhao et al., 2011, [62].	Technology payback period	-
Kilinc-Ata, 2016 [43].	Carrier prices, imports, GDP growth, electricity consumption	-
Verbruggen, 2009 [63].	Cost-effectiveness	-
Banerjee & Solomon, 2003 [64].	-	Consumer response and manufacturer/marketer response
Kim & Lee, 2012 [65].	-	Consumer choice, technology, and location characteristics

The Economic-cost subcategory includes the concerns related to direct monetary influence, particularly related to costs and expenses. Both levels, macroeconomic and microeconomic, have been assessed; e.g., the energy price is included, referring to the economic aspects in a micro economical view. The gross domestic product (GDP) impacts of energy decisions refer to the macro level of an economy.

Regarding Economic-competitiveness, this subcategory brings together market orientation concerns (e.g., attractiveness of investment, market growth and market maturity aspects). The market characteristics provide information that allows the comparison among different markets, informing about their national and international competitiveness.

4.1.3. Social Concerns

Regarding the Social category, all concerns with a social orientation have been gathered in Table 9, including the concerns in which equity is recognized. Two sub-categories were defined. Firstly, Social-equity include concerns related to social sustainability factors, as the acceptance of certain energy technologies and the impacts of subsidies on energy bills. It can be noticed how equity has been considered in several documents, being a key component of distributive energy justice, and that should contribute to new policy initiatives. Some other concerns include energy democracy, energy equity and consumer surplus. This last is an important reference to measure the welfare of consumers [29].

Table 9. Energy Policy Social concerns found.

Document	Equity	Employment
Andrew J. Chapman et al., 2016a [16].	Health, subsidy allocation, costumer's participation	Employment
Rogge et al., 2017 [66].	Distribution of energy costs (equity)	-
IRENA et al., 2014 [67].	Equity (Consumer´s impact)	-
Abotah, 2015 [15].	Customer interaction and satisfaction	-
Andre John Chapman, 2016b [17].	Impact of subsidies in energy prices, distribution of costs and benefits	Employment
(W. Li et al., 2016) [68].	Perceptions of importance and satisfaction	-
Nikolaev & Konidari, 2017 [21].	Energy equity	Employment
Wang et al., 2017 [69].	Consumers' preferences, vehicle acceptance, and WTP	-
Hafeznia et al. [22].	Social acceptance of RE technologies	Knowledge and skill of workforce (expertise)
Enzensberger [70].	Stakeholder´s consideration	-
Edwards, 2018 [71].	Social acceptance	-
Burke & Stephens, 2017 [27].	Energy democracy	-
Chang & Fang, 2017 [28].	Equity among households: Energy elasticity	-
Liu et al., 2018 [52].	Consumer surplus	-
Mirjat et al., 2017 [30].	Social context	-
Atalla et al., 2018 [31].	Electricity price elasticity (Household behavior)	-
Delmas & Montes-sancho, 2011 [32].	Customers, Renewable preferences	Unemployment
Arbolino et al., 2019 [34].	Health	-
Harmelink et al., 2006 [72].	Acceptance by the actors	-
Yao et al., 2014 [73].	Demographic Information	-
Sancho, 2010 [38].	-	Unemployment
Qiu et al., 2019 [56].	Demographic information	-
Ling-yun & Jia-jia, 2016 [74].	Public Health and people´s welfare	-
Delmas & Montes-sancho, 2011 [32].	Customers' preferences, Investors	-
Bürer & Rolf, 2009 [75].	Investors	-
J. Zhao et al., 2011 [62].	Household income; neighborhood	-
Stagl, 2006 [40].	Warm spaces	-
Neij & Kerstin, 2006 [61].	Participation and Commitment of actors	-

The second subcategory, Social-employment, as can be inferred from its name, is the subcategory to account for the employment rates affected by the energy policies. The knowledge and skills (expertise) of the workforce have also been included in this subcategory.

4.1.4. Institutional Concerns

Finally, under the institutional concerns assessment (Table 10), the variety of concerns found through the reviewed papers was larger than in the previous categories; however, no subcategories were created, due to a lack of a clear structure. The identified concerns describe the policy institutional framework, including concerns as flexibility with policy instruments, actors, and institutional environment, among others.

Table 10. Energy Policy Institutional concerns found.

Document	Concern
Rogge et al., 2017 [66].	Actor´s and institution´s participation
IRENA et al., 2014 [67].	Political accountability, source of finance and regulatory simplicity
Andrew J. Chapman et al., 2016c [76].	Governance, policy processes and priorities
Theel & Westgaard, 2017 [13].	Policy Environment
Abotah, 2015 [15].	Institutional and Government support
Bukarica & Tomšić, 2017 [46].	Political will, public participation, public dialogue, regulatory framework
Yoon & Sim, 2015 [77].	Policy framework (policy environment and policy process)
Nikolaev & Konidari, 2017 [21].	Administrative and financial capacity, Feasibility of implementation
Arababadi et al., 2017 [26].	Policy budget
Burke & Stephens, 2017 [27].	Policy implementation mechanisms
Liu et al., 2018 [52].	Government fiscal revenue
García-Álvarez et al., 2018 [29].	Policy Environment
Mirjat et al., 2017 [30].	Policy Environment
Delmas & Montes-sancho, 2011 [32].	Type of Governor
Menz & Vachon, 2006 [33].	Policy regime
Dong, 2012 [37].	Other policies (number)
Agnolucci, 2007 [54].	Policy certainty and planning constraints
Qiu et al., 2019 [56].	Policy options
Rosenow et al., 2016 [58].	Policy characteristics
Carley, 2009 [42].	Political institutions, other policies (presence)
Harmelink et al., 2008 [78].	Policy process
J. Zhao et al., 2011 [62].	Advertisement
S. Ma et al., 2020 [50].	Subsidies
Verbruggen, 2009 [63].	Robustness, efficiency

There is a recent recognition of the importance of the institutional context. Concerns like political commitment, policy choice, and citizen engagement characterize policies from an institutional perspective.

Even if the institutional performance does not measure policy success itself, it contributes to explain the policy environment. Institutional context tends to be assessed qualitatively rather than quantitatively. The lack of quantitative procedures to evaluate some dimensions of the institutional context provides the opportunity to develop scientific-based methodology proposals that enable to evaluate the concerns found under this category.

Table 11 shows the number of appearances per concern by subcategory. It is noted that Economic-cost is the subcategory that was found more often in the reviewed documents, reflecting the importance given to it by the consulted sources. The second most popular subcategories are Governance and Social equity, which regards the energy impacts on society and energy distributional impacts.

Table 12 shows an analysis of the categories. It was noted that most documents mention more than one category (42 documents). From the remaining 24 documents (single category approach), 12 documents report an Economic orientation, being the largest one.

4.2. Identification of Gaps

After the review of the concerns that have been used to characterize the effectiveness of the energy policies, some areas have been identified as not fully covered regarding an integral review of EP effectiveness characterization.

Table 11. Number of appearances per concern by subcategory.

Category	Quantity of Documents	Documents Numbers (from Tables in Appendix A)
Environmental-others	14	2, 7, 15, 17, 18, 21, 24, 25–29, 31, 33
Environmental GHG	14	1, 5, 9, 15, 18, 21–23, 36, 38, 39, 44, 47, 49
Economic-cost	36	1, 2, 5, 8–10, 14–19, 22, 24-29, 31, 32–36, 39, 40, 42–49, 60
Economic-competitiveness	15	1, 3, 5, 7, 9, 10, 14, 18, 31, 34–37, 49, 62
Social Equity	25	1, 6, 9, 11, 12, 15, 18, 20, 24–26, 28, 29, 31, 36, 37, 39, 40, 42, 48, 52, 53, 58, 60, 66
Social-Employment	7	1, 9, 10, 18, 31, 36, 47
Governance	26	3, 4, 9–15, 23, 24, 26–28, 36, 38, 40, 44, 45, 48, 51, 56, 57, 60, 62, 63, 65

Table 12. Single vs. multi-category orientation.

Category	Quantity of Documents	Documents numbers (from Tables in Appendix A)
Single	24	4, 6–8, 13, 16, 19–21, 30, 32, 34, 37, 41, 42, 46, 50, 52, 54, 57, 58, 61, 64, 66
Multi-category	42	1–3, 5, 9, 10–12, 14, 15, 17, 18, 22, 23, 24–29, 31, 33, 35, 36, 38–40, 43–45, 47–49, 51–53, 55, 56, 59, 60, 62, 63, 65
Environmental	2	21, 41
Economic	12	7, 8, 16, 19, 32, 34, 46, 50, 52, 54, 61, 64
Social	7	6, 20, 30, 37, 42, 58, 66
Institutional	3	4, 13, 57

A list of missing concerns has been elaborated (Table 13) in which additional concerns have been included. The criterion followed to add concerns was if the outcomes of the concern could be modified by a policy and their relevance for the ep effectiveness characterization.

Table 13. Additional concerns by category.

Environmental Concerns	Economic Concern	Institutional Concern
Water resources	Externalities pricing	Amount and quality of Staff
Land resources	Grid state	-
Waste Management	-	-

Not all the subcategories needed extra concerns, e.g., the Social subcategories (Social-Equity and Social-Employment) have been considered as complete, as the concerns identified through the review do provide a wide perspective of the social approach.

Under the environmental category, the concern of environmental protection has been often found through the revised documents. However, it has been considered relevant to detail it through more specific concerns (as water and land resources) to be able to better measure them. They are relevant concerns for the environmental evaluation of policy and closely related to the impacts of energy on the environment.

The externalities pricing is a concern that can be reported under the environmental or economic category, due to their interrelationship in the concern itself. For the monetarization process that externalities involve, it has been decided to keep this concern under the economic category.

It was found a lack of consideration of some aspects that may be relevant for the policy appraisal, as available funds (for technology, innovation, and infrastructure investment), the economic situation of the private sector, and the economic situation by regions, economic and fiscal parameters (like inflation, escalation, exchange rates, and taxes). All these concerns can be gathered as economic background. Additionally, the grid state is a relevant concern for its relation to the energy service provision.

Finally, the concern under the institutional category (Amount and quality of Staff) intends to involve a relevant aspect of the institutional framework: the governing human resources. They differ from Knowledge and skill of the workforce (reported in the Social concerns), as this last regards the technical workforce involved in the policy implementation, while the second one, regards the governance-related workforce.

4.3. Inferred Energy Policy Objectives and Condensed Concerns

With the intention of better organizing the concerns and facilitating the identification of their evaluation metrics, a condensation process was taken to eliminate redundancies, conglomerating them into condensed concerns. The condensed concerns have been identified after questioning where there could be an impact if the extracted concern does not comply. In addition, for every condensed concern, its end objective has been identified (as shown in Figure 1). The identification of the ends objective aims to put the review more in line with the typical multi-criteria decision making (MCDM) language and to ease the further identification of indicators.

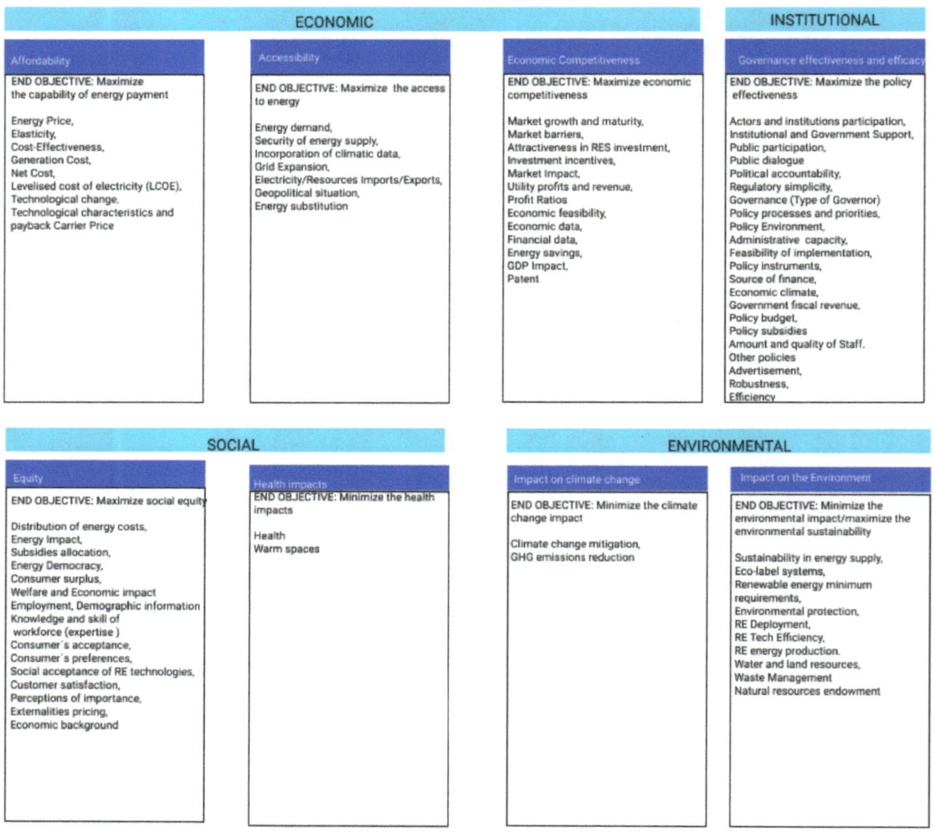

Figure 1. Extracted and condensed concerns per ends objective.

The economic extracted concerns are the largest in number when compared with other categories. It was possible to narrow down the initial list to three condensed concerns: affordability, accessibility, and economic competitiveness.

Affordability includes all the concerns that may affect the state of being cheap enough for people to be able to pay for electricity or other forms of energy [79]. Accessibility refers to the fact of energy or electricity being able to be reached or obtained easily [80].

Competitiveness brings together the concerns that relate to being able to compete successfully with other companies, countries, organizations, etc. [81].

The concerns under the environmental spectrum have been condensed through two concerns. The first one reflects the policy's impact on the environment. Its ends objective relates to environmental sustainability, defined as the quality of causing little or no damage to the environment and therefore able to continue for a long time [82]. The second condensed concern denotes the impacts on climate change.

In the social category, Equity, as the situation in which everyone is treated fairly and equally [83], gathers the majority of the social concerns. Health is however kept as a separated condensed concern.

Governance effectiveness and efficacy as a single condensed concern included all the concerns under the institutional perspective. Governance is defined as how an organization is managed at the highest level and the systems for doing so [84] and assessed through its effectiveness and efficacy.

The organization of the extracted and condensed concerns is documented in Figure 1 The economic, environmental, and social categories include more than one condensed concern.

5. Indicators for Energy Policy Effectiveness

5.1. Indicators' Relevance and Types

The relevance of indicators has been recognized in the energy field [85–89]. Agents have created ways to assess different dimensions of the energy industry through indicators, giving them several functions and purposes [90–93]. They can be used as assessment metrics, providing a comparison between the defined objectives and their achievement, therefore assessing the effectiveness of the state of systems, which can be applied to assess the success of policies.

Indicators can be differentiated by function, serving as monitoring, guideline, or communication instrument. With a monitoring purpose, indicators evaluate the policy progress towards their objective's achievement. When indicators are used to monitor the status of an energy policy, they can be helpful for the improvement of the policy. An in progress evaluation enables the identification of the need for modifying the track to achieve the settled objectives in case the progress is not fully satisfactory.

At the beginning of the policy cycle, indicators can provide a deeper orientation of where to add more resources or where to focus while designing a policy. They may allow a deeper understanding of some elements and even help to identify relations that may not be shown through basic statistics or data. Indicators, as communication instruments, can be useful tools for policymakers to express elaborated ideas or concepts to other relevant stakeholders of the sector or out of it; e.g., the indicators issued by national governments to inform regional or local authorities.

Indicators can also be differentiated by whether they are individual or composite (also known as aggregated). The first ones are commonly used to show the presence or state of a situation or condition individually for a topic [94]. Despite the usefulness of individual indicators, sometimes the information provided may not be enough for the purposes of the evaluation or an integrative assessment of more than one element. Hence, the condition of the state of a topic may require the evaluation of more than one criterion, requiring the use of composite indicators. For some interrelated information, composite indicators are a feasible alternative. They are the result of individual indicators being compiled into a single index to summarize complex and multi-dimensional situations [95]. Some well-known examples of composite indicators in the energy field are the energy poverty and energy access indicators. For energy deployment decisions, policymakers require to know the state of the energy sources, which involves different concerns. That may require an aggregated indicator. For this example, aggregated and individual indicators may be required to contribute to providing information about the energy sources, including why and how they were successful, the social perceptions around them, together with the economic details and other relevant concerns [24].

5.2. Indicators Review Methodology

The indicators' documents search was initially done in the open browser Google. The first 50 results of the search under energy policy indicators were reviewed in detail. Many of them were about energy or indicators, however, they either analyze or propose indicators related to buildings, utilities or other very specific topics of the energy sector or are not related to energy policies. Therefore, only the documents with general energy policy goals were considered, resulting in 23 documents presented in Appendix B.

Subsequently, a complementary search was done on the academic database Sciencedirect. From the first 50 documents retrieved under Energy Policy Indicators, most of the documents retrieved, do not propose new sets of indicators (from the ones previously identified). The majority of the related documents focus on analyzing the indicators proposals done by international organizations or governments (identified through thé open sources search) or documents that focus on the analysis of the implementation of the indicators in a particular country or region.

When searching the corresponding indicators for every condensed concern through the retrieved documents, some concerns could not be matched through the recovered documents. A detailed search was done to find at least one indicator for the concerns. This led to finding additional documents listed in Appendix B. For this search, the concerns themselves were used as keywords, together with the term "indicator", "assessment" and "evaluation".

5.3. Indicator Identification

In Section 4 all the extracted concerns were organized into condensed concerns, for which the indicator identification process was followed, identifying, at least one indicator for every condensed concern. The end objective of each concern has been enlisted, and at least one indicator as an assessment metric of each objective has been identified.

Table 14 shows the indicators for the condensed concerns with an economic orientation. For all the concerns, more than one indicator has been identified.

Table 14. Identified indicators for the condensed concerns—Economic.

Concern	End Objective	Indicator	Units	Source	Definition
Affordability	Max. The capability of energy payment	Willingness to Pay	Numerical value	Measuring Willingness to Pay for Electricity [96]	Indicator of how much personal satisfaction or well-being individuals derive from different outcomes
		Spending on energy as a proportion of the household income	Percentage	Assessing the performance of renewable energy support policies with quantitative indicators [97].	Percentage of income destined to pay electric energy
Accessibility	Max. The capability of energy provision	Energy access	Aggregated value	Energy Access Score [98].	Aggregated indicator of eight subindicators
		Energy security	Vary by subindicator	Energy Trilemma Index [99].	Indicator part of the Energy Trilemma Index, assessment of Security of Supply and Energy Delivery
Economic competitiveness	Max. The economic competitiveness	Harmonized Competitiveness	Numerical value	Harmonised competitiveness indicators [100].	Indicators that include price and cost competitiveness
		Competitiveness Index	Index	The Global Competitiveness Report [101].	Index that analyzes competitiveness along 12 pillars by country
		Country Context	Vary by subindicator	Energy Trilemma Index [99].	Part of the Energy Trilemma Index. Includes Investability, Stable Regulatory environment, Air, land and water impact, RD&D and innovation, Coherent and predictable framework.

Table 15 shows the proposed indicators for the condensed concern with environmental orientation, environmental sustainability, for which three alternatives were identified; the proposed indicators for the two condensed concerns with social orientation are shown in Table 16; the indicators for the Governance effectiveness and efficacy condensed concern are shown in Table 17.

Table 15. Condensed Environmental Energy Policy Effectiveness concerns and identified indicators.

Concern	End Objective	Indicator	Units	Source	Definition
Environmental sustainability	Min. The environmental impacts	Environmental Sustainability Indicator	Vary by indicator	Environmental Sustainability Indicator [102].	Indicators on air, climate, water, nature, and human influence.
		Environmental Performance Index	Index	Environmental Performance Index [103].	Ranks 180 countries on 24 performance indicators across ten issue categories covering environmental health and ecosystem vitality
		Environmental Sustainability	Vary by subindicator	Energy Trilemma Index [99].	Part of the Energy Trilemma Index. Includes the assessment of Energy resource productivity, GHG Emission and C02 emissions
Impact on climate change	Minimize the impact on climate change	GHG emissions from energy production per unit of GDP	GHG/GDP (C02/economic unit)	Indicators for Energy Sustainable Development [89].	Composed by GHG emissions from energy production over GDP

Table 16. Condensed Social Energy Policy Effectiveness concerns and identified indicators.

Concern	End Objective	Indicator	Units	Source	Definition
Equity	Max. the consumer's welfare	Energy Equity Indicators	Vary by subindicator	Energy Equity Indicators [104].	Set of indicators that evaluates Access, investment and Resilience
		Energy Equity	Vary by subindicator	Energy Trilemma Index [99].	Includes the assessment of Access, Quality of Supply and Affordability and competitiveness
Health's impacts	Min. The energy impact on health	Accident fatalities per energy produced by fuel chain	Number of fatalities by fuel chains per energy or electricity produced annually	Energy indicators for sustainable development [87].	Annual fatalities by fuel chain

Table 17. Condensed Governance Energy Policy Effectiveness concerns and identified indicators.

Concern	End Objective	Indicator	Units	Source	Definition
Governance effectiveness and efficacy	Max. the policy effectiveness	Institutional Feasibility indicators	Vary by subindicator	Institutional Feasibility indicators [24].	Aggregated and individual indicators for six dimensions of governance
		Worldwide Governance Indicators	Percentile Rank (0 to 100)	Worldwide Governance Indicators [105].	Indicators that Summarize dataset of the quality of governance

6. Discussion

6.1. Contributions of the Review

Wide variety of documents (66 documents) that meet the eligibility criteria have been analyzed. Each of them presents an alternative to characterize energy policies. The data gathered derives from multiple sources that attempt to give a solution to the same question: how to evaluate energy policy effectiveness?

The results generated a wider perspective on the different concerns that assess energy policy effectiveness. Four core categories (Economic, Environmental, Social, and Institutional) were organized and synthesized through more specific categories (eight condensed concerns organized by their ends objective), presented in Figure 2.

Figure 2. Effectiveness characterizing condensed concerns.

It was found that five documents of the review [21,27,30,32,52] consider at least one extracted concern under each of the four core categories, being the most complete documents presented in this review. Nevertheless, no single document covered the seven condensed concerns that have been identified. Therefore, our results are an expansion of those approaches, giving the possibility to apply a more integral evaluation.

The identification of the core metrics of the concerns was also done, leading to the exploration of the energy policy indicators spectrum. The main contribution of this review is the organization of attributes focused on assessing energy policy effectiveness in a more detailed way. The Energy Trilemma [99] metrics have been very useful to identify options to assess some of the identified condensed concerns; however, the Trilemma lacks the consideration of one of the core categories resulting from this review, the Institutional one.

The potential of the data from this review gives the possibility to construct more complete and integral assessment procedures and to reframe already existing energy policy evaluation methodologies with the addition of new attributes (and its metrics).

This review can be directly useful for the national and international current governance situation, in which different levels of authority are implementing approaches for climate change mitigation, whereas EP evaluation is required, e.g., for the SDGs achievement assessment, a set of indicators currently exists [106]. However, a policy effectiveness evaluation tool or methodology (in coherence with the SDGs) may contribute to assess in more detail every policy. Therefore, this report contributes to establishing a base for the future development of a policy effectiveness evaluation methodology that supports the alignment of policies to achieve the SDGs or any other international, national, or local frameworks.

6.2. Limitations of the Review

The search considered the key search engines Elservier, Google, Google scholar, and JSTOR, so in principle, all major publications obeying the criteria were covered. However, theoretically, it could happen that the consideration of additional databases (e.g.,CORE, Bielefeld Academic Search Engine (BASE), Directory of Open Access Journals (DOAJ), Social Science Research Network (SSRN), World Wide Science, Semantic Scholar, etc.) could provide some additional results. In addition, the non-published documents (studies non-accepted for publication) have not been reviewed; however, this situation does not pose a serious threat to the validity of the review [107]. The search protocol applied has a reasonable degree of easiness and reachability, resulting in a sound number of documents that could be gathered.

The data generated is suitable to be analyzed with statistical methods or other quantitative techniques as HOMA [108] or MASEM [109]. Meta-analysis techniques have not been applied in this study, which has a qualitative nature. However, useful information may be retrieved in the future by such kind of analysis.

7. Conclusions

This review aimed to characterize the policy evaluation practices in the Energy field, with a focus on the metrics: concerns, objectives, and indicators. In addition to efficacy and efficiency, an emphasis was put into trying to find indicators that can assess effectiveness. Effectiveness enables a more holistic, field-tested, and modulated assessment of the achievements.

It was found that from the different categories (Environmental, Economic, Social, and Institutional), the Economic category is the one most often included, being present in nearly all documents (36 out of 66). Ex post is the most frequent evaluation type in terms of the moment (34 out of 66). Regarding the level of disaggregation reported for EP evaluations, local studies were found in three documents, while the national ones had the lead (40 out of 66). Considerations like the stakeholder's participation, evaluation motivation, and methodological approach varied with less focus across the gathered documents.

The extracted concerns were condensed in a more compact list that eliminated redundancies, organizing them by their end objective. This resulted in seven condensed concerns: accessibility; affordability; competitiveness; impact on climate change; other impacts on the environment; equity; and health.

A total of 15 indicators were identified as possible alternatives to fully characterize the concerns: 2 for affordability, 2 for accessibility, 3 for economic competitiveness, 3 for environmental impacts, 2 for equity, 1 for health, and 2 for governance effectiveness and efficacy. This list is a considerable expansion from what could be found in any single previous document. It is proposed as a holistic assessment of the effectiveness of energy policies. The information organized through this review could be the object of further future works, e.g., on the computation of every single indicator—either ex ante or ex post, but with the goal of enabling ex ante estimation as much as possible. This would further support the elaboration of solid policies, adapted to the local institutional and geographical contexts.

Author Contributions: Conceptualization, V.L. and D.O.; formal analysis, D.O.; funding acquisition, D.O.; investigation, D.O.; methodology, D.O. and V.L.; supervision, V.L.; validation, D.O. and V.L. visualization, D.O.; writing—original draft, D.O.; writing—review and editing, D.O. and V.L. All authors have read and agreed to the published version of the manuscript.

Funding: This research was funded by the Consejo Nacional para la Ciencia y la Tecnología or Mexico, grant number 695470 within MIT SES Portugal Doctoral Program.

Conflicts of Interest: The authors declare no conflict of interest.

Abbreviations

EP	Energy Policy
GHG	Greenhouse Effect Gases
CBA	Cost-Benefit Analysis
ROA	Return on Assets
RE	Renewable Energy
GDP	Gross Domestic Product
MCDM	Multi-criteria Decision Making

Appendix A Identification of Documents by Search Terms

Table A1. Energy policy evaluation documents adopted (Sciencedirect).

#	Name of Document	Authors
1	Proposing an evaluation framework for energy policy making incorporating equity: Applications in Australia	(Chapman et al., 2016a) [16].
2	An analytical framework for energy policy evaluation	(Chen, 2011) [18].
3	Energy efficiency policy evaluation by moving from techno-economic	(Bukarica & Tomšić, 2017) [46].
4	Why is South Korea's renewable energy policy failing? A qualitative evaluation	(Yoon & Sim, 2015) [77].
5	A multiple perspective modeling and simulation approach for renewable energy policy evaluation	(Alyamani, Damgacioglu, Celik, Asfour, & Feiock, 2016) [19].
6	Consumers' evaluation of national new energy vehicle policy in China: An analysis based on a four paradigm model perceptions of importance and satisfaction	(W. Li, Long, & Chen, 2016) [68].
7	Renewable energy policy evaluation using real option model—The case of Taiwan	(Lee & Shih, 2010) [20].
8	Analysis of energy end-use efficiency policy in Spain	(Collado & Díaz, 2017) [47].
9	Development and assessment of renewable energy policy scenarios by 2030 for Bulgaria	(Nikolaev & Konidari, 2017) [21].

Table A2. Energy policy evaluation documents adopted (Google).

#	Number of Document	Number of Document
10	Residential solar PV policy: An analysis of impacts, successes, and failures in the Australian case	(Chapman et al., 2016b) [17].
11	Conceptual and empirical advances in analysing policy mixes for energy	(Rogge, Kern, & Howlett, 2017) [66].
12	IRENA: Evaluating Renewable Energy Policy: A Review of Criteria and Indicators for Assessment	(IRENA, Nicholls, Mawhood, Gross, & Castillo-Castillo, 2014) [67].
13	Strengthening the Energy Policy Making Process and Sustainability Outcomes in the OECD through Policy Design	(Chapman et al, 2016c) [76].
14	Moving Toward Energy Efficiency: A Results-Driven Analysis of Utility Based Energy Efficiency Policies	(Theel & Westgaard, 2017) [13].
15	Evaluation of Energy Policy Instruments for the Adoption of Renewable Energy: Case of Wind Energy in the Pacific Northwest U.S.	(Abotah, 2015) [15].
16	Climate and energy policy solutions for China	(Qiang et al., 2016) [44].
17	Evaluating Consistency in Environmental Policy Mixes through Policy, Stakeholder, and Contextual Interactions	(Lieu et al., 2018) [14].
18	A Framework for Energy Policy Evaluation and Improvement Incorporating Quantified Social Equity	(A.J. Chapman, 2016) [48].
19	Evaluation of Hawaii's Renewable Energy Policy and Procurement	(Energy and Environmental Economics, 2014) [45].

Table A3. Energy Policy Assessment documents adopted (Sciencedirect).

#	Name of Document	Authors
20	Public perceptions of energy policies: Predicting support, opposition, and non-substantive responses	(Edwards, 2018) [71].
21	Ex-post assessment of China's industrial energy efficiency policies during the 11th Five-Year Plan	(Yu, Wang, Li, Qi, & Tamura, 2015) [24].
22	Energy security and renewable energy policy analysis of Pakistan	(Aized, Shahid, Bhatti, Saleem, & Anandarajah, 2018) [25].
23	Energy policy assessment at strategic, tactical, and operational levels: Case studies of EU 20-20-20 and U.S. Executive Order 13514	(Arababadi, Moslehi, El Asmar, Haavaldsen, & Parrish, 2017) [26].
24	Energy democracy: Goals and policy instruments for sociotechnical transitions	(Matthew J. Burke & Stephens, 2017) [27].
25	Efficient, equitable and sustainable energy policy in a small open economy Concepts and assessments	(Chang & Fang, 2017) [28].
26	Comprehensive effectiveness assessment of renewable energy generation policy: A partial equilibrium analysis in China	(Liu et al., 2018) [52].
27	Assessment of energy policies to promote photovoltaic generation in the European Union	(García-Álvarez, García, & Soares, 2018) [29].
28	A review of energy and power planning and policies of Pakistan	(Mirjat et al., 2017) [30].
29	An Alternative Assessment of Global Climate Policies	(Atalla, Bigerna, Bollino, & Polinori, 2018) [31].

Table A4. Energy Policy Effectiveness documents adopted (Sciencedirect).

#	Name of Document	Authors
30	Effectiveness of policy incentives on electric vehicle acceptance in China: A discrete choice analysis	(Wang, Tang, & Pan, 2017) [69].
31	Analysis of the effectiveness of national renewable energy policies: A case of photovoltaic policies	(Hafeznia, Aslani, Anwar, & Yousefjamali, 2017) [22].
32	The effectiveness of China's wind power policy: An empirical analysis	(X. Zhao, Li, Zhang, Yang, & Liu, 2016) [49].
33	The policy effectiveness of economic instruments for the photovoltaic and wind power development in the European Union	(S. J. Li, Chang, & Chang, 2017) [23].
34	The effectiveness of federal renewable policies in India	(Shrimali, Srinivasan, Goel, & Nelson, 2017) [51].

Table A5. Energy Policy Effectiveness documents adopted (Google scholar).

#	Name of Document	Authors
35	Effectiveness of policy measures in transforming the energy system	(Lund, 2007) [53].
36	U.S. state policies for renewable energy: Context and effectiveness	(Delmas & Montes-sancho, 2011) [32].
37	Which renewable energy policy is a venture capitalist's best friend? Empirical evidence from a survey of international cleantech investors	(B & Rolf, 2009) [75].
38	The effectiveness of different policy regimes for promoting wind power: Experiences from the states	(Menz & Vachon, 2006) [33].
39	The effectiveness of European energy policy on the Italian system: Regional evidences from a hierarchical cluster analysis approach	(Arbolino, Boffardi, & Ioppolo, 2019) [34].
40	Analysing the effectiveness of renewable energy supporting policies in the European Union	(Harmelink, Voogt, & Crème,2006) [72].
41	The effectiveness of policy instruments in promoting bioenergy	(Thornley & Cooper, 2008) [35].
42	A quantile approach to assess the effectiveness of the subsidy policy for energy-efficient home appliances: Evidence from Rizhao, China	(Yao, Liu, & Yan, 2014) [73].
43	Assessing the strength and effectiveness of renewable electricity feed-in tariffs in European Union countries	(Jenner, Groba, & Indvik, 2013) [36].
44	Feed-in tariff vs. renewable portfolio standard: An empirical test of their relative effectiveness in promoting wind capacity development	(Dong, 2012) [37].
45	Wind electricity in Denmark: A survey of policies, their effectiveness and factors motivating their introduction	(Agnolucci, 2007) [54].
46	Effectiveness through risk reduction: a comparison of the renewable obligation in England and Wales and the feed-in system in Germany	(Mitchell, Bauknecht, & Connor, 2006) [55].
47	Double dividend effectiveness of energy tax policies and the elasticity of substitution: A CGE appraisal	(Sancho, 2010) [38].
48	Assessing the effectiveness of city-level electric vehicle policies in China	(Qiu, Zhou, & Sun, 2019) [56].
49	Renewable energy policy effectiveness. A panel data analysis across Europe and Latin America	(Menanteau & El-methni, 2020) [39].

Table A6. Energy Policy Effectiveness documents adopted (JSTOR).

#	Name of Document	Authors
50	Policy Effectiveness in Energy Conservation and Emission Reduction	(Yuan et al, 2011) [57].
51	An ex-ante evaluation of the EU Energy Efficiency Directive—Article 7	(Rosenow, Leguijt, Pato, & Eyre, 2016) [58].
52	Effectiveness, Implementation, and Policy Diffusion: Or "Can We Make That Work for Us?	(Nicholson-crotty & Carley, 2016) [59].

Table A7. Energy Policy Evaluation documents adopted (Google Scholar).

#	Name of Document	Authors
53	Multicriteria evaluation and public participation: the case of UK energy policy	(Stagl, 2006) [40].
54	A real option model for renewable energy policy evaluation with application to solar PV power generation in China	(Zhang, Zhou, & Zhou, 2014) [60].
55	Economic and Technological Models for Evaluation of Energy Policy	(Hoffman et al., 1977) [41].

Table A7. *Cont.*

#	Name of Document	Authors
56	State renewable energy electricity policies: An empirical evaluation of effectiveness	(Carley, 2009) [42].
57	Theory-based policy evaluation of 20 energy efficiency instruments	(Harmelink, Nilsson, & Harmsen, 2008) [76].
58	Policy instruments fostering wind energy projects multi-perspective evaluation approach	(Enzensberger, Wietschel, & Rentz, 2002) [70].
59	Outcome indicators for the evaluation of energy policy instruments and technical change	(Neij & Kerstin, 2006) [61].
60	Hybrid agent-based simulation for policy evaluation of solar power generation systems	(Zhao, Mazhari, Celik, & Son, 2011) [62].
61	The evaluation of renewable energy policies across EU countries and US states: An econometric approach	(Kilinc-ata, 2016) [43].
62	An evaluation of government incentives for new energy vehicles in China focusing on vehicle purchasing restrictions	(Ma, Fan, & Feng, 2020) [72].
63	Performance evaluation of renewable energy support policies, applied on Flanders' tradable certificates system	(Verbruggen, 2009) [63].
64	Eco-labeling for energy efficiency and sustainability: a meta-evaluation of US programs	(Banerjee & Solomon, 2003) [64].
65	Evaluation and optimization of feed-in tariffs	(Kim & Lee, 2012) [65].

Table A8. Energy Policy Evaluation documents adopted (JSTOR).

#	Name of Document	Authors
66	Taxing sulphur dioxide emissions: A policy evaluation from public health perspective in China	(Ling-yun & Jia-jia, 2016) [74].

Appendix B

The indicators' documents with general energy policy goals from which the indicator identification analysis was initially done is presented in Table 1. The additional energy policy indicators' documents are registered in Table 2.

Table A9. List of documents for the Energy Policy Indicators analysis.

Name	Issued by	Source
Energy Indicators	Sustainable Energy Authority of Ireland	(Statistics, 2012) [85].
Energy Indicators for Sustainable Development: Country Studies	International Atomic Energy Agency & United Nations Department of Economic and Social Affairs	(IAEA et al., 2007) [86].
Indicators for Energy Sustainable Development Guidelines and Methodologies	International Atomic Energy Agency & World Bank and Sustainable Energy for All	(IAEA et al., 2007) [87].
Eurostat Energy indicators	Eurostat	(Eurostat, 2017) [88].
Energy Indicators	Organization of the Petroleum Exporting Countries	(OPEC, 2017) [89].
Energy indicators for electricity production	Centre for Environmental Design of Renewable Energy	(CEDREN, 2012) [90].
Renewable energy indicators 2013	The Global Carbon Capture and Storage Institute	(Global CCS Institute, 2013) [91].
Energy Efficiency Indicators	World Energy Council	(World Energy Council, 2016) [92].
Energy Efficiency Indicators: Fundamentals on Statistics	International Energy Agency	(International Energy Agency, 2014) [93].
India Energy Indicators Index	Ycharts, cloud-based investment decision-making platform.	(Ycharts, 2018) [94].
New Zealand Energy Indicators	Ministry of Business, Innovation and Employment	(Business & Employment, 2016) [95].
EEA energy and environment indicators	European Environmental Agency	(European Environmental Agency, 2005) [110].
UK Energy Sector Indicators 2015	Department of Energy and Climate change	(DECC, 2015) [111].

Table A9. *Cont.*

Name	Issued by	Source
Canada Energy Statistics	Statistics Canada	(Government of Canada, 2016) [112].
Odysee Project Database(Indicators)	Enerdata	(Enerdata, 2016) [113].
Indicators and potentials in buildings, communities, and energy systems	Consultancy Group	(Forsström et al., 2011) [114].
MENA-Energy Indicators (2017)	World Bank Group	(World Bank Group, 2017) [115].
Sweden Energy Indicators	Sweden Energy Agency	(Swedish Energy Agency, 2016) [116].
Regulatory Indicators for Sustainable Energy	Energy Sector Management Assistance Program, World Bank	(World Bank Group, 2018) [117].
Nordic Countries Energy Indicators	Nordic Energy Research	(Nordic Energy Research, 2017) [118].
Evaluating Renewable Energy Policy: Indicators	International Renewable Energy Agency	(IRENA et al. 2014) [24].
Energy efficiency and conservation indicators in Yemen	Ali M. Al-Ashwal	(Al-Ashwal, 2016) [119].
Measuring Economic Policy Uncertainty	Baker et al.	(Scott R. Baker, Nicholas Bloom, 2016) [120].

Table A10. Additional Energy Policy Indicators Documents list.

Name	Issued by	Source
Government at a Glance—2017 edition	Organisation for Economic Co-operation and Development	(OECD, 2017) [105].
Indicators of Effective Policy Development & Implementation	Schoolwide Integrated Framework for Transformation	(Stonemeier, Trader, Kaloi, & Williams, 2016) [121].
Indicators for policy management	Interagency (leaded by the United Nations Development Programme)	(UNDP, 2005) [122].
Climate policy confidence indicator	Grantham Research Institute on Climate Change and indicators	(Mcdowall, Zenghelis, & Drummond, 2016) [123].
Measuring Willingness to Pay for Electricity	Asian Development Bank	(Choynowski, 2002) [96].
World Energy Trilemma Index 2018	World Energy Council	(World Energy Council, &Wyman 2018) [99].
World Bank Data	The World Bank	(The World Bank, 2019) [124].
Harmonized Competitiveness Indicators (HCIs)	European Central Bank	(European Central Bank, 2019) [100].
Global Competitiveness Index	World Economic Forum	(World Economic Forum, 2018) [101].
Environmental Sustainability Indicator	Government of Canada	(Government of Canada, 2019) [102].
Electricity Access Data	World Bank	(The World Bank, 2019) [98].
Environmental Performance Index (EPI)	Yale, Center for Environmental Law and Policy and other int. organizations	(Yale, 2019) [103].
Energy Equity Indicators	California Energy Commission	(California Energy Commission, 2019) [125].
Greenhouse gas inventories	European Commission	(Maria, Dg, & Eea, 2018) [126].
Mitigation Goal Standard	World Resource Institute	(Levin, Owen-, Dickinson, & Barth, 2014) [127].
Policy and Action Standard	World Resource Institute	(Rich et al., 2015) [128].
Climate Action Tracker	New Climate Institute, Ecofys and Climate Analytic	(Climate Action Tracker, 2019) [129].
Baseline Emission Inventory	Covenant of Mayors	(Covenant of Mayors, 2010) [130].
Environmental Protection Indicators for California (EPIC)	California Environmental Protection Agency, the Resources Agency, the Department of Health Services	(California Environmental Protection Agency (Cal/EPA), the Resources Agency, 2004) [131].
Renewables 2018 global status report	REN21 (a multi-stakeholder network)	(Hales, 2018) [132].
Cost-Effectiveness and Productivity KPIs	KPI Drafting Group	(KPI Drafting Group, 2001) [133].
New and Renewable Technologies for Sustainable Development Book	Naim Hamdia Afgan and Maria Cristina Ramos de Carvalho	(Afgan & Carvalho, 2002) [134].
New Energy Externalities Developments for Sustainability	Paul Scherrer Institut	(Stefan Hirschberg, Christian Bauer, Peter Burgherr, Roberto Dones, Andrew Simons, 2008) [135].
Sustainability and Financial Performance of Companies in the Energy Sector in Romania	Dragos Paun	(Paun, 2017) [136].
Consumers' willingness to pay for renewable energy: A meta-regression analysis	Chunbo Ma et al.	(C. Ma, Rogers, Kragt, & Zhang, 2016) [137].
Assessing the performance of renewable energy support policies with quantitative indicators	DIA-CORE Project	(Held et al., 2014) [97].
Energy Intensity Indicator	U.S. Department of Energy's Office of Energy Efficiency and Renewable Energy	(Office of Energy Efficiency & Renewable Energy, 2019) [138].
Worldwide Governance Indicators	Kaufmann et al.	(Kaufmann et al., 2010) [104].

References

1. UNFCCC. Reporting and Review under the Paris Agreement. 2020. Available online: https://unfccc.int/process-and-meetings/transparency-and-reporting/reporting-and-review-under-the-paris-agreement (accessed on 11 November 2020).
2. UNDRR. What is the Sendai Framework? 2020. Available online: https://www.undrr.org/implementing-sendai-framework/what-sendai-framework (accessed on 9 November 2020).
3. Berlin.de. Climate-Neutral Berlin 2050. 2020. Available online: https://www.berlin.de/sen/uvk/en/climate-protection/climate-neutral-berlin-2050/ (accessed on 26 October 2020).
4. The City of Ann Arbor. *Ann Arbor's Living Carbon Neutrality Plan*; Ann Arbor City Council: Ann Arbor, MI, USA, 2020.
5. Bayside City Council. *Carbon Neutrality Action Plan 2018–2020*; Bayside City Council: Bayside City, VIC, Australia, 2017.
6. Umwelt und Natur/Green City/Climate Protection/Climate-Neutral City. Climate-Neutral City. 2020. Available online: https://www.freiburg.de/pb/Len/642986.html (accessed on 26 October 2020).
7. UNFCCC. Action on Climate and SDGs. 2020. Available online: https://unfccc.int/topics/action-on-climate-and-sdgs/action-on-climate-and-sdgs (accessed on 18 September 2020).
8. Parsons, J. *Indicators of Inputs, Activities, Outputs, Outcomes and Impacts in Security and Justice Programming*; Practice Products for the CCVRI; Conflict, Crime, and Violence Results Initiative: London, UK, 2013; pp. 1–29.
9. Cambridge Dictionary. Success. 2018. Available online: https://dictionary.cambridge.org/es-LA/dictionary/essential-british-english/success (accessed on 12 September 2018).
10. Cambridge Dictionary. Efficacy. 2018. Available online: https://dictionary.cambridge.org/es-LA/dictionary/essential-british-english/efficacy (accessed on 12 September 2018).
11. Cambridge Dictionary. Effectiveness. 2018. Available online: https://dictionary.cambridge.org/esLA/dictionary/essential-british-english/effectiveness (accessed on 12 September 2018).
12. Ernst, E.; Pittler, M.H. Efficacy or effectiveness? *J. Intern. Med.* **2006**, *260*, 488–490. [CrossRef]
13. Theel, S.; Westgaard, A. Moving Toward Energy Efficiency: A Results-Driven Analysis of Utility-Based Energy Efficiency Policies. 2017. Available online: http://search.proquest.com.ezproxy.psz.utm.my/docview/272082138?pqorigsite=summon (accessed on 3 February 2018).
14. Lieu, J.; Spyridaki, N.A.; Alvarez-Tinoco, R.; van der Gaast, W.; Tuerk, A.; van Vliet, O. Evaluating consistency in environmental policy mixes through policy, stakeholder, and contextual interactions. *Sustainability* **2018**, *10*, 1896. [CrossRef]
15. Abotah, R. *Evaluation of Energy Policy Instruments for the Adoption of Renewable Energy: Case of Wind Energy in the Pacific Northwest US*; Portland State University: Portland, OR, USA, 2015; p. 282.
16. Chapman, A.J.; McLellan, B.; Tezuka, T. Residential solar PV policy: An analysis of impacts, successes and failures in the Australian case. *Renew. Energy* **2016**, *86*, 1265–1279. [CrossRef]
17. Chapman, A.J.; McLellan, B.; Tezuka, T. Proposing an evaluation framework for energy policy making incorporating equity: Applications in Australia. *Energy Res. Soc. Sci.* **2016**, *21*, 54–69. [CrossRef]
18. Chen, C.C. An analytical framework for energy policy evaluation. *Renew. Energy* **2011**, *36*, 2694–2702. [CrossRef]
19. Alyamani, T.; Damgacioglu, H.; Celik, N.; Asfour, S.; Feiock, R. A multiple perspective modeling and simulation approach for renewable energy policy evaluation. *Comput. Ind. Eng.* **2016**, *102*, 280–293. [CrossRef]
20. Lee, S.C.; Shih, L.H. Renewable energy policy evaluation using real option model—The case of Taiwan. *Energy Econ.* **2010**, *32* (Suppl. S1), S67–S78. [CrossRef]
21. Nikolaev, A.; Konidari, P. Development and assessment of renewable energy policy scenarios by 2030 for Bulgaria. *Renew. Energy* **2017**, *111*, 792–802. [CrossRef]
22. Hafeznia, H.; Aslani, A.; Anwar, S.; Yousefjamali, M. Analysis of the effectiveness of national renewable energy policies: A case of photovoltaic policies. *Renew. Sustain. Energy Rev.* **2017**, *79*, 669–680. [CrossRef]
23. Li, S.J.; Chang, T.H.; Chang, S.L. The policy effectiveness of economic instruments for the photovoltaic and wind power development in the European Union. *Renew. Energy* **2017**, *101*, 660–666. [CrossRef]
24. Yu, Y.; Wang, X.; Li, H.; Qi, Y.; Tamura, K. Ex-post assessment of China's industrial energy efficiency policies during the 11th Five-Year Plan. *Energy Policy* **2015**, *76*, 132–145. [CrossRef]

25. Aized, T.; Shahid, M.; Bhatti, A.A.; Saleem, M.; Anandarajah, G. Energy security and renewable energy policy analysis of Pakistan. *Renew. Sustain. Energy Rev.* **2018**, *84*, 155–169. [CrossRef]
26. Arababadi, R.; Moslehi, S.; El Asmar, M.; Haavaldsen, T.; Parrish, K. Energy policy assessment at strategic, tactical, and operational levels: Case studies of EU 20-20-20 and U.S. Executive Order 13514. *Energy Policy* **2017**, *109*, 530–538. [CrossRef]
27. Burke Matthew, J.; Stephens, J.C. Energy democracy: Goals and policy instruments for sociotechnical transitions. *Energy Res. Soc. Sci.* **2017**, *33*, 35–48. [CrossRef]
28. Chang, Y.; Fang, Z. Efficient, equitable and sustainable energy policy in a small open economy: Concepts and assessments. *Energy Policy* **2017**, *105*, 493–501. [CrossRef]
29. García-Álvarez, M.T.; García, L.C.; Soares, I. Assessment of energy policies to promote photovoltaic generation in the European Union. *Energy* **2018**. [CrossRef]
30. Mirjat, N.H.; Uqaili, M.A.; Harijan, K.; Valasai, G.; Das Shaikh, F.; Waris, M. A review of energy and power planning and policies of Pakistan. *Renew. Sustain. Energy Rev.* **2017**, *79*, 110–127. [CrossRef]
31. Atalla, T.; Bigerna, S.; Bollino, C.A.; Polinori, P. An alternative assessment of global climate policies. *J. Policy Model.* **2018**. [CrossRef]
32. Delmas, M.A.; Montes-sancho, M.J. US state policies for renewable energy: Context and effectiveness. *Energy Policy* **2011**, *39*, 2273–2288. [CrossRef]
33. Menz, F.C.; Vachon, S. The effectiveness of different policy regimes for promoting wind power: Experiences from the states. *Energy Policy* **2006**, *34*, 1786–1796. [CrossRef]
34. Arbolino, R.; Boffardi, R.; Ioppolo, G. The effectiveness of European energy policy on the Italian system: Regional evidences from a hierarchical cluster analysis approach. *Energy Policy* **2019**, *132*, 47–61. [CrossRef]
35. Thornley, P.; Cooper, D. The effectiveness of policy instruments in promoting bioenergy. *Biomass Bioenergy* **2008**, *32*, 903–913. [CrossRef]
36. Jenner, S.; Groba, F.; Indvik, J. Assessing the strength and effectiveness of renewable electricity feed-in tariffs in European Union countries. *Energy Policy* **2013**, *52*, 385–401. [CrossRef]
37. Dong, C.G. Feed-in tariff vs. renewable portfolio standard: An empirical test of their relative effectiveness in promoting wind capacity development. *Energy Policy* **2012**, *42*, 476–485. [CrossRef]
38. Sancho, F. Double dividend effectiveness of energy tax policies and the elasticity of substitution: A CGE appraisal. *Energy Policy* **2010**, *38*, 2927–2933. [CrossRef]
39. Bersalli, G.; Menanteau, P.; El-methni, J. Renewable energy policy effectiveness: A panel data analysis across Europe and Latin America. *Renew. Sustain. Energy Rev.* **2020**, *133*. [CrossRef]
40. Stagl, S. Multicriteria evaluation and public participation: The case of UK energy policy. *Land Use Policy* **2006**, *23*, 53–62. [CrossRef]
41. Hoffman, K.C.; Jorgenson, D.W.; The, S.; Journal, B.; Autumn, N.; Hoffman, K.C. Economic and technological models for evaluation of energy policy published. *Bell. J. Econ.* **1977**, *8*, 444–466. [CrossRef]
42. Carley, S. State renewable energy electricity policies: An empirical evaluation of effectiveness. *Energy Policy* **2009**, *37*, 3071–3081. [CrossRef]
43. Kilinc-ata, N. Energy for sustainable development the evaluation of renewable energy policies across EU countries and US states: An econometric approach. *Energy Sustain. Dev.* **2016**, *31*, 83–90. [CrossRef]
44. Qiang, L.; Chuan, T.; Xiaoqi, Z.; Kejun, J.; Chenmin, H.; Harvey, H.; Aggarwal, S.; Busch, C.; Rissman, J.; Orvis, R.; et al. Climate and Energy Policy Solutions for China Quantitative Analysis and Policy Recommendations for the 13Th Five-Year Plan. 2016. Available online: https://energyinnovation.org/wp-content/uploads/2016/07/ChinaPolicySolutions_ExecutiveSummary_EN.pdf (accessed on 22 February 2018).
45. Energy and Environmental Economics. *Evaluation of Hawaii's Renewable Energy Policy and Procurement*; Energy and Environmental Economics, Inc.: San Francisco, CA, USA, 2014.
46. Bukarica, V.; Tomšić, Ž. Energy efficiency policy evaluation by moving from techno-economic towards whole society perspective on energy efficiency market. *Renew. Sustain. Energy Rev.* **2017**, *70*, 968–975. [CrossRef]
47. Collado, R.R.; Díaz, M.T.S. Analysis of energy end-use efficiency policy in Spain. *Energy Policy* **2017**, *101*, 436–446. [CrossRef]
48. Chapman, A.J. A Framework for Energy Policy Evaluation and Improvement Incorporating Quantified Social Equity. Ph.D. Thesis, Kyoto University, Kyoto, Japan, 2016.
49. Zhao, X.; Li, S.; Zhang, S.; Yang, R.; Liu, S. The effectiveness of China's wind power policy: An empirical analysis. *Energy Policy* **2016**, *95*, 269–279. [CrossRef]

50. Ma, S.; Fan, Y.; Feng, L. An evaluation of government incentives for new energy vehicles in China focusing on vehicle purchasing restrictions. *Energy Policy* **2020**, *110*, 609–618. [CrossRef]
51. Shrimali, G.; Srinivasan, S.; Goel, S.; Nelson, D. The effectiveness of federal renewable policies in India. *Renew. Sustain. Energy Rev.* **2017**, *70*, 538–550. [CrossRef]
52. Liu, D.; Liu, M.; Xu, E.; Pang, B.; Guo, X.; Xiao, B.; Niu, D. Comprehensive effectiveness assessment of renewable energy generation policy: A partial equilibrium analysis in China. *Energy Policy* **2018**, *115*, 330–341. [CrossRef]
53. Lund, P.D. Effectiveness of policy measures in transforming the energy system. *Energy Policy* **2007**, 627–639. [CrossRef]
54. Agnolucci, P. Wind electricity in Denmark: A survey of policies, their effectiveness and factors motivating their introduction. *Renew. Sustain. Energy Rev.* **2007**, *11*, 951–963. [CrossRef]
55. Mitchell, C.; Bauknecht, D.; Connor, P.M. Effectiveness through risk reduction: A comparison of the renewable obligation in England and Wales and the feed-in system in Germany. *Energy Policy* **2006**, *34*, 297–305. [CrossRef]
56. Qiu, Y.Q.; Zhou, P.; Sun, H.C. Assessing the effectiveness of city-level electric vehicle policies in China. *Energy Policy* **2019**, *130*, 22–31. [CrossRef]
57. Yuan, M.; Tuladhar, S.; Bernstein, P.; Lane, L. Policy Effectiveness in Energy Conservation and Emission Reduction. Strategies for Mitigating Climate Chang. *Energy J.* **2011**, *32*, 153–172. [CrossRef]
58. Rosenow, J.A.N.; Leguijt, C.O.R.; Pato, Z.; Eyre, N. An ex-ante evaluation of the EU Energy Efficiency Directive—Article 7. *Econ. Energy Environ. Policy* **2016**, *5*, 45–64. [CrossRef]
59. Nicholson-Crotty, A.S.; Carley, S. Effectiveness, Implementation, and Policy Diffusion: Or "Can We Make That Work for Us?". *State Polit. Policy Q.* **2016**, *16*, 78–97. [CrossRef]
60. Zhang, M.; Zhou, D.; Zhou, P. A real option model for renewable energy policy evaluation with application to solar PV power generation in China. *Renew. Sustain. Energy Rev.* **2014**, *40*, 944–955. [CrossRef]
61. Neij, L.; Kerstin, A. Outcome indicators for the evaluation of energy policy instruments and technical change. *Energy Policy* **2006**, *34*, 2662–2676. [CrossRef]
62. Zhao, J.; Mazhari, E.; Celik, N.; Son, Y. Simulation Modelling Practice and Theory Hybrid agent-based simulation for policy evaluation of solar power generation systems. *Simul. Model. Pract. Theory* **2011**, *19*, 2189–2205. [CrossRef]
63. Verbruggen, A. Performance evaluation of renewable energy support policies, applied on. *Energy Policy* **2009**, *37*, 1385–1394. [CrossRef]
64. Banerjee, A.; Solomon, B.D. Eco-labeling for energy efficiency and sustainability: A meta-evaluation of US programs. *Energy Policy* **2003**, *31*, 109–123. [CrossRef]
65. Kim, K.; Lee, C. Evaluation and optimization of feed-in tariffs. *Energy Policy* **2012**, *49*, 192–203. [CrossRef]
66. Rogge, K.S.; Kern, F.; Howlett, M. Conceptual and empirical advances in analysing policy mixes for energy transitions. *Energy Res. Soc. Sci.* **2017**, *33*, 1–10. [CrossRef]
67. Nicholls, J.; Mawhood, R.; Gross, R.; Castillo-Castillo, A.; IRENA. Evaluating Renewable Energy Policy: A Review of Criteria and Indicators for Assessment. 2014. Available online: http://www.irena.org/documentdownloads/publications/evaluating_re_policy.pdf (accessed on 18 January 2019).
68. Li, W.; Long, R.; Chen, H. Consumers' evaluation of national new energy vehicle policy in China: An analysis based on a four paradigm model. *Energy Policy* **2016**, *99*, 33–41. [CrossRef]
69. Wang, N.; Tang, L.; Pan, H. Effectiveness of policy incentives on electric vehicle acceptance in China: A discrete choice analysis. *Transp. Res. Part. A Policy Pract.* **2017**, *105*, 210–218. [CrossRef]
70. Enzensberger, N.; Wietschel, M.; Rentz, O. Policy instruments fostering wind energy projects a multi-perspective evaluation approach. *Energy Policy* **2002**, *30*, 793–801. [CrossRef]
71. Edwards, M.L. Public perceptions of energy policies: Predicting support, opposition, and nonsubstantive responses. *Energy Policy* **2018**, *117*, 348–357. [CrossRef]
72. Harmelink, M.; Voogt, M.; Cremer, C. Analysing the effectiveness of renewable energy supporting policies in the European Union. *Energy Policy* **2006**, *34*, 343–351. [CrossRef]
73. Yao, X.; Liu, Y.; Yan, X. Short Communication A quantile approach to assess the effectiveness of the subsidy policy for energy-ef fi cient home appliances: Evidence from Rizhao, China. *Energy Policy* **2014**, *73*, 512–518. [CrossRef]

74. Ling-Yun, H.E.; Jia-Jia, O.U. Taxing sulphur dioxide emissions: A policy evaluation from public health perspective in China. *Energy Environ.* **2016**, *27*, 755–764. [CrossRef]
75. Bürer, M.J.; Wüstenhagen, R. Which renewable energy policy is a venture capitalist's best friend? Empirical evidence from a survey of international cleantech investors. *Energy Policy* **2009**, *37*, 4997–5006. [CrossRef]
76. Chapman, A.; Mclellan, B.; Tezuka, T. Administrative sciences Strengthening the Energy Policy Making Process and Sustainability Outcomes in the OECD through Policy Design. *Adm. Sci.* **2016**, *6*, 9. [CrossRef]
77. Yoon, J.H.; Sim, K. Why is South Korea's renewable energy policy failing? A qualitative evaluation. *Energy Policy* **2015**, *86*, 369–379. [CrossRef]
78. Harmelink, M.; Nilsson, L.; Harmsen, R. Theory-based policy evaluation of 20 energy efficiency instruments. *Energy Effic.* **2008**, 131–148. [CrossRef]
79. Cambridge Dictionary. Affordability. 2019. Available online: https://dictionary.cambridge.org/es-LA/dictionary/english/affordability (accessed on 12 September 2019).
80. Cambridge Dictionary. Accessibility. 2019. Available online: https://dictionary.cambridge.org/es-LA/dictionary/english/accessibility (accessed on 12 September 2019).
81. Cambridge Dictionary. Competitiveness. 2019. Available online: https://dictionary.cambridge.org/es-LA/dictionary/english/competitiveness (accessed on 12 September 2019).
82. Cambridge Dictionary. Environmental Sustainability. 2019. Available online: https://dictionary.cambridge.org/es-LA/dictionary/english/environmentalsustainability (accessed on 12 September 2019).
83. Cambridge Dictionary. Equity. 2019. Available online: https://dictionary.cambridge.org/es-LA/dictionary/english/equity (accessed on 12 September 2019).
84. Cambridge Dictionary. Governance. 2019. Available online: https://dictionary.cambridge.org/es-LA/dictionary/english/goverance (accessed on 12 September 2019).
85. Sustainable Energy Authority of Ireland. *Energy in Ireland*; Sustainable Energy Authority of Ireland: Dublin, Ireland, 2018.
86. IAEA; UNDESA; IEA; EUROSTAT; EEA. Energy indicators for sustainable development: Guidelines and Methodologies. *Energy* **2007**, *32*, 875–882. [CrossRef]
87. International Atomic Energy Agency; United Nations Department of Economic and Social Affairs. *Energy Indicators for Sustainable Development: Country Studies*; International Atomic Energy Agency: Paris, France; United Nations Department of Economic and Social Affairs: Rome, Italy, 2007; p. 463.
88. Eurostat. *Energy, Transport and Environment Indicators*, 2017 ed.; Eurostat Pocketbook; Eurostat: Luxembourg, 2017; Volume 25; ISBN 978-92-79-74192-0. [CrossRef]
89. OPEC. Energy Indicators. 2017. Available online: http://www.ofid.org/DATA/Energy-indicators (accessed on 12 September 2018).
90. CEDREN. *Energy Indicators for Electricity Production Comparing Technologies*; Centre for Environmental Design of Renewable Energy: Trondheim, Norway; DN (Directorate for Nature Management): Trondheim, Norway, 2012.
91. Global CCS Institute. Renewable Energy Indicators 2013. Global CCS Institute: 2013. Available online: https://hub.globalccsinstitute.com/publications/renewables-2014-global-status-report/renewable-energy-indicators-2013 (accessed on 25 May 2018).
92. World Energy Council. Energy Efficiency Indicators. 2016. Available online: https://www.worldenergy.org/data/efficiency-indicators/ (accessed on 28 May 2018).
93. International Energy Agency. *Energy Efficiency Indicators: Fundamentals on Statistics*; IEA: Paris, France, 2014.
94. Ycharts. India Energy Indicators. 2018. Available online: https://ycharts.com/indicators/regions/countries/IND/indicators?category=energy (accessed on 7 June 2018).
95. Ministry of Business, Innovation & Employment in New Zealand 2016, 69. 2016. Available online: http://ci.nii.ac.jp/naid/40020762305/ (accessed on 8 September 2018).
96. Choynowski, P. *Measuring Willingness to Pay for Electricity*; Asian Development Bank: Manila, Philippines, 2002.
97. Held, A.; Ragwitz, M.; Boie, I.; Wigand, F.; Janeiro, L.; Klessmann, C.; Nabe, C.; Hussy, C.; Neuhoff, K.; Grau, T.; et al. *Assessing the Performance of Renewable Energy Support Policies with Quantitative Indicators*; Fraunhofer ISI: Karlshöhe, Germany, 2014.
98. World Bank. World Bank Data. 2019. Available online: https://data.worldbank.org/indicator (accessed on 30 January 2019).
99. Wyman, O.; World Energy Council. *World Energy Trilemma Index*; World Energy Council: London, UK, 2018.

100. European Central Bank. Harmonised Competitiveness Indicators. 2019. Available online: https://data.europa.eu/euodp/en/data/dataset/ecb-competitiveness-indicators (accessed on 30 January 2019).
101. Schwab, K.; World Economic Forum. *The Global Competitiveness Report*; World Economic Forum: Geneva, Switzerland, 2018.
102. Government of Canada. Environmental Indicators-Canada.Ca. 2019. Available online: https://www.canada.ca/en/environment-climate-change/services/environmental-indicators.html (accessed on 25 January 2019).
103. Yale. Introduction|Environmental Performance Index. 2019. Available online: https://epi.envirocenter.yale.edu/2018-epi-report/introduction (accessed on 8 January 2019).
104. Kaufmann, D.; Kraay, A.; Mastruzzi, M. *The Worldwide Governance Indicators: Methodology and Analytical Issues*; The Worlwide Governance Indicators: Washington, DC, USA; Geneva, Switzerland, 2010.
105. OECD. *Goverment at Glance 2017*; OECD Publishing: Paris, France, 2017.
106. SDG Indicators. SDG Indicators. 2020. Available online: https://unstats.un.org/sdgs/indicators/indicators-list/ (accessed on 18 September 2020).
107. Dalton, D.R.; Aguinis, H.; Dalton, C.M.; Bosco, F.A.; Pierce, C.A. Revisiting the file drawer problem in meta-analysis: An Assessment of published and non-published correlation matrices. *Pers. Psychol.* **2012**, *65*, 221–249. [CrossRef]
108. Hedges, L.V.; Olkin, I. Statistical Methods for Meta-Analysis. *J. Educ. Stat.* **1985**, *13*, 75. [CrossRef]
109. Bergh, D.D.; Aguinis, H.; Heavey, C.; Ketchen, D.J.; Boyd, B.K.; Su, P.; Joo, H. Using meta-analytic structural equation modeling to advance strategic management research: Guidelines and an empirical illustration via the strategic leadership-performance relationship. *Strateg. Manag. J.* **2016**, *37*, 477–497. [CrossRef]
110. European Environmental Agency. *EEA Core Set of Indicators*; European Environmental Agency: Copenhagen, Denmark, 2005. [CrossRef]
111. DECC. UK Energy Sector Indicators 2015, 102. 2015. Available online: https://www.gov.uk/government/statistics/uk-energy-sector-indicators-2015 (accessed on 25 September 2018).
112. Government of Canada. About Environmental Sustainability Indicators-Canada.ca. 2016. Available online: https://www.canada.ca/en/environment-climate-change/services/environmental-indicators/about-sustainability.html (accessed on 7 January 2019).
113. Enerdata. Free Energy Indicators|ODYSSEE. 2016. Available online: http://www.indicators.odyssee-mure.eu/online-indicators.html (accessed on 12 January 2019).
114. Forsström, J.; Lahti, P.; Pursiheimo, E.; Rämä, M.; Shemeikka, J.; Sipilä, K.; Tuominen, P.; Wahlgren, I. *Measuring Energy Efficiency: Indicators and Potentials in Buildings, Communities and Energy Systems*; VTT Tiedotteita-Valtion Teknillinen Tutkimuskesku: Espoo, Finland, 2011.
115. World Bank Group. MENA-Energy Indicators-Datasets-Energydata.Info. 2017. Available online: https://energydata.info/dataset/mena-energy-indicators-2017 (accessed on 15 January 2019).
116. Swedish Energy Agency. Energy Indicators in Figures 2016-Follow-up of Sweden's Energy Policy Goals. 2016. Available online: http://www.energimyndigheten.se/en/news/2016/energy-indicators-in-figures-2016---follow-up-of-swedens-energy-policy-goals/ (accessed on 25 January 2019).
117. World Bank Group. Indicators|RISE. 2018. Available online: http://rise.esmap.org/indicators (accessed on 14 January 2019).
118. Nordic Energy Research. Indicators-Nordic Energy Research. 2017. Available online: http://www.nordicenergy.org/indicators/ (accessed on 18 January 2019).
119. Al-Ashwal, A.M. *Future Cities and Environment*; Springer Open: Luxembourg, 2016. [CrossRef]
120. Baker, S.R.; Bloom, N.; Davis, S.J. Measuring Economic Policy Uncertainty. *Q. J. Econ.* **2016**, *131*, 1593–1636. [CrossRef]
121. Stonemeier, B.J.; Trader, B.; Kaloi, L.; Williams, G. *Indicators of Effective Policy Development & Implementation*; SWIFT Center: Lawrence, NY, USA, 2016; pp. 1–7.
122. UNDP. *Indicators for Policy Management*; United Nations: New York, NY, USA, 2005.
123. Mcdowall, W.; Zenghelis, D.; Drummond, P. *Climate Policy Confidence Indicator: Final Report to CCCEP*; Grantham Research Institute on Climate Change and the Environment: London, UK, 2016.
124. The World Bank. Access to Electricity (% of Population)|Data. 2019. Available online: https://data.worldbank.org/indicator/eg.elc.accs.zs (accessed on 29 January 2019).
125. California Energy Commission. Energy Equity Indicators. 2019. Available online: https://www.energy.ca.gov/sb350/barriers_report/equity-indicators.html (accessed on 10 January 2019).

126. Maria, A.; Dg, D.; Eea, R.F. *EEA Report/2018 Annual European Union Greenhouse Gas Inventory 1990–2016 and Inventory Report 2018 Submission to the UNFCCC Secretariat*; European Environment Agency: Luxembourg, 2018.
127. Levin, K.; Owen, R.; Dickinson, J.; Barth, K.-K. *Mitigation Goal Standard*; World Resources Institute: Washington, DC, USA, 2014.
128. Rich, D.; Finnegan, J.; Owen-Jones, R.; García-Guerrero, A.; Dickinson, J.; Barth, K. *Policy and Action Standard*; World Resources Institute: Washington, DC, USA, 2015.
129. Climate Action Tracker. Scaling Up Climate Action Series|Climate Action Tracker. 2019. Available online: https://climateactiontracker.org/publications/scalingup/ (accessed on 24 January 2019).
130. Covenant of Mayors. *How to Develop a Sustainable Energy Action Plan*; Covenant of Mayors: Brussels, Belgium, 2010.
131. California Environmental Protection Agency (Cal/EPA) the Resources Agency; Department of Health Services. *Environmental Protection Indicators for California (EPIC)*; California Environmental Protection Agency: Sacramento, CA, USA, 2004.
132. Hales, D. *Renewables 2018 Global Status Report*; Renewable Energy Policy Network for the 21st Century: Paris, France, 2018.
133. KPI Drafting Group. *Working Paper Cost Effectiveness and Productivity KPIs*; Version 1; EUROCONTROL: Brussels, Belgium, 2001.
134. Afgan, N.H.; de Carvalho, M.C.R. *New and Renewable Technologies for Sustainable Development*; Springer Science & Business: New York, NY, USA, 2002.
135. Hirschberg, S.; Bauer, C.; Burgherr, P.; Dones, R.; Simons, A.; Bachmann, T.; Carrera, D.G. *New Energy Externalities Developments for Sustainability. Deliverable n° D3. 2-RS 2b Final Set of Sustainability Criteria and Indicators for Assessment of Electricity Supply Options*; Paul Scherrer Institut: Villigen, Switzerland, 2008.
136. Paun, D. Sustainability and Financial Performance of Companies in the Energy Sector in Romania. *Sustainability* **2017**, *9*, 1722. [CrossRef]
137. Ma, C.; Rogers, A.A.; Kragt, M.E.; Zhang, F.F.; Polyakov, M.; Gibson, F.; Chalak, M.; Pandit, R.; Tapsuwan, S. Consumer Willingness to Pay for Renewable Energy: A Meta Analysis. Master's Thesis, Norwegian University of Life Sciences, Ås, Norway, 2015.
138. Office of Energy Efficiency & Renewable Energy. Energy Intensity Indicators|Department of Energy. 2019. Available online: https://www.energy.gov/eere/analysis/energy-intensity-indicators (accessed on 13 January 2019).

energies

MDPI

Article

A Study of Design Variables in Daylight and Energy Performance in Residential Buildings under Hot Climates

Ali Mohammed AL-Dossary and Daeung Danny Kim *

Architectural Engineering Department, King Fahd University of Petroleum and Minerals (KFUPM), Dhahran 31261, Saudi Arabia; Ali.moh.aldossary@gmail.com
* Correspondence: dkim@kfupm.edu.sa

Received: 19 September 2020; Accepted: 4 November 2020; Published: 9 November 2020

check for updates

Abstract: In Saudi Arabia, residential buildings are one of the major contributors to total energy consumption. Even though there are abundant natural resources, it is somewhat difficult to apply them to building designs, as design variables, due to slow progress and private issues in Saudi Arabia. Thus, the present study demonstrated the development of sustainable residential building design by examining the daylighting and energy performance with design variables. Focusing on the daylighting system, the design variables were chosen, including window-to-wall ratios (WWR), external shading devices, and types of glazing. The illuminance level by these design variables in a building was evaluated by using daylight metrics, such as spatial daylight autonomy and annual sunlight exposure. Moreover, the building energy consumption with these design variables was analyzed by using energy simulation. As a result, the daylighting was improved with the increase in WWRs and the tinted double glazing, while these design options can cause overheating in a residential building. Among types of glazing, the double pane windows with a low-E coating showed better energy performance. Based on the results, it is necessary to find the proper design variables that can balance the daylighting and energy performance in residential buildings in hot climates.

Keywords: daylighting; energy consumption; design variables; residential building; hot climates

1. Introduction

Energy has become a global concern in developed, as well as developing, countries. The International Energy Agency (IEA) indicated that 81% of the world's total energy was primarily supplied by fossil fuels, which are depletable resources [1]. Buildings have consumed a large share of global energy and they have contributed to about 33% of the greenhouse gas (GHG) emissions [2]. The GHGs emitted by buildings have a detrimental impact on the environment. Subsequently, this issue has attracted the attention of scientists as well as public attention [3,4]. Energy conservation has also received great attention in Saudi Arabia. The building industry in Saudi Arabia has experienced major developments with a rapidly escalating population [5]. In addition, a high level of economic growth in Saudi Arabia has caused a vigorous expansion of infrastructure, in that buildings, including residential, commercial, and governmental buildings over the last two decades, were highly demanded [6]. Eventually, Saudi Arabia needs to pay more attention to energy, both in terms of energy resources and energy usage in buildings.

In Saudi Arabia, the building sector comprised 79% of the total electricity consumption [7]. Specifically, about 50% of the total building energy was consumed by residential sectors [8]. As the main contributor to the total energy consumption in Saudi Arabia, the energy efficiency in residential buildings has become a main concern. Several studies have been conducted to improve energy efficiency in buildings, in that a number of strategies have proposed. These strategies have mainly dealt with the different building systems, including envelope, HVAC (Heating, ventilation, and air- conditioning), and lighting. These were: (1) improvement of thermal resistance in building envelopes [9–11]; (2) application of advanced window systems [12,13]; (3) application of more efficient lighting bulbs [14,15]; (4) and the use of energy-efficient mechanical systems [16]. Other strategies included the utilization of renewable energy technology [17], daylight systems [18,19], and natural ventilation techniques [20]. Since most buildings in Saudi Arabia are heavily dependent on mechanical systems to maintain thermal comfort, as well as to provide lighting, it is necessary to find a proper strategy to reduce building energy consumption.

In addition to the strategies above, several studies have performed investigations to improve energy efficiency in residential buildings in Saudi Arabia. In a study by Taleb and Sharples, energy use patterns in apartment buildings were analyzed, and energy efficiency was improved by applying various measures, such as improved building thermal insulation and external shading devices [5]. In the case of the retrofitting project for villas by Mejjaouli and Alzahrani, they applied various types of mechanical systems and building envelope components as retrofitting measures to improve the energy efficiency of villas [21]. In addition, Waleed et al. performed experiments and simulations to investigate the thermal performance of building materials for walls of residential buildings [22]. A similar study was conducted by Alaidroos and Krarti [23]. In their study, the impact on energy consumption by various components of building envelopes, including thermal insulation, window shading, and glazing types, was investigated to develop an optimum building envelope system for residential buildings. Moreover, the advanced air-conditioning system was applied to residential buildings to reduce the energy consumed by mechanical systems [24]. As can be shown, most studies focused on the improvement of building envelopes or high energy-efficient mechanical systems for energy-efficient residential buildings. Considering the reduction of greenhouse gas emission, passive design solutions should be implemented more than active design strategies.

Among passive design strategies, daylighting is one of the effective design solutions for improving building energy efficiency. According to the study by Li and Liam, daylighting in a building can satisfy the human visual response, and it can make a more attractive and pleasing environment [25]. By installing daylight responsive control systems in an office building, considerable energy savings were achieved [26]. Furthermore, Do et al. proposed the use of semi-transparent solar cell window systems with daylight dimming systems for residential buildings to obtain the opportunity of energy saving [27]. A similar study for utilizing daylighting systems was conducted by Reffat and Ahmad [28]. In their study, energy performance by various daylighting systems in an office building was extensively investigated to reduce energy consumption for cooling. To summarize, the use of a daylighting system in buildings can provide benefits regarding visual comfort as well as potential for energy consumption reduction.

Daylighting is one of the most abundant natural resources in Saudi Arabia. If the admitted daylighting in the building is well-controlled, it can provide great potential for energy-saving by reducing energy consumption for artificial lighting, heating, etc., while uncontrolled daylighting can cause overheating and glare [29–31]. Considering the substantial advantage of daylighting, there were a few studies for the investigation of both daylighting and energy performance in residential buildings in Saudi Arabia. For the present study, the daylighting and energy performance by the design variables of residential buildings were investigated. The outcomes of the study were used to develop an energy-efficient residential building design considering daylighting performance.

2. Principal Design Variables for Daylighting in a Residential Building

There have been many design factors influencing daylighting performance in buildings. The daylighting performance is highly influenced by the geometries of buildings and rooms. One of the most important parameters is a window-to-wall ratio (WWR) that is designed to increase daylight admission. In general, higher WWRs can increase the quantity of daylight in the buildings, while it can cause unwanted heat gains. According to the study by Abel et al., WWRs should be properly designed in the early design stage to prevent overheating in buildings [32]. In addition to their study, the function of WWRs can be varied by the shading design. A similar point was observed in the study of Li et al [33]. In their study, daylighting and energy performance of key design variables, including shading devices, window areas, and glass types in a residential building, were investigated. Another study for the daylighting performance with a shading device, such as a solar screen, was conducted by Sherif et al. [30]. The illuminance levels by the solar screen were measured and some design suggestions were proposed. Based on the literature, the design variables, such as WWRs, shading devices, and windows, are the most influential key features of the daylighting system. Moreover, they can play a significant role in thermal behaviors in buildings. To prevent overheating through the daylighting system, as well as improve energy efficiency in a residential building under hot climate conditions, it was suggested to conduct both daylighting and energy simulations to optimize the design of WWRs, shadings, and the glazing types [34,35]. Thus, the daylighting and energy performance by design variables for the present study were evaluated by the simulation tools.

As mentioned above, it is imperative to mention the importance of the selection of glazing since the quality of daylighting and thermal behaviors in a building can be significantly altered by glazing types of window systems. In general, the application of energy-efficient glazing, such as double-pane windows, has been used in buildings under hot climate conditions, to reduce heat gain by about 15–20% [36]. In addition, a higher reduction for heat gain can be achieved by low-E coated glass. Focusing on daylighting and thermal performance, several glazing types were tested by Taleb and Antony [37]. Based on their results, tinted glazing can lower the cooling load by about 20%. Special gases, such as krypton and xenon between the layers of double glazing, can also lower the additional cooling load. In case of the study of Liu et al., thermal and daylight performance of triple glazing was also investigated [38]. Even though the energy consumption by triple pane windows is always lower than that by double glazing units, the triple-pane windows are not practically used due to the high initial costs. Thus, it is crucial to select the proper glazing for window systems, considering a balance between daylighting and energy aspects [12]. Based on the previous studies, various types of glazing were selected for the present study.

3. Methodology

3.1. Building Description

In Saudi Arabia, there are three types of residential buildings: apartments, villas, and traditional houses. Among these buildings, apartment buildings and villas have become the most popular residential buildings, accounting for about 80% of total residential buildings. For the present study, a typical villa was chosen as a reference building in the eastern region of Saud Arabia, in which the latitude and longitude are 26.2361° N and 50.0393° E, respectively. The highest temperature was observed in July (about 44 °C). In addition, the average global horizontal irradiance in the eastern area was 5874 W/m^2. For the reference building, the total floor area was 590 m^2 and the window-to-wall area ratio (WWR) was 15%. The floor to floor height was 3.5 m and the building area for the ground floor and the first floor were 300 m^2 and 290 m^2, respectively. Moreover, the reference building had 14 rooms and two kitchens. This reference building was occupied for a whole year, from 8 a.m. to 6 p.m. on all weekdays. The specification of the building envelopes is presented in Table 1. For the air conditioning system, a single packaged rooftop electric cooling unit was used to provide 15 tons of nominal cooling. For the building envelopes system, the lower thermal transmittance value was considered for the roof system than the wall system to reduce heat gain by the sun.

Table 1. The specification of building envelopes and the systems for the reference building.

Component and System	Specification
Walls	U-value: 0.54 W/m^2K
Roof	U-value: 0.24 W/m^2K
Window systems	Single clear glazing (U-value: 6.08 W/m^2K, SHGC:0.8, visible transmittance: 0.9)
Shading	No external shading
Air infiltration	0.57 ACH @ 50 PA
Internal heat gain	8 occupants Lighting: fluorescent lamps (lighting power density: 27.3 W/m^2) Equipment: 13 W/m^2
Design temperature	19 °C for cooling and 20 °C for heating
HVAC	Packaged unit air-conditioning system (Capacity: 15 tons, COP: 3.28)

3.2. Design Variables

Since the admitted sunlight is highly influenced by the design features of window systems, the design variables of window systems were chosen for the analysis of daylighting and energy performance in a residential building. Table 2 presents three different design variables of window systems. The first variable is the WWR. Since the WWR of the reference building was about 15%, three different WWRs (25%, 50%, and 70%) were used to assess the daylighting and energy performance. Another important design variable is a shading device. To admit sunlight effectively, it is imperative to use external shading devices. Focusing on external shadings, the impact on the daylighting and energy performance by overhangs and fins were investigated. As can be shown in Table 2, two different projection lengths of overhangs, and overhangs with fins, were designed for the analysis. For the reference building, a single clear glazing was used and the basic properties of the single clear glazing are presented in Table 1. The third design variable was the glazing types To figure out the improvement of daylighting and energy use in a residential building, four different types of glazing were selected.

Table 2. The design variables.

Design Variable	Specification		
WWR	(1) 25%,	(2) 50%,	(3) 70%
External shading device	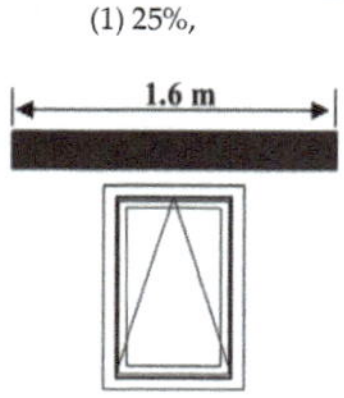 Overhang	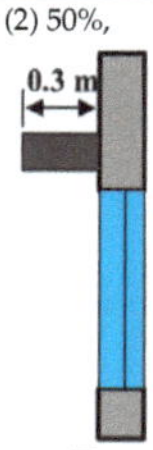 a. Type A (Projection length:0.3 m)	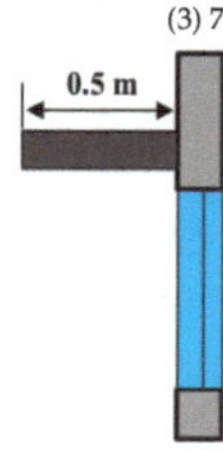 b. Type B (Projection length:0.5 m)
	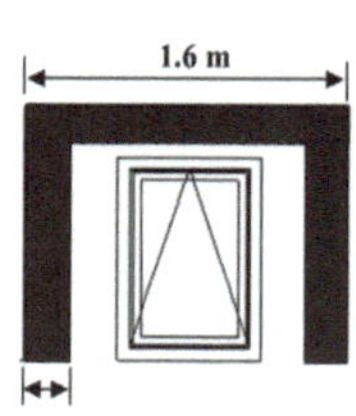 Overhang + Fin	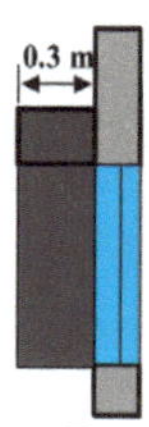 c. Type C (Projection length:0.3 m)	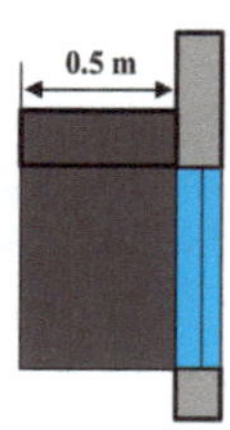 d. Type D (Projection length:0.5 m)

Table 2. *Cont.*

Glazing	(1) Tinted single glazing (2) Air-filled double glazing (3) Air-filled double glazing with a low-E coating (4) Tinted double glazing

3.3. Assessment of Daylighting and Energy Performance

For the present study, the design variables were tested by using the building performance analysis software provided by AUTODESK (Version 2019, New York, NY, USA) for a residential building satisfying both daylighting and energy performance [39]. This software is a plug-in module for Revit, enabling the assessment of both daylighting and energy consumption in a building [39]. Figure 1 shows a model of the reference building created by Revit.

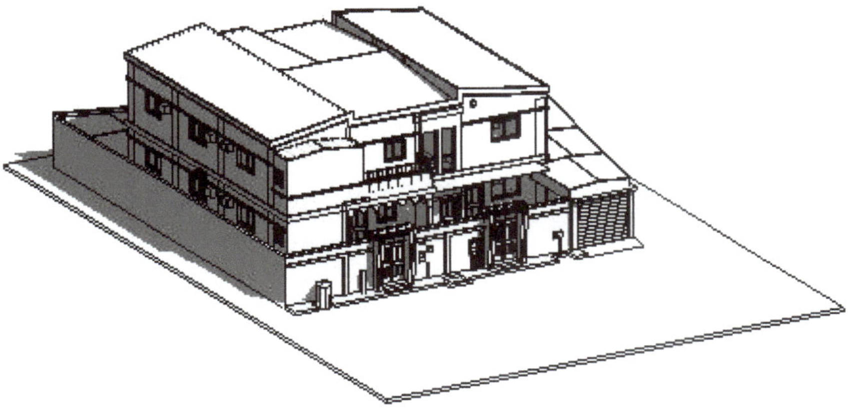

Figure 1. The reference building created by Revit.

To evaluate the daylighting performance, two daylighting metrics were implemented: annual Sunlight Exposure (aSE) and spatial Daylight Autonomy (sDA). Specifically, aSE is the percentage of the area in the space where the direct sunlight illuminance is greater than a specified level, while sDA is the percentage of the area in the space where the daylight illuminance is greater than the target level for more than a specified number of occupied hours in a year [40]. The illuminance threshold for the aSE was set to 1000 lux, while the number of annual operation hours exceeding 1000 lux should be lower than 250 h to prevent glare [41]. For the sDA metrics, the target illuminance was set to 300 lux, in which the value should be retained about 50% of the occupied hours (8 a.m. to 6 p.m.) from January 1 to December 31 [40]. For the present study, the simulation for the daylighting performance was evaluated by using these two measures. In addition, the suggested values for these two measures were presented in Table 3.

Table 3. The suggested values for $aSE_{1000,250h}$ and $sDA_{300/50\%}$.

Metrics	Value	Ref.
$aSE_{1000,250h}$	<10%: Accepted <7%: Neutrality <3%: Preferred	[40–42]
$sDA_{300/50\%}$	>75%: Preferred 55–74%: Nominally accepted	[40,42,43]

For the evaluation of the energy performance by design variables, the energy analysis module of the Revit was employed. Before the analysis, it is essential to validate the computational conditions of energy simulation. Using the weather file of Saudi Arabia provided by EnergyPlus and the specified building condition in Table 1 [44], the energy simulation with the reference villa was conducted through the energy analysis module of Revit. The monthly energy consumption obtained from the simulation was compared with the energy consumption of the reference building by using the coefficient of variation of the root mean squared error (CV(RMSE)) provided by ASHRAE Guideline 14 [45]. The models will be declared calibrated if they produce CV(RMSE)s within ± 15% with monthly energy data.

$$RMSE = \sqrt{\frac{\sum_{i=1}^{n}(M_i - S_i)^2}{n}} \quad (1)$$

$$C_V(RMSE) = \frac{RMSE}{M_{avg}} \times 100 \quad (2)$$

where M_i is the energy consumption of the residential building, while S_i is the monthly energy consumption by energy simulation. N is the period and M_{avg} is the average for the energy consumption of the residential building. After the validation, the annual and artificial lighting energy consumption of each design variable was compared.

4. Result

4.1. Assessment of Daylighting Performance

Using the values for $aSE_{1000,250h}$ and $sDA_{300/50\%}$, the daylighting performance of design variables was assessed. For this assessment, three WWRs, four different external shading designs, and four different types of glazing were applied.

4.1.1. WWRs

The reference building had a WWR of 15% with a sing pane glass. As with the WWR (15%) of the reference case, it was difficult to reach the preferred or accepted levels of daylight metrics, in which $sDA_{300/50\%}$ was less than 20%, while the value of $aSE_{1000,250h}$ was about 5% (Figure 2). When the WWR was increase by 25%, 50%, and 70%, the values of $sDA_{300/50\%}$ were increased to 38%, 58%, and 75%, respectively. Among them, only the WWR of 70% reached the preferred range of $sDA_{300/50\%}$. In addition, the WWR of 50% reached the nominally accepted range of $sDA_{300/50\%}$. For the values of $aSE_{1000,250h}$, it increased gradually, as with the increase in the WWRs. However, the WWR of 70% exceeded 10% of the value of $aSE_{1000,250h}$. Even though the WWR of 70% can provide a reliable daylight level, there is also a risk of glare or overheat due to the excessive admitted daylight in a building. Considering the results, the WWR of 50% can thus provide reliable daylight by preventing glare. While the increase in the WWRs showed improvement by satisfying with the accepted or preferred ranges of $sDA_{300/50\%}$., it also has potential for glare.

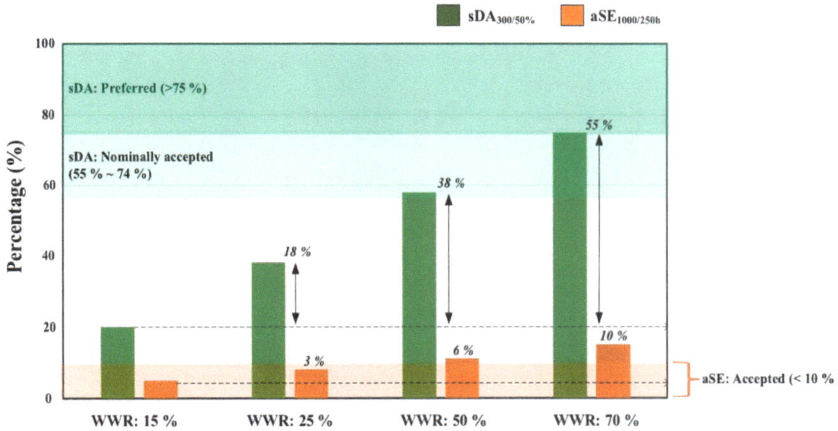

Figure 2. Daylighting performance by WWRs.

4.1.2. External Shadings

Generally, shading devices have been used to control daylighting in a building to prevent glare and overheat by blocking direct sunlight. It is designed externally for building design in the early design stage as an essential design variable. In this view, the daylighting performance of four different designs of external shading devices was evaluated, as shown in Figure 3. For this analysis, the WWR of 50% with a single pane glass was used as a reference case to figure out whether the daylighting level by the shading device was satisfied with the accepted values of $sDA_{300/50\%}$ and $aSE_{1000,250h}$. Largely, the values of $sDA_{300/50\%}$ were decreased when applying four shading devices. Consequently, all the cases were not satisfied with the accepted range of $sDA_{300/50\%}$. Specifically, about 8% to 10% of the decrease in $sDA_{300/50\%}$ was observed when two horizontal overhang designs were applied (Type A and Type B). When the overhang with fins was used, more than 15% of the $sDA_{300/50\%}$ was decreased compared with the reference case. For the $aSE_{1000,250h}$, all external shading designs showed that the illuminance level was within the accepted ranges. Moreover, the best performance regarding the $aSE_{1000,250h}$ was observed with the shading designs of Type C and Type D. Even though the use of external shading devices can decrease the illuminance level, it can prevent glare and overheat. Considering the results, Type A showed the best overall performance.

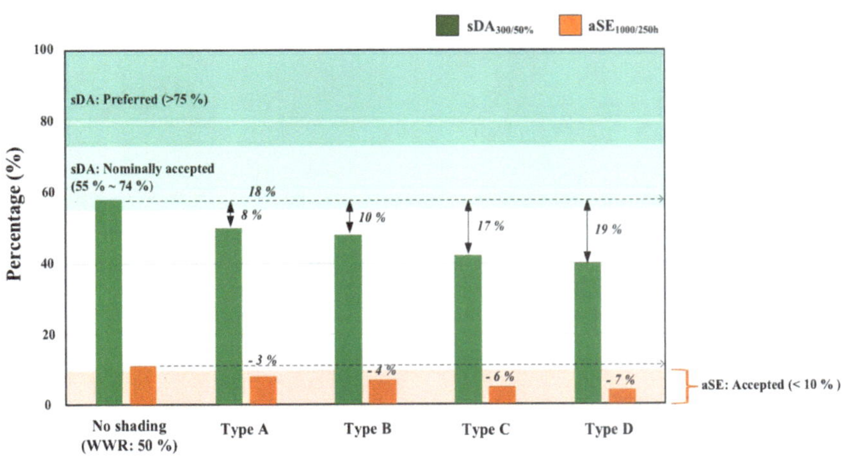

Figure 3. Daylighting performance by different designs of overhangs and overhangs with fins.

4.1.3. Types of Glazing

In general, the double-glazed windows with low-E coating have been applied to buildings to reject heat in buildings. However, most residential buildings in Saudi Arabia have equipped with a single pane of glass. For the present study, the daylighting performance of four different types of glazing was compared with that of a single pane glass in the reference villa, where the WWR was 50%.

As shown in Figure 4, each glazing showed different values of $sDA_{300/50\%}$ and $aSE_{1000,250h}$. Among them, about 5% of the decrease in $sDA_{300/50\%}$ was observed, when a tinted single glazing was used. In addition, a similar trend was observed, when a low-E coated double glazing was used. In the case of the air-filled double-glazed clear windows, there was a little impact on $sDA_{300/50\%}$. The lowest illuminance level was observed by the air-filled tinted double glazing among all the cases, which was about 10% decrease in $sDA_{300/50\%}$. A similar trend was observed among all the cases regarding the values of $aSE_{1000,250h}$. In sum, a single tinted glass and the double-glazed clear windows provided a somewhat accepted illuminance level for the $sDA_{300/50\%}$, while the illuminance by all the cases was within the accepted range of the $aSE_{1000,250h}$.

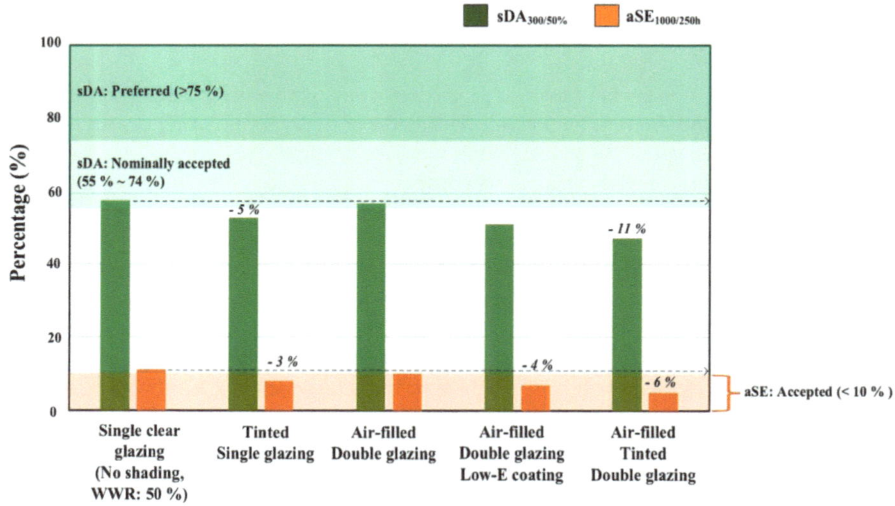

Figure 4. Daylighting performance by different types of glazing.

4.2. Energy Performance Assessment of Design Variables

4.2.1. The Monthly Energy Consumption Comparison between the Data of the Reference Villa and Energy Simulation

Before the energy performance assessment of the design variables, the computational conditions of energy simulation should be validated. For the validation, the monthly energy consumption of the reference villa was compared with the energy consumption prediction by the simulation. The total energy consumption of the residential building was about 76.2 MWh, while about 84 MWh of the energy consumption was predicted by the energy simulation, in which the difference between them was about 9%. As shown in Table 4, the largest difference was observed in the winter from November to February, and the root-mean-squared errors (CV(RMSEs) ranged from 0.46 to 3.64. Since these results were within the acceptable range, the predicted results by the simulation met the requirement by ASHRAE Guideline 14 [45].

Table 4. The monthly energy consumption comparison.

Month	Energy Consumption (MWh)			CV(RMSE) (%)
	The Reference Villa, 2019	Energy Simulation	Difference	
January	4.1	5.0	−0.9	2.05
February	3.9	5.5	−1.6	3.64
March	4.9	5.7	−0.8	1.82
April	6.0	6.5	−0.5	1.14
May	7.6	7.4	0.2	0.46
June	7.7	7.9	−0.2	0.46
July	8.6	8.3	0.3	0.68
August	8.9	8.6	0.3	0.68
September	7.8	8.4	−0.6	1.46
October	7.2	7.9	−0.7	1.59
November	5.1	6.5	−1.4	3.19
December	4.4	6.0	−1.6	3.64

4.2.2. The Energy Performance of Design Variables

One of the most important considerations for the design features of the daylighting system is to control thermal performance in buildings. Considering this point, energy consumption for the annual energy consumption, artificial lighting, and the HVAC system were presented.

As shown in Figure 5, the energy consumption in the reference residential building for each WWR was presented. The reference building was equipped with a single pane window without external shading devices and the WWR was 15%. The other conditions were specified in Table 1. As the WWR increased, the energy consumption for the HVAC system operation increased due to the increase in cooling demand. With the WWR of 75%, about 30% of energy for the HVAC system increased compared with the energy consumption of the reference building. Consequently, the total energy consumption also increased with the WWR increase. However, some energy consumption was offset by the energy consumption reduction of artificial lighting due to daylight. Comparing the artificial lighting energy consumption for the reference case, the energy consumption for artificial lighting was decreased to 30% and 40% for the WWR of 50% and 70%, respectively. Even though the annual energy consumption increased with the increase of the WWRs, the smallest increase in the annual energy consumption was about 5% with the WWR of 25%. In addition, only a 2% difference in the annual energy consumption was observed between the WWR of 25% and 50%.

Figure 6 showed the energy consumption comparisons in the reference building with the application of four different shading devices. For the reference building, no shading device was equipped. As can be shown, the annual energy consumption and energy consumed by artificial lighting and HVAC systems were reduced as the projection lengths of the overhang was increased. In addition, better energy performance was observed when the overhang with fins was applied (Type C and Type D), which was about a 30% decrease in annual energy consumption. Among the cases, Type D showed the best overall energy performance, in which about 25% of the HVAC system load was reduced, compared with the reference case. Moreover, the highest reduction for the annual energy consumption was observed. For all of the cases, a slight energy consumption difference was observed for artificial lighting energy consumption.

Moreover, the energy performance by the different types of glazing was also investigated as shown in Figure 7. For this analysis, the reference building equipped a single pane glass with a WWR of 15%. With the use of the double-glazed windows with low-E coating, the annual energy consumption decreased by about 35%, which was the lowest annual energy consumption among the cases. The second-lowest annual energy consumption was observed when the tinted double glazing was used (about 27% decrease). However, only a 5% difference in the annual energy consumption was observed between the tinted double glazing and the air-filled double glazing. Considering this point, it can be seen that the low-E coating has a significant impact on thermal performance. A similar trend was observed for HVAC energy consumption by different types of glazing. In addition, there was little difference in artificial lighting energy consumption among the cases.

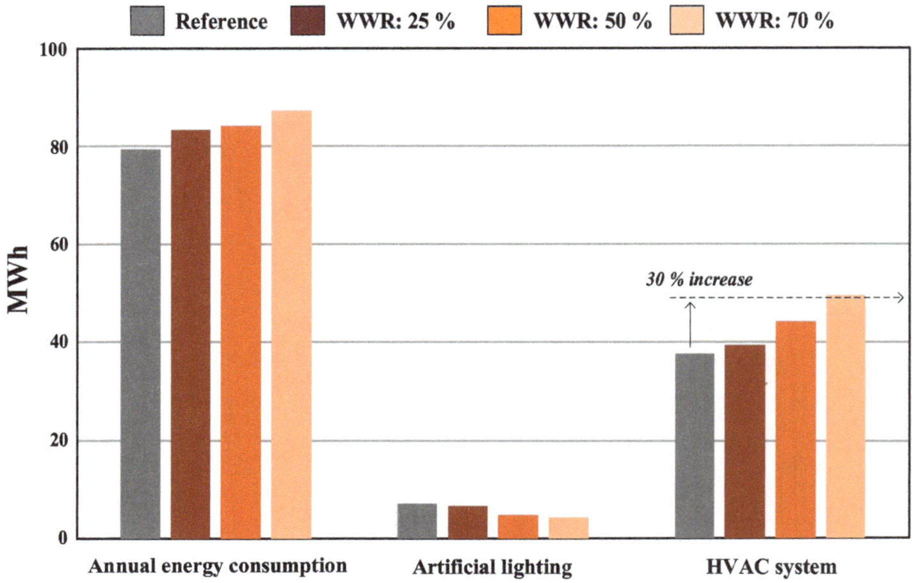

Figure 5. The energy consumption by different WWRs.

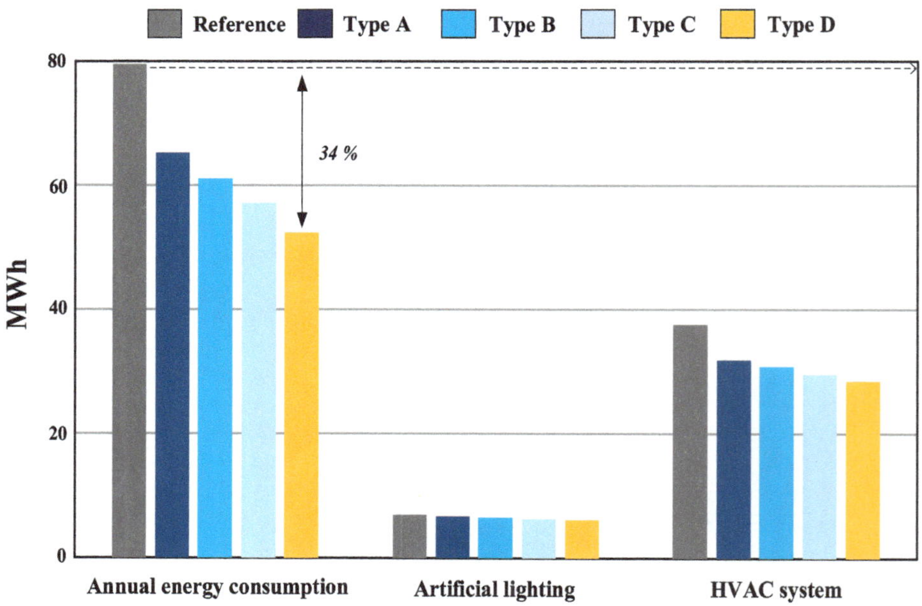

Figure 6. The energy consumption by different designs of external shadings.

In sum, more than 30% of the annual energy consumption was reduced by applying the overhang with fins and the air-filled double low-E glazing. For the WWRs, the annual energy consumption was increased as with the increase in the WWRs. Based on the result of the daylighting performance analysis, the WWR of 50% provided the acceptable range of both sDA and aSE. Thus, the WWR of 50% was chosen. By applying the selected design variables, the energy consumption was compared with that of the reference building (Figure 8). As a result, about 35% of annual energy consumption was reduced compared with that of the reference villa. In addition, 17–20% of the energy consumption was reduced for the artificial lighting and HVAC system.

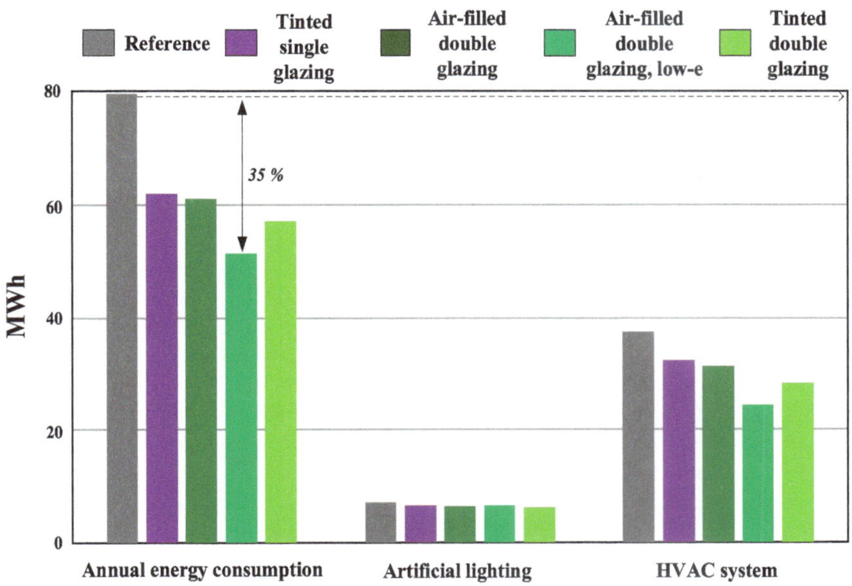

Figure 7. The energy consumption by different types of glazing.

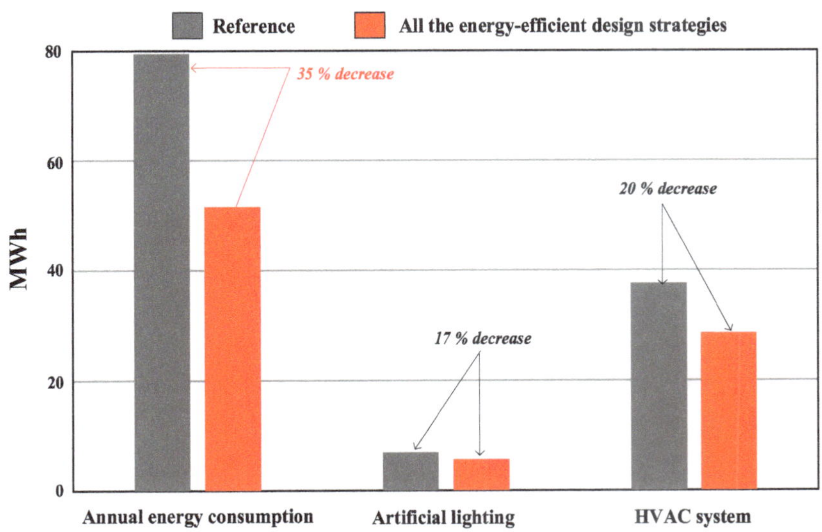

Figure 8. The energy consumption comparison between the reference villa and the villa with energy-efficient design strategies.

5. Discussion

For the present study, the daylighting and energy performance of design variables for residential buildings under hot climates were investigated. Contrary to typical residential buildings in other countries, the residential buildings in Saudi Arabia have relatively small WWRs to reduce cooling loads, and for other reasons, such as cultural characteristics [46]. Subsequently, it is, thus, difficult to find out daylighting systems in current residential buildings. As mentioned previously, the important functions of daylighting systems are to control the quality of admitted daylight as well as reduce heat gain in buildings. In this view, the daylighting performance of various sizes of WWRs, shading devices, and types of glazing was examined by using daylighting metrics. In addition, the energy performance of these design variables was evaluated. Among these design variables, windows with higher WWR can provide more illuminance and more daylighting distribution in the room [47]. Thus, higher WWRs have shown more acceptable daylighting performance, while the potential for glare or overheating was also increased. Similarly, the result of the present study showed that the daylighting performance was improved as the WWR increased. However, the energy consumption also increased. A similar result was observed in the study of Alghoul [48] et al. An increase in WWRs can cause significant cooling loads. In the case of the study of Xue et al., they investigated the cooling energy performance by increasing WWRs [49]. Their result showed that the cooling load was increased as with the increase in WWRs. According to the study of Mangkuto et al., larger WWR can cause increasing cooling loads and rising glare in a building [50]. They pointed out that different energy consumption and daylighting performance can be caused by the same WWR through different window configurations. For the present study, a similar trend was observed that there was a difference in daylighting and energy performance by same design variables. For example, the annual energy consumption increased as the WWR increased, while the increased WWR can provide better daylighting performance. It can be seen that the evaluation result of either daylighting or energy performance can cause visual or thermal discomfort for occupants. The daylighting performance in the current study was only analyzed by two daylighting metrics: sDA and aSE. Due to insufficient information on daylighting performance, it is difficult to find the proper design variables satisfying both energy and daylighting performance. Moreover, the other difficulty was from the limitation of the range of design variables. Based on the obtained results, the daylighting and energy performance was sensitive to the design variables. Therefore, it is necessary to apply various daylighting metrics into the daylighting analysis. It also requires to have more configurations of design variables for further study to design more proper daylight systems.

Regarding the assessment results for various types of glazing, the windows with low-E coating showed better performance for preventing heat gain than other windows. This was also investigated by the study of Huang et al. [36]. In their study, low-E glazing was the best design option for reducing cooling loads, while maintaining a satisfactory daylighting level compared with double-pane clear glazing. According to the study of Leung et al., the aerogel glazing showed better performance than the windows with low-E coating regarding the thermal performance in buildings [51]. Moreover, krypton or xenon gas-filled double-glazing systems can improve lighting performance in building under hot climate conditions [37]. Some studies mentioned the importance of the selection of the color of the glazing since it affects significantly overall comfort in buildings [52]. According to the study by Chen et al., the color of the tinted glazing had an impact on occupant behavior and green glazing can improve working performance [53]. For the present study, only bronze glazing and double low-E glazing were used. Considering the current building industry in Saudi Arabia, it is still difficult to apply advanced glazing systems for residential buildings. Based on the obtained results, the tinted double glazing has shown the second-best performance in energy consumption. For more accurate analysis, it is thus imperative to analyze the economic impact of the tinted and the low-E coated double glazing.

6. Conclusions

In Saudi Arabia, residential buildings account for more than half of total buildings, and the energy consumed by residential buildings has become one of the most significant concerns. To improve energy efficiency in residential buildings, several studies have mainly investigated the performance of mechanical systems, because of high cooling demand due to hot climates in Saudi Arabia. In addition, privacy issues in Saudi Arabia have caused small WWRs in residential buildings. Even though there are abundant natural resources in Saudi Arabia, it is still difficult to apply passive design strategies for residential buildings. Focusing on the use of natural resources, such as daylight, the present study has investigated the daylight and energy performance of design variables for developing energy-efficient residential buildings.

For the design variables, three different WWRs, four designs of the external shading device, and four types of glazing were selected. By using two different simulations, the daylighting and energy performance of these design variables in a typical villa were examined. Moreover, two daylight metrics—sDA and aSE—were used to evaluate the daylighting performance of design variables. As a result, the daylighting performance was satisfied with the accepted ranges of $sDA_{300/50\%}$ and $aSE_{1000,250h}$, when the WWR of 50% and the double-glazed clear windows were applied into the reference building. The illuminance levels by two overhang designs (Type A and B) were somewhat satisfied with the accepted values of sDA and aSE, while the other shading designs were not.

For the energy performance assessment, the annual, artificial, and HVAC energy consumption were analyzed. While the daylighting performance was improved with the increased WWRs, the energy consumption also increased as the WWR increased. This was caused by the increase in cooling demand by the excessive sunlight through the increased WWRs. Similar to the daylighting performance, the energy consumption decreased, as with the increase of the projection lengths and fins for the external shading devices. A villa with the double-pane windows with a low-E coating showed a better energy performance. In summary, the lowest energy consumption was achieved by using the external shading device.

Based on the results, the energy-saving potential for residential buildings under hot climates with the implementation of daylighting was observed. In addition, the daylighting performance can be improved by using the proper design variables. However, the design variables for the present study seemed to be insufficient to draw out more valuable outcomes. Thus, it is necessary to examine the performance of more various design variables, such as triple-glazed windows with various inert gases, and various types of internal/external shading devices for further study. For the present study, the daylighting performance by design variables was evaluated by only using simulation tools. To figure out the illuminance level practically, it requires further field measurements.

Author Contributions: A.M.A.-D. designed and performed the simulation and collected data; D.D.K. wrote the manuscript and analyzed the data. All authors have read and agreed to the published version of the manuscript.

Funding: This research received no external funding.

Conflicts of Interest: The authors declare no conflict of interest.

References

1. Energy Technology Perspectives 2017—Catalysing Energy Technology Transformations, International Energy Agency. Available online: https://www.Iea.Org/buildings/ (accessed on 1 August 2020).
2. Radhi, H. Evaluating the potential impact of global warming on the uae residential buildings—A contribution to reduce the CO_2 emissions. *Build. Environ.* **2009**, *44*, 2451–2462. [CrossRef]
3. Abdul-Wahab, S.A.; Charabi, Y.; Al-Maamari, R.; Al-Rawas, G.A.; Gastli, A.; Chan, K. CO_2 greenhouse emissions in oman over the last forty-two years: Review. *Renew. Sustain. Energy Rev.* **2015**, *52*, 1702–1712. [CrossRef]
4. Fahmy, M.; Mahdy, M.M.; Nikolopoulou, M. Prediction of future energy consumption reduction using GRC envelope optimization for residential buildings in Egypt. *Energy Build.* **2014**, *70*, 186–193. [CrossRef]

5. Taleb, H.M.; Sharples, S. Developing sustainable residential buildings in Saudi Arabia: A case study. *Appl. Energy* **2011**, *88*, 383–391. [CrossRef]
6. Al-Sulaiman, F.A.; Zubair, S.M. A survey of energy consumption and failure patterns of residential air-conditioning units in eastern Saudi Arabia. *Energy* **1996**, *21*, 967–975. [CrossRef]
7. Almutairi, K.; Thoma, G.; Burek, J.; Algarni, S.; Nutter, D. Life cycle assessment and economic analysis of residential air conditioning in Saudi Arabia. *Energy Build.* **2015**, *102*, 370–379. [CrossRef]
8. The Annual Report of 2011, Electricity & Cogeneration Regulatory Authority, Saudi Arabia. Available online: https://www.ecra.gov.sa/en-us/MediaCenter/DocLib2/Pages/SubCategoryList.aspx?categoryID=4 (accessed on 1 September 2020).
9. Levy, J.I.; Woo, M.K.; Tambouret, Y. Energy savings and emissions reductions associated with increased insulation for new homes in the United States. *Build. Environ.* **2016**, *96*, 72–79. [CrossRef]
10. Evin, D.; Ucar, A. Energy impact and eco-efficiency of the envelope insulation in residential buildings in Turkey. *Appl. Therm. Eng.* **2019**, *154*, 573–584. [CrossRef]
11. Friess, W.A.; Rakhshan, K.; Hendawi, T.A.; Tajerzadeh, S. Wall insulation measures for residential villas in Dubai: A case study in energy efficiency. *Energy Build.* **2012**, *44*, 26–32. [CrossRef]
12. Hee, W.J.; Alghoul, M.A.; Bakhtyar, B.; Elayeb, O.; Shameri, M.A.; Alrubaih, M.S.; Sopian, K. The role of window glazing on daylighting and energy saving in buildings. *Renew. Sustain. Energy Rev.* **2015**, *42*, 323–343. [CrossRef]
13. Fasi, M.A.; Budaiwi, I.M. Energy performance of windows in office buildings considering daylight integration and visual comfort in hot climates. *Energy Build.* **2015**, *108*, 307–316. [CrossRef]
14. Dubois, M.-C.; Blomsterberg, Å. Energy saving potential and strategies for electric lighting in future north european, low energy office buildings: A literature review. *Energy Build.* **2011**, *43*, 2572–2582. [CrossRef]
15. Ahn, B.-L.; Jang, C.-Y.; Leigh, S.-B.; Yoo, S.; Jeong, H. Effect of led lighting on the cooling and heating loads in office buildings. *Appl. Energy* **2014**, *113*, 1484–1489. [CrossRef]
16. Homod, R.Z.; Sahari, K.S.M. Energy savings by smart utilization of mechanical and natural ventilation for hybrid residential building model in passive climate. *Energy Build.* **2013**, *60*, 310–329. [CrossRef]
17. Kim, J.; Choi, H.; Kim, S.; Yu, J. Feasibility analysis of introducing renewable energy systems in environmental basic facilities: A case study in Busan, South Korea. *Energy* **2018**, *150*, 702–708. [CrossRef]
18. Kunwar, N.; Cetin, K.S.; Passe, U.; Zhou, X.; Li, Y. Energy savings and daylighting evaluation of dynamic venetian blinds and lighting through full-scale experimental testing. *Energy* **2020**, *197*, 117190. [CrossRef]
19. Srisamranrungruang, T.; Hiyama, K. Balancing of natural ventilation, daylight, thermal effect for a building with double-skin perforated facade (DSPF). *Energy Build.* **2020**, *210*, 109765. [CrossRef]
20. Souza, L.C.; Souza, H.A.; Rodrigues, E.F. Experimental and numerical analysis of a naturally ventilated double-skin façade. *Energy Build.* **2018**, *165*, 328–339. [CrossRef]
21. Mejjaouli, S.; Alzahrani, M. Decision-making model for optimum energy retrofitting strategies in residential buildings. *Sustain. Prod. Consum.* **2020**, *24*, 211–218. [CrossRef]
22. Al-Awsh, W.A.; Qasem, N.A.A.; Al-Amoudi, O.S.B.; Al-Osta, M.A. Experimental and numerical investigation on innovative masonry walls for industrial and residential buildings. *Appl. Energy* **2020**, *276*, 115496. [CrossRef]
23. Alaidroos, A.; Krarti, M. Optimal design of residential building envelope systems in the kingdom of Saudi Arabia. *Energy Build.* **2015**, *86*, 104–117. [CrossRef]
24. Krarti, M.; Howarth, N. Transitioning to high efficiency air conditioning in Saudi Arabia: A benefit cost analysis for residential buildings. *J. Build. Eng.* **2020**, *31*, 101457. [CrossRef]
25. Li, D.; Lam, J. Predicting vertical luminous efficacy using horizontal solar data. *Lighting Res. Technol.* **2001**, *33*, 25–42. [CrossRef]
26. Shishegar, N.; Boubekri, M. Quantifying electrical energy savings in offices through installing daylight responsive control systems in hot climates. *Energy Build.* **2017**, *153*, 87–98. [CrossRef]
27. Do, S.L.; Shin, M.; Baltazar, J.-C.; Kim, J. Energy benefits from semi-transparent bipv window and daylight-dimming systems for IECC code-compliance residential buildings in hot and humid climates. *Sol. Energy* **2017**, *155*, 291–303. [CrossRef]
28. Reffat, R.M.; Ahmad, R.M. Determination of optimal energy-efficient integrated daylighting systems into building windows. *Sol. Energy* **2020**, *209*, 258–277. [CrossRef]

29. Sabry, H.; Sherif, A.; Gadelhak, M.; Aly, M. Balancing the daylighting and energy performance of solar screens in residential desert buildings: Examination of screen axial rotation and opening aspect ratio. *Sol. Energy* **2014**, *103*, 364–377. [CrossRef]
30. Sherif, A.; Sabry, H.; Rakha, T. External perforated solar screens for daylighting in residential desert buildings: Identification of minimum perforation percentages. *Sol. Energy* **2012**, *86*, 1929–1940. [CrossRef]
31. Sherif, A.H.; Sabry, H.M.; Gadelhak, M.I. The impact of changing solar screen rotation angle and its opening aspect ratios on daylight availability in residential desert buildings. *Sol. Energy* **2012**, *86*, 3353–3363. [CrossRef]
32. Sepúlveda, A.; De Luca, F.; Thalfeldt, M.; Kurnitski, J. Analyzing the fulfillment of daylight and overheating requirements in residential and office buildings in Estonia. *Build. Environ.* **2020**, *180*, 107036. [CrossRef]
33. Li, D.H.W.; Wong, S.L.; Tsang, C.L.; Cheung, G.H.W. A study of the daylighting performance and energy use in heavily obstructed residential buildings via computer simulation techniques. *Energy Build.* **2006**, *38*, 1343–1348. [CrossRef]
34. Toutou, A.; Fikry, M.; Mohamed, W. The parametric based optimization framework daylighting and energy performance in residential buildings in hot arid zone. *Alex. Eng. J.* **2018**, *57*, 3595–3608. [CrossRef]
35. Dabe, T.J.; Adane, V.S. The impact of building profiles on the performance of daylight and indoor temperatures in low-rise residential building for the hot and dry climatic zones. *Build. Environ.* **2018**, *140*, 173–183. [CrossRef]
36. Huang, Y.; Niu, J.-l.; Chung, T.-m. Comprehensive analysis on thermal and daylighting performance of glazing and shading designs on office building envelope in cooling-dominant climates. *Appl. Energy* **2014**, *134*, 215–228. [CrossRef]
37. Taleb, H.M.; Antony, A.G. Assessing different glazing to achieve better lighting performance of office buildings in the United Arab Emirates (Uae). *J. Build. Eng.* **2020**, *28*, 101034. [CrossRef]
38. Liu, M.; Heiselberg, P.K.; Antonov, Y.I.; Mikkelsen, F.S. Parametric analysis on the heat transfer, daylight and thermal comfort for a sustainable roof window with triple glazing and external shutter. *Energy Build.* **2019**, *183*, 209–221. [CrossRef]
39. Building Performance Analysis Software, Autodes. Available online: https://www.Autodesk.Com/products/insight/overview (accessed on 1 August 2020).
40. IES. *IES LM-83-12 IES Spatial Daylight Autonomy (SDA) and Annual Sunlight Exposure (ASE)*; Illuminating Engineering Society: New York, NY, USA, 2013.
41. Costanzo, V.; Gianpiero, E.; Marletta, L.; Pistone Nascone, F. Application of climate based daylight modelling to the refurbishment of a school building in Sicily. *Sustainability* **2018**, *10*, 2653. [CrossRef]
42. Leed v4, Leed bd+c: Healthcare, the U.S. Green Building Council. 2005. Available online: https://www.Usgbc.Org/credits/healthcare/v4-draft/eqc-0 (accessed on 24 June 2019).
43. Lee, J.; Boubekri, M.; Liang, F. Impact of building design parameters on daylighting metrics using an analysis, prediction, and optimization approach based on statistical learning technique. *Sustainability* **2019**, *11*, 1474. [CrossRef]
44. Weather Data, Energyplus. Available online: https://energyplus.Net/weather (accessed on 1 August 2020).
45. *American Society of Heating, Refrigerating and Air Conditioning Engineers, Ashrae Guideline 14-2002, Measurement of Energy and Demand Savings—Measurement of Energy, Demand and Water Savings*; ASHRAE: Atlanta, GA, USA, 2002.
46. Rashwan, A.; El Gizawi, L.; Sheta, S. Evaluation of the effect of integrating building envelopes with parametric patterns on daylighting performance in office spaces in hot-dry climate. *Alex. Eng. J.* **2019**, *58*, 551–557. [CrossRef]
47. Alhagla, K.; Mansour, A.; Elbassuoni, R. Optimizing windows for enhancing daylighting performance and energy saving. *Alex. Eng. J.* **2019**, *58*, 283–290. [CrossRef]
48. Alghoul, S.K.; Rijabo, H.G.; Mashena, M.E. Energy consumption in buildings: A correlation for the influence of window to wall ratio and window orientation in Tripoli, Libya. *J. Build. Eng.* **2017**, *11*, 82–86. [CrossRef]
49. Xue, P.; Li, Q.; Xie, J.; Zhao, M.; Liu, J. Optimization of window-to-wall ratio with sunshades in china low latitude region considering daylighting and energy saving requirements. *Appl. Energy* **2019**, *233–234*, 62–70. [CrossRef]

50. Mangkuto, R.A.; Rohmah, M.; Asri, A.D. Design optimisation for window size, orientation, and wall reflectance with regard to various daylight metrics and lighting energy demand: A case study of buildings in the tropics. *Appl. Energy* **2016**, *164*, 211–219. [CrossRef]
51. Leung, C.K.; Lu, L.; Liu, Y.; Cheng, H.S.; Tse, J.H. Optical and thermal performance analysis of aerogel glazing technology in a commercial building of Hong Kong. *Energy Built Environ.* **2020**, *1*, 215–223. [CrossRef]
52. Chinazzo, G.; Wienold, J.; Andersen, M. Combined effects of daylight transmitted through coloured glazing and indoor temperature on thermal responses and overall comfort. *Build. Environ.* **2018**, *144*, 583–597. [CrossRef]
53. Chen, X.; Zhang, X.; Du, J. Glazing type (colour and transmittance), daylighting, and human performances at a workspace: A full-scale experiment in Beijing. *Build. Environ.* **2019**, *153*, 168–185. [CrossRef]

Publisher's Note: MDPI stays neutral with regard to jurisdictional claims in published maps and institutional affiliations.

energies

MDPI

Article

Evaluation of Daylight and Cooling Performance of Shading Devices in Residential Buildings in South Korea

Taesub Lim [1], Woong Seog Yim [2,*] and Daeung Danny Kim [3,*]

1 Department of Architectural Engineering, Seoil University, Seoul 02192, Korea; francis9@seoil.ac.kr
2 Department of Industrial Management, Seoil University, Seoul 02192, Korea
3 Architectural Engineering Department, King Fahd University of Petroleum and Minerals (KFUPM), Dhahran 31261, Saudi Arabia
* Correspondence: wsyim8452@hanmail.net (W.S.Y.); dkim@kfupm.edu.sa (D.D.K.)

Received: 17 August 2020; Accepted: 10 September 2020; Published: 11 September 2020

check for updates

Abstract: Accounting for more than half of buildings in South Korea, the energy consumed by residential buildings has become a main concern and the cooing demand has rapidly increased. To reduce energy consumption, several passive and active design strategies have generally been applied. However, there has been an increasing demand for high window-to-wall ratios in residential buildings, it is imperative to block sunlight into a building effectively. Focusing on the reduction of cooling energy consumption in a residential building, the present study assessed the daylight and energy performance of shading devices. Among various types of shading devices, the Venetian blind, horizontal louver, light shelf, and egg-crate were selected. The illuminance levels in three different areas in a building were measured. In addition, the annual cooling energy consumption by these shading devices was investigated. As a result, both daylight and energy performance varied with different design options of these shading devices. Because of the slight performance difference among shading devices, the artificial loads of two best shading devices were compared. In sum, the egg-crate shading was the most proper shading device to block sunlight as well as reduce the cooling energy consumption effectively.

Keywords: daylight; cooling energy; shading device; residential building

1. Introduction

Accounting for more than half of buildings in South Korea (hereafter Korea), residential buildings have consumed a significant amount of energy and it has rapidly increased [1]. Specifically, more than half of the total building energy consumption was used by residential buildings, which was about 60% [2,3]. Focusing on energy consumption by the residential sector, many studies have been conducted to improve energy efficiency and develop energy conservation measures [4–7]. Considering the effort to reduce greenhouse gas emissions, it is necessary to find ways of reducing energy consumption by residential buildings.

Generally, most of the energy in the residential building has been used for heating and cooling to maintain thermal comfort. Thus, it is necessary to reduce the heating and cooling demand in residential buildings in that the total energy consumption in buildings can be reduced. To reduce the energy for heating and cooling, building envelope-enhancement is initially considered. In general, advanced materials have been used to improve thermal resistance in building envelopes such as phase change materials, vacuum insulation panels, various types of double pane glazing, etc. [8–12]. In addition, the use of shading devices and airtightness improvement can be also effective strategies to reduce heating or cooling in buildings [13,14].

Among components of building envelopes, window systems are the main contributor to the heat loss and their energy performance generally depends on thermal properties of glazing and frames [15]. Even though the use of advanced glazing can improve the thermal performance of window systems, there are still issues about the poor thermal performance of window systems compared to other components of building envelopes [16,17]. As mentioned above, another way of reducing heating and cooling demand is the use of internal and external shading devices that can control sunlight through window systems [13]. According to the study of Cho et al. [18], external shading devices for high-rise residential buildings reduced cooling energy demand effectively by minimizing solar heat gain during the summer. Similar studies about the energy-saving potential by using external shadings were performed [19–22]. By absorbing or reflecting the sun's radiation, shading devices play a significant role to maintain thermal comfort in buildings.

Since daylighting highly influences not only the energy performance of buildings but also occupant satisfaction, productivity, and health, it is one of the most important factors that need to be considered from the architectural design stage. In addition, daylighting should be well-controlled and appropriately used to avoid a further increase in heat gains including internal heat gains such as from artificial lighting and home appliances. Regarding the recent design trend of residential buildings, the demand for high window-to-wall ratios in residential buildings in Korea is rapidly increasing. Moreover, the cooling demand has been consistently increasing [18,23,24]. While most studies have focused on the energy performance of in buildings, the present study assesses the daylighting and cooling energy performance of both internal and external shading devices. In addition, the relationship between the daylighting of shading devices with cooling energy demand was also investigated.

2. The Control of Building Energy Consumption by Shading Devices

As essential factors influencing thermal conditions in a building via thermal gains/losses, solar radiation has been controlled with shading devices [25]. Since the solar shadings were initially introduced, the interest of this area has been growing rapidly. Researchers are focusing on diverse methods that can maximize the performance of shading devices. Marrero and Oliviera [26] showed that the installation of louver shading devices in buildings can lead to more comfortable indoor thermal conditions for the residents as well as significant savings in building energy consumption. A similar result was also obtained by Lim et al. [27]. Their result showed that window glazing coupled with shading devices enhanced tropical daylighting quantity as well as visual comfort. Solar shading devices can also provide an energy saving of 8% and 20% during the winter and summer, respectively [28].

Many types of shading devices have been developed by designers and building scientists over the years. They vary in several aspects such as shape, size, and overall performance. Several shading types were investigated by Nedhal and Fadzil [29]. Among various types, the egg-crate shading had the most significant impact on the indoor air temperature decrease in high-rise residential buildings. In addition, locations, size, and color of the internal blinds can significantly influence building energy consumption and the proper design of internal blinds can save energy about 14% of total building cooling consumption during the summer season [30]. According to the study of Uribe [31], the optimum design of shading devices can influence annual energy consumption in buildings and the design variables of the shading devices should consider the time of day and seasonal differences. Focusing on the design of shading devices, window glass types, seasonal and daily temperature change, and the operation of mechanical systems for heating and cooling should be also considered [32]. Previous studies have revealed that the use of shading devices play an important role in reducing electric lighting, while it provides a required amount of light for visual tasks.

According to the study of Datta [33], externally fixed horizontal louvers with different slat lengths and tilts on buildings have a different impact on thermal conditions in buildings. Therefore, shading strategies were developed to control daylight quantities. One of the strategies brought by Tzempelikos and Athienitis [34] was the control of building cooling and lighting demand by exterior roller shades. As a result, the shading devices reduced the energy demand for cooling and lighting by limiting 20% of solar transmittance. Horizontally installed shading devices in buildings in tropical regions controlled solar heat gains as well as reduced HVAC (Heating, Ventilation, and Air Conditioning) loads [35]. In the case of the internal blinds, the automated Venetian blind was used to limit direct sunlight while admitting diffused sunlight for visual comfort [36]. Similarly, absorptive blinds in an office building can also reduce the indoor air temperature by 1 °C during the summer [37].

3. Methodology

3.1. Building Description

To evaluate the daylighting and energy performance of a residential building with shading devices, a typical apartment building in Seoul in Korea was chosen, in which the latitude and longitude are 37.5665° N and 126.9780° E, respectively. The annual mean air temperature in Seoul is 12.5 °C, and the highest and lowest mean air temperatures are −2 °C and 26 °C, respectively [38]. In addition, the mean annual insolation is 4125 MJ/m^2 [39]. The total area of the unit space was 145 m^2 with a ceiling height of 2.3 m and located on the eighth floor of an apartment building with 17 floors. All bedrooms and the living room were headed to the south (Figure 1). Due to a high WWR of 90% of the southern wall, a large quantity of solar radiation was expected. Thermal properties and other conditions for the energy simulation were presented in Table 1. In addition, the energy model was created by SketchUp.

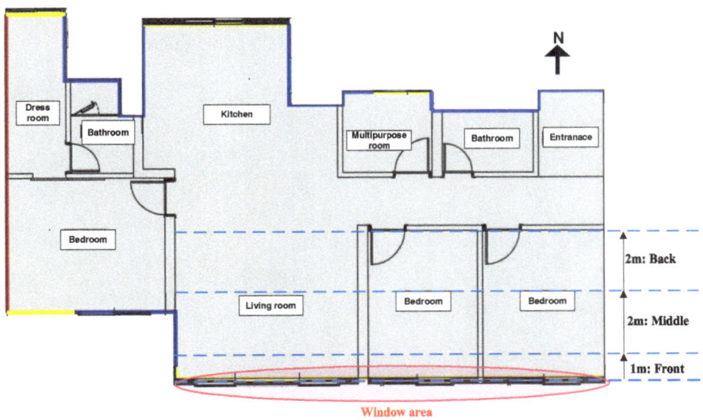

Figure 1. The plan of the selected unit of the reference residential building.

Table 1. The specification for the energy simulation.

Building System	Design Value
Walls	0.397 W/m^{2}K
Window systems	1.46 W/m^{2}K
Shading coefficient	0.6
Air infiltration	2.1 cm^2/m^2 (3.5 ACH @ 50 PA)
Internal heat gain	4 occupants Lighting: 5.4 W/m^2 Equipment: 7.0 W/m^2
HVAC system setpoint temperature	26 °C for cooling and 20 °C for heating

3.2. The Selection of Shading Devices

The main role of a shading device is to block direct sunlight as well as diffuse sunlight and reflective sunlight. Direct sunlight can be controlled simply with external shading devices, while diffuse sunlight is usually controlled with internal shading devices. For the present study, 4 different shading devices were selected as shown in Figure 2 the design variables of these shading devices were presented in Table 2.

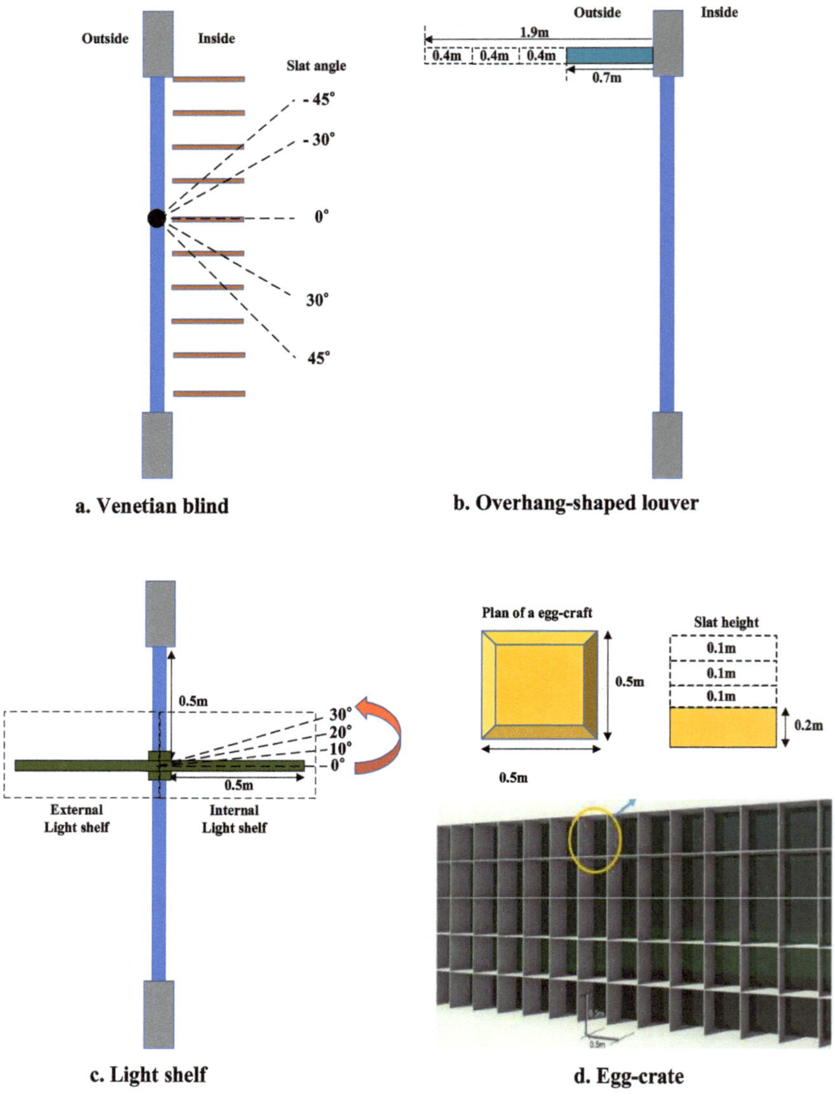

Figure 2. Design variables of the shading devices.

Table 2. The design variables of shading devices.

Shading Device	Design Variable
Venetian blind	Slat angles: −45°, −30°, 0°, 30°, and 45°.
Overhang-shaped louver	Projection length: 0.7 m, 1.1 m, 1.5 m, and 1.9 m.
Light shelf	Internal shelf angles: 0°, 10°, 20°, and 30°.
Egg-crate	Vertical and horizontal slat height: 0.2 m, 0.3 m, 0.4 m, and 0.5 m.

The first shading device is the Venetian blind. The Venetian blind is generally used to control both direct and diffuse sunlight. Slat angle adjustment is the most important variable and these slats are adjusted negatively or positively when there is a need to reduce incoming light. To assess the daylighting and cooling energy performance of the Venetian blind, 5 different slat angles were set. The second shading device was an overhang-shaped horizontal louver. Since the admitted sunlight quantity can be highly influenced by the projection lengths of the louver [26], 4 different projection lengths with 0.4 m intervals were created for the present study. The third shading device was the light-shelf modeled based on the design guideline provided by BRE [40]. To figure out the daylighting and energy performance of an angle of the internal shelf, 4 different angles were considered. The last shading device was an egg-crate shading device that refers to a shape formed by combining the vertical and horizontal shading types. The horizontal shade functions to control direct solar radiation depending on the solar altitude, while the vertical shade functions to block or control side light depending on the change of azimuth. The plan size of each egg-crate was 0.5 m by 0.5 m. For the present study, 4 different heights of a vertical and horizontal slat were considered for the assessment since the solar radiation differs by the heights of the egg-crate.

3.3. Simulation for the Assessment of Daylighting and Cooling Energy

To evaluate the daylighting and energy performance of a residential building with various types of shading devices, two simulation tools were used. For the analysis of daylighting through shading devices, the Radiance was used, which was developed by the lighting research team of the LBNL (Lawrence Berkeley National Laboratory) in the US [41]. By verifying and visualizing light rays from the light source, illuminance through shading devices was calculated. The simulation conditions were based on IES (Illuminating Engineering Society) guidelines as presented in Table 3 [42].

Table 3. Input parameters for the Radiance.

Input Parameters	Value
Orientation	South-facing window
Time	June and December 21 (noon)
Sky condition	CIE Clear sky for direct sunlight
Glazing visual Transmittance	70%: Glazing double pane Low E
Material properties (Reflectance)	20%: A standard floor made by an opaque material 70%: A standard ceiling 50%: A standard wall 50%: A standard external shadings and curtain wall frames

Based on the building description above and the specifications of the building envelopes (Table 1), the energy simulation was performed. In addition, hot water and space heating was assumed to be provided by the central heating systems fueled by natural gas. Compact Fluorescent lamps were mainly used as a lighting system. For the weather condition, the "Seoul" EPW (Energy Plus Weather format) file in South Korea was used. The energy model was created by using the SketchUp program (Figure 3). The design variables of shading devices were also modeled by using SketchUp for energy simulation. For the evaluation of the cooling energy, IES VE was selected as a building energy simulation program [43].

Figure 3. The energy model created by SketchUp.

Since the energy performance of shading devices in a building can be varied by the degree of indoor temperature in the space, the design temperature in the cooling season was set to 20 °C. For the energy consumption comparison, the annual energy consumption operated by each shading device was compared. By using the coefficient of variation of the root mean squared error (*CV(RMSE)*) provided by ASHRAE Guideline 14, the monthly energy consumption of the reference residential building without any shading devices was compared with the energy simulation [44]. The models will be declared to be calibrated if they produce *CV(RMSE)s* within ± 15% with monthly energy data

$$RMSE = \sqrt{\frac{\sum_{i=1}^{n}(M_i - S_i)^2}{n}} \tag{1}$$

$$C_V(RMSE) = \frac{RMSE}{M_{avg}} \times 100 \tag{2}$$

where M_i is the energy consumption of the residential building, while S_i is the monthly energy consumption by energy simulation. N is the period and M_{avg} is the average for the energy consumption of the residential building.

4. Result

4.1. Daylighting Performance

Since the solar azimuth changes depending on the changes of time and seasons, to evaluate by season, the daylighting performance in the front, middle, and back areas were assessed for the standard time of noon during the spring and autumn equinoxes, the summer solstice, and the winter solstice.

4.1.1. The Venetian Blind

As shown in Figure 4, the illuminance was decreased when the slats of the blind were parallel to the X-axis of the window (0°) about 90% in the front area during the summer. When the slat angles were set to −30° and −45°, the illuminance was also significantly decreased in the front area, while the positive slat angles of the blind exceeded 12,000 lux during the summer. According to the lighting handbook of IESNA, the recommended light level ranges 200–300 lux for residential buildings [45]. The observed illuminance levels in the front area during the summer through the slat angles of −30° and −45° were satisfied with the recommendation of IESNA. While the illuminance with the slat angle of 0° was somewhat higher than the recommendation of IESNA, this slat angle can be applied to the summer season because the illuminance level was decreased in the middle and back areas. The other two positive slat angles may cause discomfort such as glare. During the winter, the positive angles of the blind admitted more illuminance than the negative angles of the blind in the middle area of the space. Considering the obtained results, it can be seen that the illuminance level can be highly influenced by the slat angles of the blind.

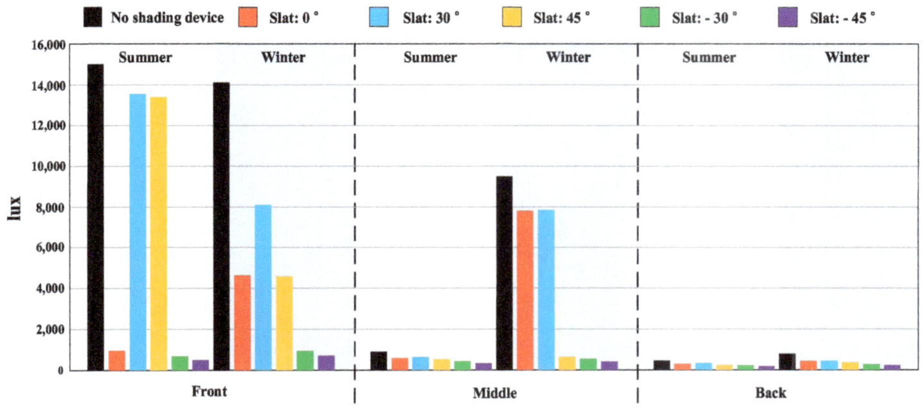

Figure 4. Daylighting performance through the Venetian blind.

4.1.2. The Overhang-Shaped Louver

When the projection length was ranged from 0.7 m to 1.1 m, about 50% of the illuminance was reduced compared to the building without a shading device in the front area during the summer as shown in Figure 5. As the projection length was increased from 1.1 m to 1.9 m, the illuminance was significantly reduced to below 1000 lux. Even though this illuminance level was somewhat higher than the IESNA recommendation, the projection length from 1.1 m to 1.9 m can be applied because the illuminance level in the middle and back areas was close to the recommendation. During the winter, there was a slight illuminance difference among the cases with various projection lengths in the front area.

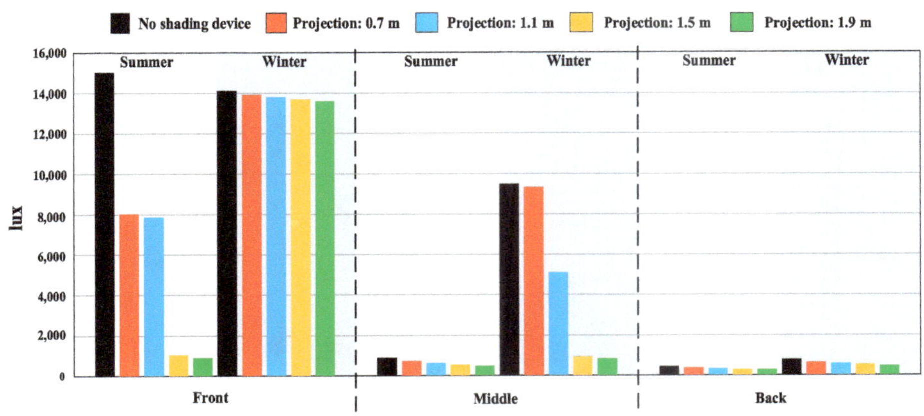

Figure 5. Daylighting performance through the overhang-shaped louver.

4.1.3. The Light Shelf

In the case of the light shelf, the illuminance level was measured when applying four different internal shelf angles in a building (Figure 6). Comparing the illuminance in the reference building without a shading device, the illuminance of all the cases was significantly reduced to below 1000 lux in the front area during the summer. In addition, there was a little illuminance difference among the cases with the angles of the internal shelf from 0° to 30°. A similar trend was also observed in the front area during the winter. Thus, it can be seen that the angle of the internal shelf had little impact on the illuminance in a building.

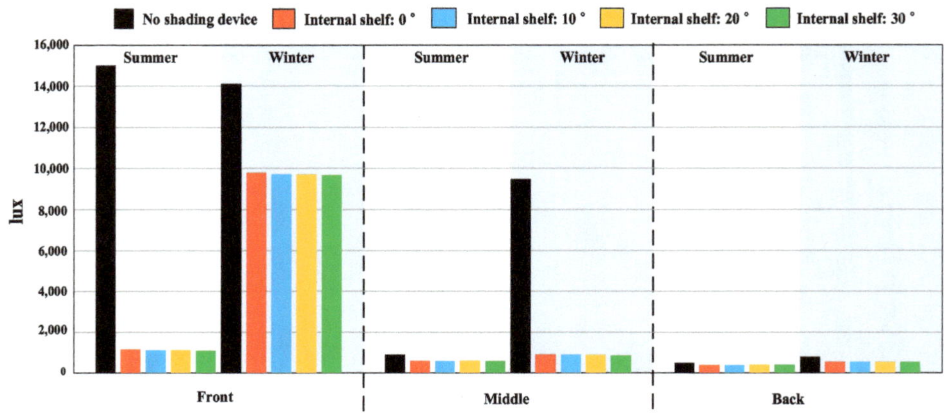

Figure 6. Daylighting performance through the light shelf.

4.1.4. The Egg-Crate

During the summer, the illuminance was reduced to below 2000 lux in the front area when the egg-crate shading was applied (Shown in Figure 7). While there was a little illuminance increase about 10% from with the vertical and horizontal slat height increase from 0.2 m to 0.3 m, the illuminance was significantly reduced as with the increase in the vertical and horizontal slat height during the summer in the front area. During the winter, 30% of the illuminance was reduced from the building without a shading device to the cases with the vertical and horizontal slat heights of 0.2 m ~0.3 m in the front area. As the vertical and horizontal slat height was increased, about 45% of the illuminance was additionally decreased. Based on the result, the use of the egg-crate shading can effectively reduce the illuminance in a building during the summer while admitting sunlight during the winter.

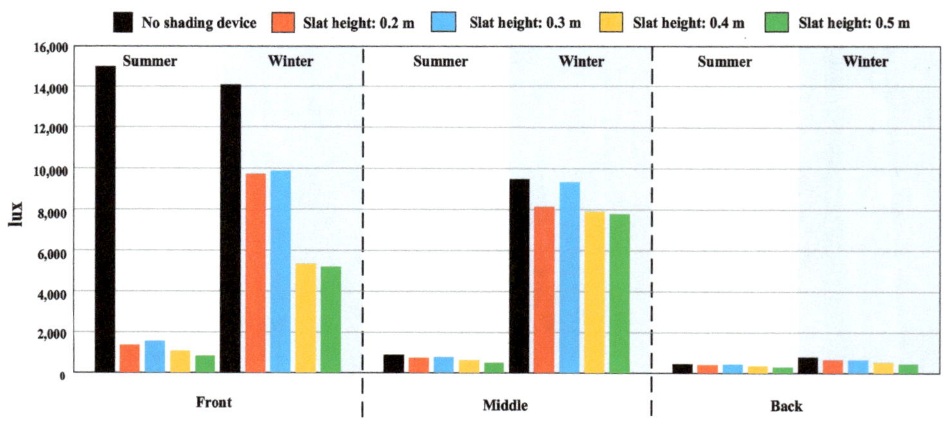

Figure 7. Daylighting performance through the egg-crate.

4.2. The Analysis of Annual Cooling Energy Consumption by Shading Devices

4.2.1. The Comparison between Energy Simulation and the Monthly Energy Consumption

To validate the energy simulation, the monthly energy consumption of the reference residential building was compared with the energy prediction by the simulation. The total energy consumption of the residential building was about 76.4 MWh, while there was a 5% decrease for the energy use predicted by the energy simulation. As shown in Figure 8, the largest difference was observed in

April, which about 30%. Specifically, the root-mean-squared errors (CV(RMSEs) were calculated and these were ranged from 0.16 to 3.5 (Table 4). Since these results were within the acceptable range, the predicted results by the simulation met the requirement by ASHRAE Guideline 14 [44].

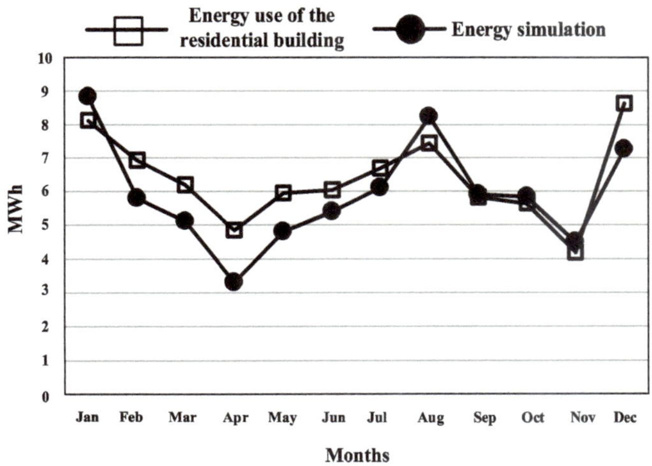

Figure 8. Energy consumption comparison between the reference residential building and the energy simulation.

Table 4. The monthly energy consumption comparison for the reference case.

Month	Energy Consumption (MWh)			CV(RMSE) (%)
	The Reference Residential Building, 2017	Energy Simulation	Difference	
January	8.1	8.8	−0.7	1.59
February	6.9	5.8	1.1	2.54
March	6.2	5.1	1.1	2.46
April	4.8	3.3	1.5	3.50
May	5.9	4.8	1.1	2.61
June	6.0	5.4	0.6	1.46
July	6.7	6.1	0.6	1.34
August	7.4	8.2	−0.8	1.85
September	5.8	5.9	−0.1	0.16
October	5.6	5.8	−0.2	0.48
November	4.2	4.5	−0.3	0.72
December	8.6	7.3	1.4	3.10

4.2.2. The Cooling Energy Comparison by Shading Devices

To find the most effective shading device for reducing cooling demand in a residential building, the annual cooling energy consumption by four shading devices was compared (Figure 9). When applying the Venetian blind in a building, about 14% of the annual cooling energy was reduced. Among five different slat angles, the largest cooling energy was reduced by about 18%, when the slat angle was set to 45°. Conversely, the smallest amount of cooling energy was reduced, which was about 8%, when a slat angle was set to −45°. It can be seen that the reduction of cooling energy consumption was varied with the different slat angles of the blind. In the case of the overhang-shaped louver, the annual cooling energy was reduced as the projection length was increased. When the projection length was 1.9 m, the largest energy consumption reduction was observed, which was about 51% of the total annual cooling energy consumption. The average cooling energy reduction was about 32% with the louver. When the projection length was 0.7 m, only 6% of the cooling energy was reduced. As can be shown, the cooling energy in a building with the overhang-shaped louver was highly influenced by the projection lengths of the louver. Moreover, about 30% of annual cooling energy saving was made when the light shelf was used. However, there was little difference in the cooling energy consumption reduction among the cases with different angles of the internal shelf. Therefore, the angles of the internal shelf have a less impact on cooling energy consumption in a building. Lastly, about 50% of the annual cooling energy consumption was reduced with the egg-crate shading in a building. As the length of vertical and horizontal slats was increased, the cooling energy consumption reduction was also increased. The largest energy-saving was obtained, when the vertical and horizontal slats were 0.5 m, which was about 54%.

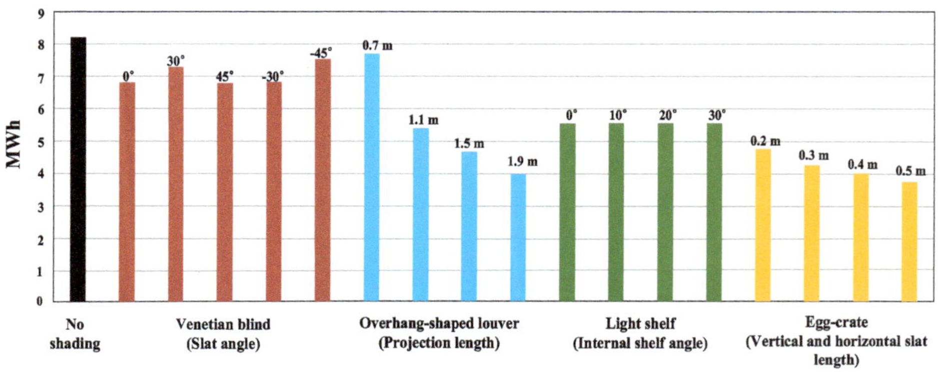

Figure 9. The reduction of annual cooling energy consumption by shading devices.

In sum, the use of an overhang-shaped louver with a projection length of 1.9 m and the egg-crate shading with a vertical and horizontal slat height of 0.5 m can effectively reduce the annual cooling energy consumption in a residential building among various shading devices.

5. Discussion

For the present study, the daylighting and energy performance of shading devices in a residential building was assessed. As can be shown above, each design variable had a different impact on the illuminance level and cooling energy demand in the building. Considering the daylighting performance, the light shelf, and the egg-crate effectively blocked sunlight in the front area during the summer more effectively than the others, while the overhang-shaped louver and the egg-crate shading had more impact on reducing the cooling energy consumption than the others. A similar result was obtained through the study of Alhuwayil et al. [20]. According to their result, the energy-saving was increased as the fin widths and the projection lengths of the overhang were increased. This was also mentioned in the present study. In addition, Alzoubi and Al-Zoubi [46] have investigated the energy performance of three types of external shadings including vertical, horizontal, and horizontal 45°. They concluded that the most effective shading for the reduction of energy consumption was the vertical shading device. This showed a similar result to the present study. Even though the egg-crate in the present study is not the vertical shading device, it showed the best overall performance among other shading devices by blocking the sun both horizontally and vertically.

In the present study, the annual energy consumption was calculated for the IES VE simulation. However, the time and date for the Radiance simulation program were set on only two days such as June and December 21st at noon to find the most effective shading devices during the summer and winter as well as reduce the computational resources, while other studies have considered the annual daylighting performance [20,47–49]. Considering the outcome of the present study, the obtained illuminance levels can be overestimated or underestimated. Thus, it is necessary to consider annual daylighting performance for further study.

Moreover, it can be seen that the performance difference between daylighting and cooling energy consumption was caused by the illuminance in the middle and back areas. This can influence other loads in buildings. Thus, it is necessary to evaluate other building energy use such as artificial lighting. As mentioned previously, the best shading device for reducing the cooling energy consumption was the egg-crate shading with a vertical and horizontal slat height of 0.5 m, and the second one was the overhang-shaped louver with a projection length of 1.9 m. However, there was only a 3% difference in annual cooling energy consumption between the two strategies. To find out the shading device with the best overall performance, the artificial lighting loads by these strategies were compared. As can be shown in Figure 10, about 15% of energy saving for the artificial load was obtained by using the egg-crate shading with a vertical and horizontal slat height of 0.5 m, while 5% of artificial lighting energy was reduced by the overhang-shaped louver. Therefore, the use of the eff-crate was a more effective shading device for blocking sunlight as well as reducing cooling energy demand in a residential building than the overhang-shaped louver.

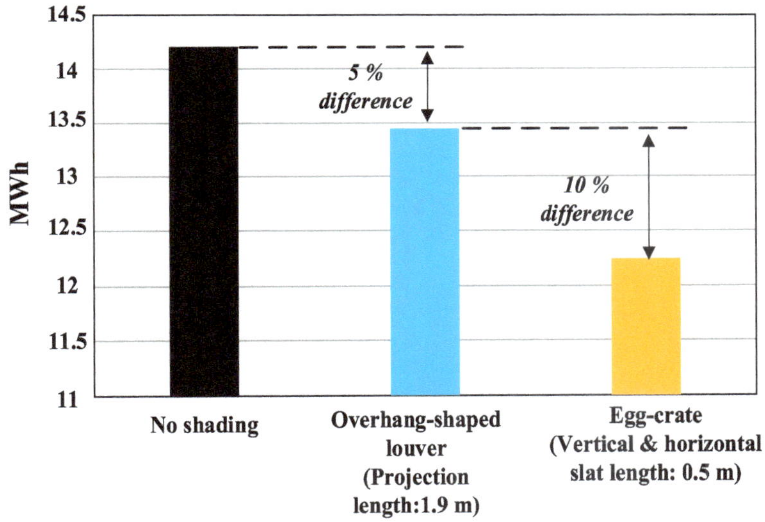

Figure 10. The reduction of artificial lighting energy consumption by two shading devices.

6. Conclusions

As one of the main contributors to energy consumption in South Korea, the number of residential building has gradually increased and the energy consumed by residential buildings has become the main concern. Thus, many studies have been conducted to reduce energy consumption in residential buildings by applying passive and active design solutions. In addition, several renewable energy systems were used to improve energy efficiency in residential buildings. In South Korea, the energy demand for cooling in residential buildings has been increasing rapidly. Focusing on possible passive design strategies, the present study investigated the daylight and cooling performance of several shading devices in a residential building.

Among various types of shading devices, four shading devices were selected including the Venetian blind, horizontal louver, light shelf, and egg-crate. Using two different simulations, the illuminance level and annual cooling energy reduction by various design variables of these shading devices were analyzed. For both assessments, the illuminance and cooling energy performance were varied by the design options of shading devices. For the daylight assessment, all the shading devices blocked the sunlight effectively. Specifically, the light shelf and egg-crate were able to block a significant amount of sunlight in a building during the summer, while admitting sunlight during the winter. Regarding the cooling energy consumption, the overhang-shaped louver and egg-crate were able to reduce the cooling demand more effectively than the other shading devices. Among various design options of these two devices, the overhang-shaped louver with a projection length of 1.9 m and the egg-crate shading with a vertical and horizontal slat height of 0.5 m reduced about 51% to 54% of the annual cooling energy consumption. Because of only a 3% difference in the cooling energy consumption, the artificial lighting loads of these two design options were compared. As a result, about 15% of the artificial lighting load was reduced by the use of the egg-crate shading with a vertical and horizontal slat height of 0.5 m, while the overhang-shaped louver only reduced 5% of the load. Therefore, it is necessary to consider both daylight and energy performance for developing energy reduction strategies by using shading devices.

Considering the outcome of the present study, the proper use of shading devices has the potential for energy saving by reducing the cooling and artificial lighting loads in a building. Moreover, the obtained result can be used for the development of energy-efficient building design. Finally, it can reduce greenhouse gas emissions. For further study, more various design options and shading devices will be included in the analysis of energy efficiency and daylighting performance. In addition, it is necessary to consider the annual daylighting performance for more accurate analysis.

Author Contributions: T.L. and W.S.Y. designed and performed the simulation and collected the data. D.D.K. wrote the manuscript and analyzed the data. All authors have read and agreed to the published version of the manuscript.

Funding: This research received no external funding

Conflicts of Interest: The authors declare no conflict of interest.

References

1. Kim, M.J.; Cho, M.E.; Kim, J.T. Energy use of households in apartment complexes with different service life. *Energy Build.* **2013**, *66*, 591–598. [CrossRef]
2. Ihm, P.; Park, L.; Krarti, M.; Seo, D. Impact of window selection on the energy performance of residential buildings in south korea. *Energy Policy* **2012**, *44*, 1–9. [CrossRef]
3. The Annual Energy Consumption in 2017. Korea Energy Agency. Available online: http://www.Energy.Or.Kr/renew_eng/main/main.Aspx (accessed on 10 January 2019).
4. Choi, I.Y.; Cho, S.H.; Kim, J.T. Energy consumption characteristics of high-rise apartment buildings according to building shape and mixed-use development. *Energy Build.* **2012**, *46*, 123–131. [CrossRef]
5. Seo, Y.-K.; Hong, W.-H. Constructing electricity load profile and formulating load pattern for urban apartment in korea. *Energy Build.* **2014**, *78*, 222–230. [CrossRef]
6. Lee, S.; Kim, S.; Na, Y. Comparative analysis of energy related performance and construction cost of the external walls in high-rise residential buildings. *Energy Build.* **2015**, *99*, 67–74. [CrossRef]
7. Park, J.S.; Lee, S.J.; Kim, K.H.; Kwon, K.W.; Jeong, J.-W. Estimating thermal performance and energy saving potential of residential buildings using utility bills. *Energy Build.* **2016**, *110*, 23–30. [CrossRef]
8. Sharma, V.; Rai, A.C. Performance assessment of residential building envelopes enhanced with phase change materials. *Energy Build.* **2020**, *208*, 109664. [CrossRef]
9. Yaşar, Y.; Kalfa, S.M. The effects of window alternatives on energy efficiency and building economy in high-rise residential buildings in moderate to humid climates. *Energy Convers. Manag.* **2012**, *64*, 170–181. [CrossRef]
10. Tetlow, D.; De Simon, L.; Liew, S.Y.; Hewakandamby, B.; Mack, D.; Thielemans, W.; Riffat, S. Cellulosic-crystals as a fumed-silica substitute in vacuum insulated panel technology used in building construction and retrofit applications. *Energy Build.* **2017**, *156*, 187–196. [CrossRef]

11. Park, J.H.; Wi, S.; Chang, S.J.; Kim, S. Analysis of energy retrofit system using latent heat storage materials applied to residential buildings considering climate impacts. *Appl. Therm. Eng.* **2020**, *169*, 114904. [CrossRef]
12. De Gracia, A. Dynamic building envelope with pcm for cooling purposes–proof of concept. *Appl. Energy* **2019**, *235*, 1245–1253. [CrossRef]
13. Sghiouri, H.; Mezrhab, A.; Karkri, M.; Naji, H. Shading devices optimization to enhance thermal comfort and energy performance of a residential building in morocco. *J. Build. Eng.* **2018**, *18*, 292–302. [CrossRef]
14. Fine, J.P.; Gray, J.; Tian, X.; Touchie, M.F. An investigation of alternative methods for determining envelope airtightness from suite-based testing in multi-unit residential buildings. *Energy Build.* **2020**, *214*, 109845. [CrossRef]
15. Gasparella, A.; Pernigotto, G.; Cappelletti, F.; Romagnoni, P.; Baggio, P. Analysis and modelling of window and glazing systems energy performance for a well insulated residential building. *Energy Build.* **2011**, *43*, 1030–1037. [CrossRef]
16. Berardi, U.; Kisilewicz, T.; Kim, S.; Lechowska, A.; Paulos, J.; Schnotale, J. Experimental and numerical investigation of the thermal transmittance of pvc window frames with silica aerogel. *J. Build. Eng.* **2020**, *2020*, 101665. [CrossRef]
17. Paulos, J.; Berardi, U. Optimizing the thermal performance of window frames through aerogel-enhancements. *Appl. Energy* **2020**, *266*, 114776. [CrossRef]
18. Cho, J.; Yoo, C.; Kim, Y. Viability of exterior shading devices for high-rise residential buildings: Case study for cooling energy saving and economic feasibility analysis. *Energy Build.* **2014**, *82*, 771–785. [CrossRef]
19. Liu, S.; Kwok, Y.T.; Lau, K.K.-L.; Chan, P.W.; Ng, E. Investigating the energy saving potential of applying shading panels on opaque façades: A case study for residential buildings in hong kong. *Energy Build.* **2019**, *193*, 78–91. [CrossRef]
20. Alhuwayil, W.K.; Abdul Mujeebu, M.; Algarny, A.M.M. Impact of external shading strategy on energy performance of multi-story hotel building in hot-humid climate. *Energy* **2019**, *169*, 1166–1174. [CrossRef]
21. Huo, H.; Xu, W.; Li, A.; Cui, G.; Wu, Y.; Liu, C. Field comparison test study of external shading effect on thermal-optical performance of ultralow-energy buildings in cold regions of china. *Build. Environ.* **2020**, *180*, 106926. [CrossRef]
22. Chan, A.L.S. Effect of adjacent shading on the energy and environmental performance of photovoltaic glazing system in building application. *Energy* **2019**, *187*, 115939. [CrossRef]
23. Mun, S.-H.; Kwak, Y.; Huh, J.-H. A case-centered behavior analysis and operation prediction of ac use in residential buildings. *Energy Build.* **2019**, *188–189*, 137–148. [CrossRef]
24. Song, D.; Choi, Y.-J. Effect of building regulation on energy consumption in residential buildings in korea. *Renew. Sustain. Energy Rev.* **2012**, *16*, 1074–1081. [CrossRef]
25. Kristl, Ž.; Košir, M.; Trobec Lah, M.; Krainer, A. Fuzzy control system for thermal and visual comfort in building. *Renew. Energy* **2008**, *33*, 694–702. [CrossRef]
26. Palmero-Marrero, A.I.; Oliveira, A.C. Effect of louver shading devices on building energy requirements. *Appl. Energy* **2010**, *87*, 2040–2049. [CrossRef]
27. Lim, Y.-W.; Kandar, M.Z.; Ahmad, M.H.; Ossen, D.R.; Abdullah, A.M. Building façade design for daylighting quality in typical government office building. *Build. Environ.* **2012**, *57*, 194–204. [CrossRef]
28. Bellia, L.; De Falco, F.; Minichiello, F. Effects of solar shading devices on energy requirements of standalone office buildings for italian climates. *Appl. Therm. Eng.* **2013**, *54*, 190–201. [CrossRef]
29. Al-Tamimi, N.A.; Fadzil, S.F.S. The potential of shading devices for temperature reduction in high-rise residential buildings in the tropics. *Procedia Eng.* **2011**, *21*, 273–282. [CrossRef]
30. Gratia, E.; De Herde, A. The most efficient position of shading devices in a double-skin facade. *Energy Build.* **2007**, *39*, 364–373. [CrossRef]
31. Uribe, D.; Vera, S.; Bustamante, W.; McNeil, A.; Flamant, G. Impact of different control strategies of perforated curved louvers on the visual comfort and energy consumption of office buildings in different climates. *Sol. Energy* **2019**, *190*, 495–510. [CrossRef]
32. Bessoudo, M.; Tzempelikos, A.; Athienitis, A.K.; Zmeureanu, R. Indoor thermal environmental conditions near glazed facades with shading devices – part i: Experiments and building thermal model. *Build. Environ.* **2010**, *45*, 2506–2516. [CrossRef]
33. Datta, G. Effect of fixed horizontal louver shading devices on thermal perfomance of building by trnsys simulation. *Renew. Energy* **2001**, *23*, 497–507. [CrossRef]

34. Tzempelikos, A.; Athienitis, A.K. The impact of shading design and control on building cooling and lighting demand. *Sol. Energy* **2007**, *81*, 369–382. [CrossRef]
35. Cheng, C.-L.; Liao, L.-M.; Chou, C.-P. A study of summarized correlation with shading performance for horizontal shading devices in taiwan. *Sol. Energy* **2013**, *90*, 1–16. [CrossRef]
36. Koo, S.Y.; Yeo, M.S.; Kim, K.W. Automated blind control to maximize the benefits of daylight in buildings. *Build. Environ.* **2010**, *45*, 1508–1520. [CrossRef]
37. Frontini, F.; Kuhn, T.E. The influence of various internal blinds on thermal comfort: A new method for calculating the mean radiant temperature in office spaces. *Energy Build.* **2012**, *54*, 527–533. [CrossRef]
38. Korea Meteorological Administration. Available online: https://www.Weather.Go.Kr/weather/main.Jsp (accessed on 10 January 2019).
39. Korean Statistical Information Service. Available online: http://kosis.Kr/index/index.Do (accessed on 10 January 2019).
40. Rennie, D.; Parand, F. *Environmental Design Guide for Naturally Ventilated and Daylit Offices*; Building Research Communications Ltd.: London, UK, 1998.
41. Radiane. Available online: https://windows.Lbl.Gov/software/radiance (accessed on 24 May 2019).
42. Illuminating Engineering Society of North America (IESNA). *Ies Spatial Daylight Autonomy (SDA) and Annual Sunlight Exposure (ASE)*; lm-83-12; Iesna Lighting Measurement: New York, NY, USA, 2012.
43. IES VE. Available online: https://www.Iesve.Com/software/virtual-environment (accessed on 24 May 2019).
44. American Society of Heating, Refrigerating and Air Conditioning Engineers. *Ashrae Guideline 14-2002, Measurement of Energy and Demand Savings–Measurement of Energy, Demand and Water Savings*; American Society of Heating, Refrigerating and Air Conditioning Engineers: Atlanta, GA, USA, 2002.
45. David, L.D.; Houser, K.; Mistrick, R.; Steffy, G.R. *The Lighting Handbook*, 10th ed.; Illuminating Engineering Society: New York, NY, USA, 2011. Available online: https://www.Ies.Org/ (accessed on 24 May 2019).
46. Alzoubi, H.H.; Al-Zoubi, A.H. Assessment of building façade performance in terms of daylighting and the associated energy consumption in architectural spaces: Vertical and horizontal shading devices for southern exposure facades. *Energy Convers. Manag.* **2010**, *51*, 1592–1599. [CrossRef]
47. Nematchoua, M.K.; Noelson, J.C.V.; Saadi, I.; Kenfack, H.; Andrianaharinjaka, A.-Z.F.R.; Ngoumdoum, D.F.; Sela, J.B.; Reiter, S. Application of phase change materials, thermal insulation, and external shading for thermal comfort improvement and cooling energy demand reduction in an office building under different coastal tropical climates. *Sol. Energy* **2020**, *207*, 458–470. [CrossRef]
48. Han, Y.; Taylor, J.E.; Pisello, A.L. Exploring mutual shading and mutual reflection inter-building effects on building energy performance. *Appl. Energy* **2017**, *185*, 1556–1564. [CrossRef]
49. Li, D.H.W.; Wong, S.L. Daylighting and energy implications due to shading effects from nearby buildings. *Appl. Energy* **2007**, *84*, 1199–1209. [CrossRef]

energies

MDPI

Article

Total Solar Reflectance Optimization of the External Paint Coat in Residential Buildings Located in Mediterranean Climates

Tiago Souto [1], Margarida Almeida [2], Vítor Leal [3], João Machado [1] and Adélio Mendes [2,*]

1 Corporação Industrial do Norte, S.A., avenida Dom Mendo, n. 831, Apartado 1008, 4471-909 Maia, Portugal; tiago.souto@cin.pt (T.S.); jmachado@cin.pt (J.M.)
2 Laboratory for Process Engineering, Environment, Biotechnology and Energy, Faculty of Engineering, University of Porto, Rua Dr. Roberto Frias, 4200-465 Porto, Portugal; margarida.almeida@fe.up.pt
3 Department of Mechanical Engineering, Faculty of Engineering, University of Porto, Rua Dr. Roberto Frias, 4200-465 Porto, Portugal; vleal@fe.up.pt
* Correspondence: mendes@fe.up.pt; Tel.: +35-122-508-1695; Fax: +35-122-508-1449

Received: 17 April 2020; Accepted: 27 May 2020; Published: 29 May 2020

check for updates

Abstract: This work addresses the effect of the total solar reflectance (TSR) value of paints applied in residential buildings upon their thermal performance. A semi-detached residential building was modeled in the ESP-r software, and taken as the basis for parametric studies which assessed the effects of variations in (i) the TSR values; (ii) the thermal characteristics of the building envelope; (iii) the location/climate; and: (iv) the way how the indoor temperature is controlled. The parametric studies were used to find optimal TSR values for each combination of Location + Building envelope characteristics (mainly the existence of thermal insulation). It was concluded that paints having a carefully chosen TSR value lead to better indoor thermal temperatures if the buildings have no mechanical heating or cooling, or to energy savings of up to 32% if they do.

Keywords: energy efficiency; solar reflectance; building simulation; thermal comfort; passive architecture

1. Introduction

Climate change refers to a statistically significant variation, persisting for an extended period, in either the mean state of the climate or in its variability [1]. Climate change can be caused by anthropogenic actions, particularly by the increase of greenhouse gases (GHG) emissions in the atmosphere. The climate change has serious and severe consequences, direct and/or indirectly related with human's life, such as: extreme weather events [2], mass movements [3], temperature rises in mainland's [4] and ocean's surface [5], melting ice of the polar ice caps, rising sea level, among others [1].

According to the European Commission, the Mediterranean area "is becoming drier, making it even more vulnerable to drought and wildfires" [6]. Residential demand will be impacted by climate change due to the increase in average temperature, weather extremes, and the consequent change on space heating and cooling needs [7]. In 2017, in the EU-28, the residential sector accounted for 27.2% of the final end-use of energy [8]. The increase in GHG is mainly related to the increase in energy consumption. To decrease the amount of GHG realized to the atmosphere, the new and existing building stock needs the implementation of both renewable energy supply and energy efficiency measures.

The thermal insulation of buildings has been applied since the earliest civilizations, which used soil as an insulator. During the Industrial Revolution, mineral fibers were introduced as pipe insulation and later adapted for buildings as fiberboards [9]. In the 1910s, with the introduction of the first HVAC (heating, ventilation, and air-conditioning) systems in buildings [10], major concerns about the thermal insulation of buildings have emerged. Several approaches were considered, namely ETICS (external thermal insulation composite systems) with numerical [11], and experimental studies [12], which confirms its benefits in terms of thermal comfort and reduction of initial and operative costs in air conditioning [13]. Also, multilayer façades can be considered [14], such as a double-wall façade with and without thermal insulation [15], or a ventilated façade able to reverse the heat flux through the building envelope and reduce the indoor heating demand [16], among others. More recently, an innovative thermal comfort passive strategy became commercially available: the use of phase change materials (PCM). Contrary to the common thermal insulation (such as ETICS), PCM allows obtaining indoor thermal comfort with thinner building envelopes, because they do not require high volume, which is normally a limitation in the building sector [17].

Studies concerning thermal comfort are not limited to the vertical opaque envelope of buildings; the heat transfer through the building's roof has also been assessed. A good example was reported by Pisello et al. [18], that studied the thermal and energy performance of a historic residential building with a green roof. The simulation results showed that green roofs could decrease the indoor temperature during the summer period up to 3 °C compared to a concrete roof (CR) and up to 6 °C compared to a bitumen roof (BR); in winter, green roofs allowed to increase the indoor temperature up to 2 °C compared to CR and ca. 3 °C compared to BR, due to the high thermal insulation produced by the vegetation layer. Assuming a comfort temperature range between 20 °C–26 °C, these researchers concluded that the green roof with green grass reduced the annual primary energy requirements of about 29.4% compared to CR and approximately 42.1% compared to BR.

Although glazed areas have an important role to obtain thermal comfort, opaque façades often occupy most of the building envelope, especially in old buildings, and could account for a great part of the total heat gain or loss. For instance, it is more likely for a high-absorptivity opaque wall to achieve 60–70 °C, under strong solar radiation, rather than a transparent window with high transmissivity [19].

Solar reflectance is the ratio of solar energy reflected by a surface [20]. The reflectance of a building's coating is normally characterized by the percentage of total solar reflectance (TSR). Low values of TSR are associated with darker colors, while light colors typically have a high TSR [21]. High or low values of a surface's solar reflectance may lead, respectively, to lower or higher heat gains [22] and surface temperatures [23]. Consequently, the buildings' coating color choice is often considered a solar radiation control strategy [24]. Paints may be designed to retain the solar heat (hot paints) or to reflect most of the solar radiation (cool paints) to prevent the treated surface from heating [25].

Cool paints are high solar-reflective coatings, and a passive thermal comfort strategy easy to apply, with a short period of investment cost recovery, which can be applied to any type of building (either residential or office or industrial, either with old or new construction, either single-story or high-rise) [26]. Their application is especially useful in hot climates: with no heating needs, and without considerable winter drawbacks; and in warm climates: when no cooling system has been installed, and cool paints account for the reduction of cooling needs [27].

The Lawrence Berkeley National Laboratory (LBNL) had been conducting an extensive researcher concerning the development and the use of cool paints. Researchers of this laboratory identified and characterized cool pigments with high reflective properties and created a free access pigment database [28]. Cool pigments, similarly to hot pigments, are pigments that display a much different reflectance in the infrared IR spectrum compared with the visible one [29]. In the case of the cool pigments, they display a very high reflectance in the IR spectrum while displaying a lower reflectance in the visible spectrum.

Following, LBNL developed experimental and simulation works to study the impact of coating with cool paints residential [30] or non-residential roofs [31,32] in terms of thermal comfort and energy saving [33], and improvement of air quality in urban areas [34]. Weathering tests and studies including paint coating soiling and cleaning effects were also performed [35,36]. In one of these studies [32,37], researchers of LBNL performed an experimental test, on six California commercial buildings located in Sacramento, San Marcos, and Reedley. The energy performance and the environment parameters of each building were recorded considering a cool roof application (TSR value of about 70%). These authors observed reductions between 15.6 °C and 23.9 °C on the maximum roof surface temperature. The experimental values allowed them to calibrate a simulation model used to estimate the energy savings of similar buildings located in all sixteen California climate zones. These authors concluded that energy savings up to 6.6 $W{\cdot}m^{-2}$ could be achieved on the average peak demand.

Despite the beneficial effect of using cool paints, some studies depict that these paints may also harm the thermal comfort of a building. Namely, Dias et al. [26] simulated the impact of altering the value of TSR from 50% to 92% both on the external walls and roof of a residential building located in three different cities of Portugal. They observed a summer maximum indoor temperature drop of ca. 4 °C, and cooling energy savings up to 1.9 $MWh{\cdot}y^{-1}$. Nonetheless, these authors also observed a noteworthy increase in the heating demand of up to 2.9 $MWh{\cdot}y^{-1}$ (a penalty of about 25.9%); the overall effect was then a significant upsurge in energy consumption for reaching indoor thermal comfort.

To compensate for the heating penalty of using cool materials, thermochromic materials have been investigated for their great potential in reducing, simultaneously, the cooling and heating demand in buildings. Thermochromic paints can change the absorption of solar radiation dynamically according to external temperature [38]. Granadeiro et al. [21] performed thermal and energy simulations with thermochromic paints applied in the residential building façade located in Porto, Madrid, and Abu Dhabi. For all the climates, the application of the optimized thermochromic paint always leads to annual energy savings, up to 51% when compared with black paint and 48% when compared with white paint. However, the cost of thermochromic materials are still very expensive [39,40], and further developments on the chemical formulation of such paints to ensure longevity and cost-reductions should be encouraged. Another advanced option for adaptive façade response to the weather is the use of dynamic facades, capable of rotation and folding motion according to the external conditions. Xuepeng et al. [41] depicted energy consumption reductions by 14 21% with dynamic façades implemented, nonetheless, they also emphasized the high initial investment and maintenance cost that dynamic façades require compared to pristine façades.

As seen from the review, cool paints are cost-effective and easy to apply, while thermochromic paints and dynamic façades are costly and emergent techniques, even though their great potential. Nevertheless, cool paints typically improve performance related to summer but slightly worsen performance related to winter. There must, therefore, exist a trade-off that has an optimum for some TSR value. This work focuses precisely on the optimization of the TSR value of paints to obtain the maximum energy savings when applied both on the exterior façades and on the roof of a residential building. The study considers different cities of the Mediterranean European area: Bragança, Porto, Coimbra, Lisbon, Évora, Beja, Faro, Lajes and Funchal, in Portugal; Madrid, Barcelona, Seville, and Tenerife, in Spain; Paris and Nice, in France; Rome, in Italy; and Athens, in Greece.

Previous studies were focused on the optimization of solar reflectance to be applied to roof coatings or building exterior walls. For example, Yuan et al. [42] studied the combined optimum values of reflectivity and insulation thickness of building exterior walls for different regions of Japan, while Piselli et al. [43] reported optimum values of roof solar reflectance for six different Italian climatic zones. As far as the authors are aware, this work focuses, for the first time, on the optimization of the total solar reflectance value of paints, applied on both the external walls and the roof, to decrease the space heating and cooling needs. This article tackles the case of the European Mediterranean climate, which differs from most of the central and northern European climate zones, and shows appropriate solutions for this region, taking also into account the presence or absence of thermal insulation, as well as, the presence or absence of a thermal control system in residential buildings.

2. Case Study Description and Simulation Scenarios

2.1. Building Description

This study considers a single villa, with two floors and a not inhabited attic, with 90 m^2 of floor area per story, as shown in Figure 1. Briefly, the ground floor has a living room, a dining room, a kitchen, a toilet, a hall, and stairs. The first floor has three bedrooms, a suite, a toilet, a bathroom, a corridor, and stairs. The attic is an open space, as mentioned before not inhabited.

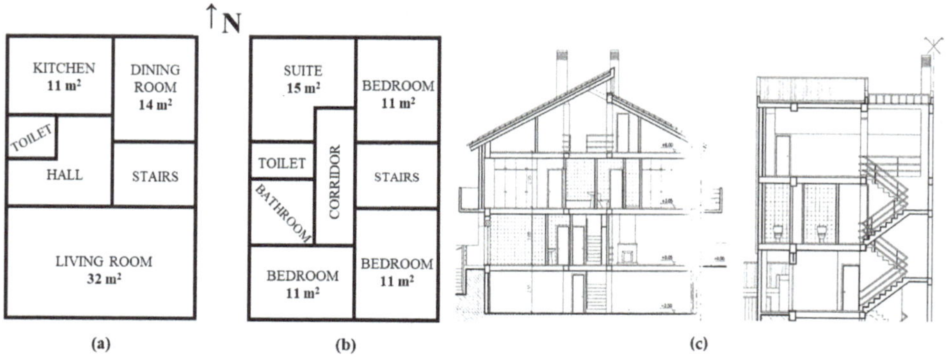

Figure 1. Plans of the residential building: (**a**) ground floor and (**b**) first floor, and (**c**) building sections.

The residential building has a total area of 26.3 m^2 of windows with a single clear glass. It was considered the existence of venetian blinds on windows that cover 50% of each window. The zones considered to obtain the building indoor temperature were the living room, dining room, and the kitchen on the ground floor and the three bedrooms and the suite on the first floor. The internal gains of the zones previously mentioned were settled to 4 $W{\cdot}m^{-2}$. On the remaining zones, it was assumed 1 $W{\cdot}m^{-2}$ of internal gains (except for the uninhabited attic that was assumed to have no internal gains). A constant value of 0.6 air changes per hour was considered.

The dynamic thermal behavior and energy demand of this residential building were simulated using the open-source software ESP-r [44]. ESP-r is a building performance tool based on finite volume formulation, in which a problem specified in terms of geometry, construction, operation, leakage distribution, is converted into a set of discretized conservation equations—energy, mass, and momentum—which are then solved at successive time-steps in response to climate, occupant and control system influences—boundary conditions. ESP-r can model thermal, visual, and acoustic performance of buildings as well as estimate heat, moisture, and electrical power needs [44–46]. ESP-r has been extensively validated [47–49].

Two building construction types were considered: (i) a building with no thermal insulation, neither on the façades nor on the roof (named case BD1); and (ii) a building with thermal insulation both on external walls and on the roof slab (named case BD2).

The construction details of BD1 and BD2 are shown in Tables 1 and 2. The specifications of each layer of the construction (e.g., conductivity, density, specific heat, emissivity, and absorption) follow the recommendations by the Portuguese national legislation [50].

Table 1. Construction details of the building with a single wall façade (BD1).

Construction	Layer	Thickness/mm	$U/W{\cdot}m^{-2}{\cdot}K^{-1}$
External wall	Coating	0.2	1.3
	Plaster	30	
	Brick	190	
	Plaster	10	
Internal wall	Plaster	10	2.1
	Brick	110	
	Plaster	10	
Ground floor	Common earth	300	0.7
	Gravel	300	
	Concrete	300	
	Asphalt	10	
	Concrete	20	
	Wood floor	20	
Window	Clear glass	8	5.6
Slab	Plaster	10	2.0
	Concrete	240	
	Plaster	10	
Roof slab	Clay tile	5	1.6
	Air	3	
	Plaster	5	
	Brick	110	
	Plaster	5	

Table 2. Construction details of BD2—in bold the differences from BD1.

Construction	Layer	Thickness /mm	$U/W{\cdot}m^{-2}{\cdot}K^{-1}$
External wall	Coating	0.2	**0.5**
	Plaster	30	
	Insulation	**60**	
	Brick	190	
	Plaster	10	
Roof slab	Clay tile	5	**0.5**
	Air	3	
	Plaster	5	
	Insulation	**60**	
	Brick	110	
	Plaster	5	

2.2. Simulation Scenarios

The study considers different cities of the Mediterranean European area (except Paris): Bragança, Porto, Coimbra, Lisbon, Évora, Beja, Faro, Lajes, and Funchal, in Portugal; Madrid, Barcelona, Seville, and Tenerife, in Spain; Paris and Nice, in France; Rome, in Italy; and Athens, in Greece. To characterize the different climates, values of minimum and maximum outdoor air temperatures—during the coldest and hottest months, are shown in Figure 2, while values of sunshine hours—during the coldest and warmest months, and rainfall—during the wettest and driest months, are shown in Figure 3.

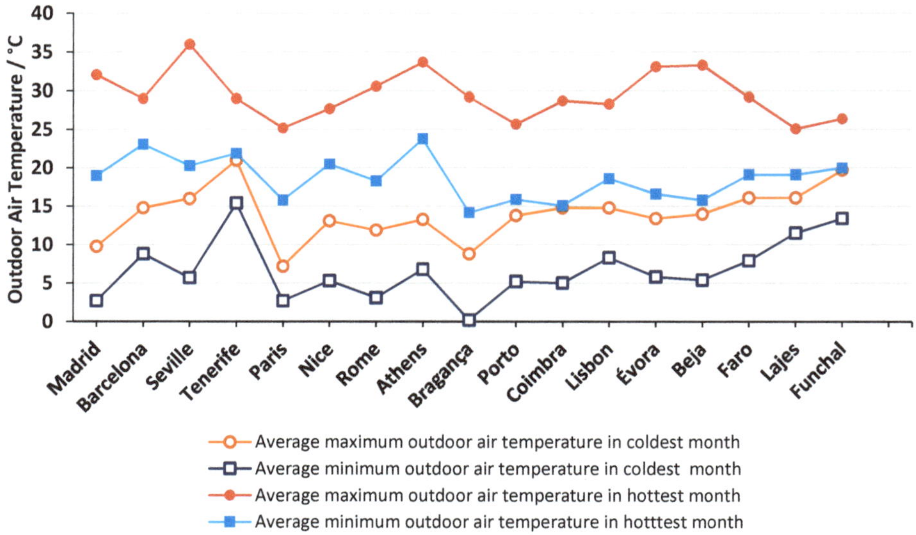

Figure 2. Average minimum and maximum outdoor air temperature during the coldest and hottest months, for all the cities under study, according to [51].

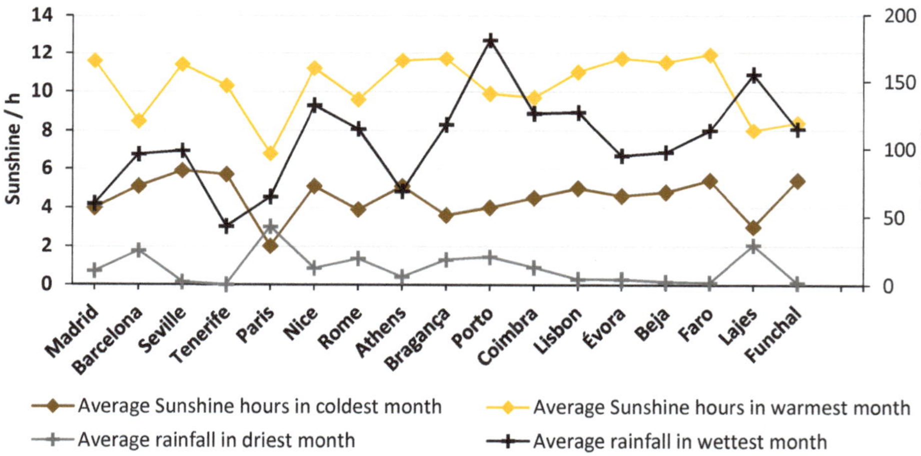

Figure 3. Daily sunshine hours during the coldest and warmest months, and rainfall during the wettest and driest months, for all the cities under study, according to [51].

There are cities in which the outdoor air temperature variation between the average minimum and maximum values can differ approximately 30 °C, such as Seville, Madrid, and Bragança. Cities with lower temperature variation are mostly islands, such as Tenerife, Funchal, and Lajes, in which the average minimum outdoor temperature during the coldest months is above 10 °C. Most of the Mediterranean cities display ca. 10 h of sunshine during the warmest months and 5 h during the coldest ones. Paris, the only non-Mediterranean city, differs from the others by displaying lower values of sunshine hours and a smaller rainfall range. Coastal cities such as Porto, Lajes, and Nice display higher rainfall values during the wettest months than inland cities, such as Madrid, Évora, and Beja.

2.2.1. Case A (Conditioned Mode)

The energy demand needed to keep the indoor temperature within the recommended range of the Portuguese legislation at the time of the calculations (between 20 °C and 25 °C, [52]) was assessed by a full year simulation. The climatized room area is shown in Figure 1, totaling 105 m^2. The cooling and heating thermal energy needs were converted into final electrical energy demand by considering as air conditioning system a reference heat pump with an average coefficient of performance (COP) of 4 for heating and 3 for cooling (also called EER instead of COP, in some literature). While practice the COP values do depend on the values of the outdoor temperature, this approach avoids having to choose a specific machine for the COP variation, which would introduce some level of arbitrariness. Given that the final heating and cooling energy demand (E_{demand}) are presented in the following chapter for both building simulation cases (BD1 and BD2).

2.2.2. Case B (Free-Floating Mode)

In this case, only a free-floating simulation (building having no mechanical cooling nor heating) was performed to evaluate the indoor temperature over the year. The weekly average daily maximum temperature (T_{max}) and the weekly average daily minimum temperature (T_{min}) were determined. Besides assessing the maximum and minimum temperatures, an indicator of time-integrated indoor temperature deviation from the comfort zone was also considered. The indicator f_E was introduced, and defined as given by Equation (1):

$$f_B = \begin{cases} 0 \ , \ T \in [20, 25] \\ \sum_{i=1}^{365} (\overline{T_M} - 25)^2 \wedge \overline{T_M} > 25 + \sum_{i=1}^{365} (20 - \overline{T_m})^2 \wedge \overline{T_m} < 20 \end{cases} \quad (1)$$

where $\overline{T_M}$ is the average maximum temperature of each day of the year (in °C) and $\overline{T_m}$ is the average minimum temperature of each day of the year (in °C).

To determine the optimized TSR value for each city and each type of building (BD1 and BD2), in cases A and B, several dynamic simulations were performed with an increment of 5% of the TSR value between simulations, from 5% to 90%. The optimized TSR value corresponds to the smallest value of annual energy consumption, in case A, and to the minimization of objective functions f_B (described above), for case B. All the weather data used were obtained from the US Department of Energy/Energy Plus database [53]. Simulations were run with 10-min time-step, with results integrated over each hour.

The reference used for assessing the optimized external coating TSR was always the corresponding building coated with 50% TSR paint, both on the façades and the roof.

3. Results and Discussion

The TSR values that achieved higher energy savings (in case A) and higher f_B reductions (in case B) are presented in Table 3 for several Mediterranean cities, and in Table 4 for Portuguese cities. The maximum and minimum indoor temperatures (T_{max} and T_{min}) are also compared between the reference case (50% TSR) and the optimized TSR value.

Table 3. Energy demand in case A (conditioned mode) and Indoor temperature indicators for case B (free-floating), as function of the Optimized TSR values for several Mediterranean cities.

			Case A			Case B				
City	Building	TSR	TSR Value	E_{demand}/kWh·m^{-2}·y^{-1}	E_{demand} Reduction	TSR Value	T_{min}/°C	T_{max}/°C	f_B	f_B Reduction
Madrid	BD1	Reference	50%	90.78	3.3%	50%	1.2	37.1	48,462	1.4%
		Optimal	25%	87.74		35%	1.5	39	47,783	
	BD2	Reference	50%	45.90	3.2%	50%	2.6	32.1	39,790	5.4%
		Optimal	5%	44.45		25%	3.2	34.2	37,634	
Barcelona	BD1	Reference	50%	67.90	4.5%	50%	4.6	31.8	27,447	2.8%
		Optimal	15%	64.85		30%	5	33.7	26,668	
	BD2	Reference	50%	33.68	3.4%	50%	5.6	28.4	22,561	8.3%
		Optimal	10%	32.52		10%	6.4	31.1	20,699	
Seville	BD1	Reference	50%	59.61	0.7%	50%	4.9	38.9	25,599	6.7%
		Optimal	60%	59.22		70%	4.5	36.4	23,883	
	BD2	Reference	50%	29.13	1.0%	50%	7.1	33.7	15,708	0.2%
		Optimal	65%	28.84		55%	7	33.3	15,670	
Tenerife	BD1	Reference	50%	18.01	22.4%	50%	14.8	33.4	8733	41.5%
		Optimal	85%	13.97		90%	14.3	31	5107	
	BD2	Reference	50%	11.09	32.2%	50%	15.4	29.9	4697	28.5%
		Optimal	90%	7.51		90%	14.9	28.5	3358	
Paris	BD1	Reference	50%	123.66	7.6%	50%	−3	28.7	71,881	6.8%
		Optimal	5%	114.29		5%	−2.5	33.3	66,982	
	BD2	Reference	50%	62.30	6.3%	50%	−0.7	27.3	64,454	7%
		Optimal	5%	58.36		5%	−0.2	28.9	59,946	
Nice	BD1	Reference	50%	71.90	2.8%	50%	6.1	32.3	28,663	1.7%
		Optimal	20%	69.90		35%	6.5	33.9	28,189	
	BD2	Reference	50%	35.19	1.7%	50%	7.1	28.7	23,933	6.0%
		Optimal	20%	34.59		15%	7.8	31.6	22,487	
Rome	BD1	Reference	50%	72.14	2.4%	50%	4.7	32.1	30,392	4.4%
		Optimal	20%	70.41		25%	5.2	34.6	29,065	
	BD2	Reference	50%	35.34	2.1%	50%	6.3	28.7	24,338	9.7%
		Optimal	15%	34.60		5%	7	31.4	21,987	
Athens	BD1	Reference	50%	69.04	0.1%	50%	6.4	36.8	27,220	0.3%
		Optimal	55%	68.94		55%	6.3	36.4	27,144	
	BD2	Reference	50%	34.04	0.3%	50%	7.7	32.8	18,890	0.4%
		Optimal	60%	33.92		45%	7.8	33	18,817	

Table 4. Energy demand, in case A (conditioned mode), and indoor temperature indicators for case B (free-floating), as function of the Optimized TSR values for several Portuguese Cities.

			Case A			Case B				
City	**Building**	**TSR**	**TSR Value**	**E_{demand}/kWh·m^{-2}·y^{-1}**	**E_{demand} Reduction**	**TSR Value**	**T_{min}/°C**	**T_{max}/°C**	**f_B**	**f_B Reduction**
Bragança	BD1	Reference	50%	98.01	9.3%	50%	−0.4	32.6	53,678	7.2%
		Optimal	5%	88.86		20%	0.3	36.6	49,812	
	BD2	Reference	50%	49.34	8.6%	50%	1.8	29.9	46,619	11.4%
		Optimal	5%	45.10		5%	2.8	31.9	41,327	
Porto	BD1	Reference	50%	59.31	11.6%	50%	6.2	29.6	24,304	6.4%
		Optimal	5%	52.45		25%	7.0	32.5	22,743	
	BD2	Reference	50%	28.81	10.4%	50%	7.6	28.3	20,539	10.6%
		Optimal	5%	25.82		5%	8.8	29.9	18,354	
Coimbra	BD1	Reference	50%	54.07	5.3%	50%	5.9	32.4	22,265	2.6%
		Optimal	25%	51.21		35%	6.3	33.8	21,676	
	BD2	Reference	50%	26.42	5.6%	50%	7.1	28.9	18,195	8.7%
		Optimal	20%	24.93		5%	8.2	31.2	16,610	
Lisbon	BD1	Reference	50%	52.50	2.7%	50%	8.0	32.1	20,337	1.4%
		Optimal	30%	51.07		40%	8.2	32.6	20,051	
	BD2	Reference	50%	24.94	1.8%	50%	8.8	30.7	16,405	2.8%
		Optimal	30%	24.49		25%	9.1	31.7	15,942	
Évora	BD1	Reference	50%	63.02	3.2%	50%	6.7	33.8	24,676	1.1%
		Optimal	25%	60.99		40%	6.9	34.6	24,401	
	BD2	Reference	50%	30.08	1.9%	50%	7.7	30.4	19,565	5.0%
		Optimal	25%	29.50		20%	8.2	31.6	18,583	
Beja	BD1	Reference	50%	63.40	0.6%	50%	7.8	36.3	26,282	0.1%
		Optimal	40%	63.05		45%	7.9	36.8	26,262	
	BD2	Reference	50%	30.82	0.2%	50%	8.5	32.3	20,251	1.0%
		Optimal	40%	30.74		35%	8.8	33.1	20,042	
Faro	BD1	Reference	50%	47.30	-	50%	9.2	34.4	18,508	0.2%
		Optimal	50%	47.30		55%	9.2	34	18,477	
	BD2	Reference	50%	22.50	0.2%	50%	10.2	31	13,670	≈0%
		Optimal	40%	22.46		45%	10.3	31.2	13,668	
Lajes	BD1	Reference	50%	31.31	6.2%	50%	10.8	28.9	9882	5.3%
		Optimal	20%	29.36		30%	11.1	30.3	9361	
	BD2	Reference	50%	14.94	3.4%	50%	12.2	27.3	7665	12.4%
		Optimal	25%	14.44		5%	12.6	28.7	6712	
Funchal	BD1	Reference	50%	29.57	0.5%	50%	11.3	31.5	11,186	0.5%
		Optimal	45%	29.42		55%	11.2	30.8	11,129	
	BD2	Reference	50%	13.73	0.1%	50%	12.4	30.2	8792	1.1%
		Optimal	45%	13.71		40%	12.6	30.7	8695	

Additionally, Figures 4–7 display examples of coating colors with the optimized TSR value for each location, for cases A and B and building constructions BD1 and BD2, according to the color catalog reported by Resines Company [54].

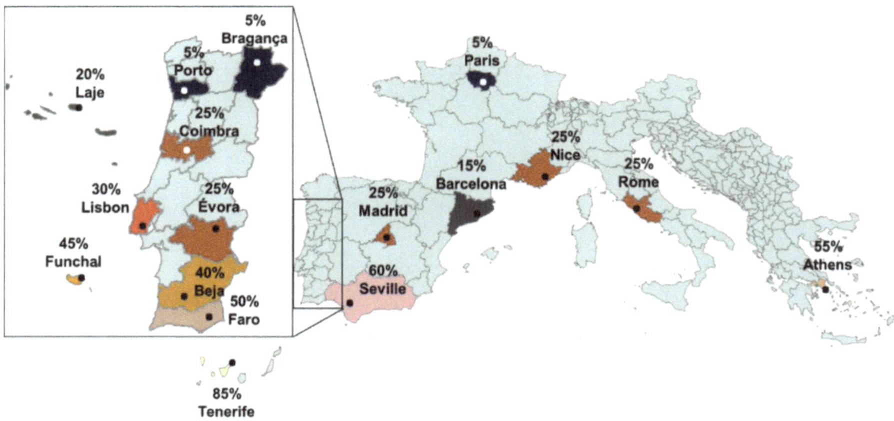

Figure 4. Optimized total solar reflectance (TSR) and pseudo-color, for regions of the Mediterranean cities under study, in case A (with heating, ventilation, and air-conditioning (HVAC) system) and type BD1 (without insulation).

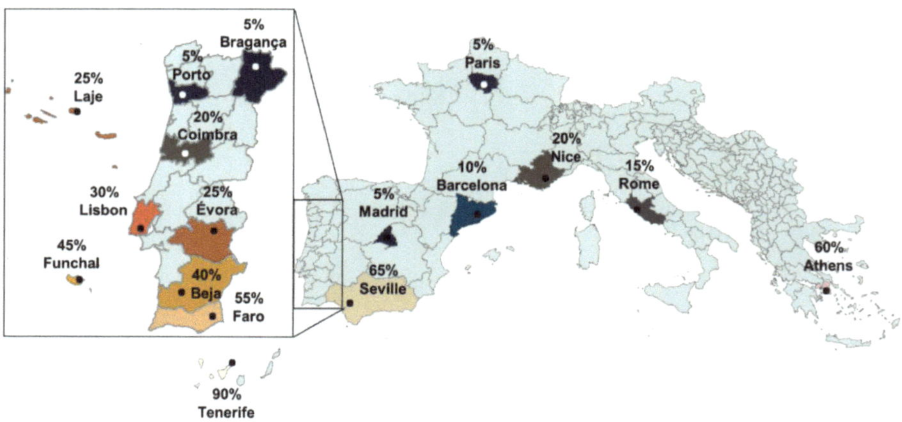

Figure 5. Optimized TSR and pseudo-color, for regions of the Mediterranean cities under study, in case A (with HVAC system) and type BD2 (with insulation).

For Case A, energy savings of up to 32.2% (in BD2, in Tenerife) are observed when the TSR coating is optimized from 50% (reference) to 90%. Since Portugal is a country with heating needs higher than cooling, the optimized TSR values obtained are mostly low, for maximizing the sunlight absorption. The Portuguese city that would take the most advantage from this TSR optimization is Porto, with 11.6% energy savings when coated with 5% TSR paint; and 9.6% if coated with the minimum established value of 25% TSR. The optimized TSR value for BD1 is mostly equal or superior to that of BD2, exceptions are Tenerife, Athens, Seville, Faro, and Lajes—Figures 4 and 5.

In Case B, the indicator f_B decreases by up to 41.5% (in BD1, in Tenerife) when the optimized TSR coatings are used. Building configuration BD2, coated with an optimized TSR paint, always displays the best indoor temperatures, which corresponds to darker paint colors, compared to BD1. Figures 6 and 7 show that generally the mainland region displays lighter colors as one moves from the inland to the coast region, as expected due to the decrease of heating needs.

Considering the studied cases, the optimized TSR values obtained in Case A are similar to those obtained in Case B. Results show that building BD2 coated with the optimized TSR paints, both on the external walls and the roof, always display the best indoor temperatures.

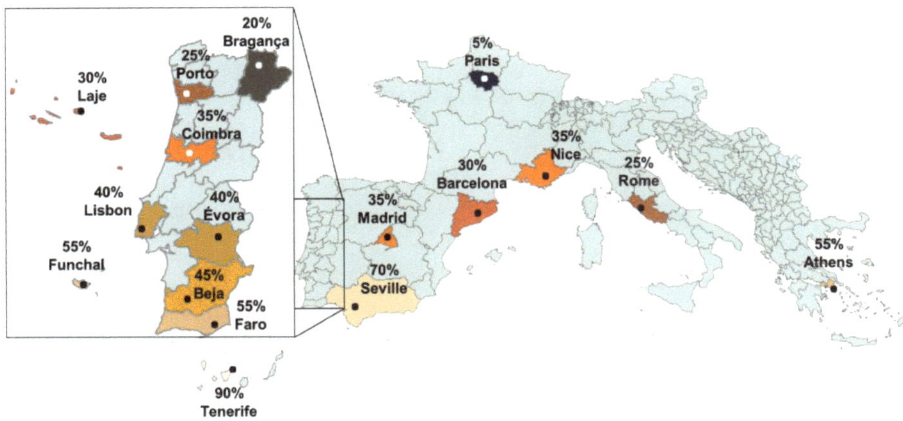

Figure 6. Optimized TSR and pseudo-color, for regions of the Mediterranean cities under study, in case B (without air conditioning) and type BD1 (without insulation).

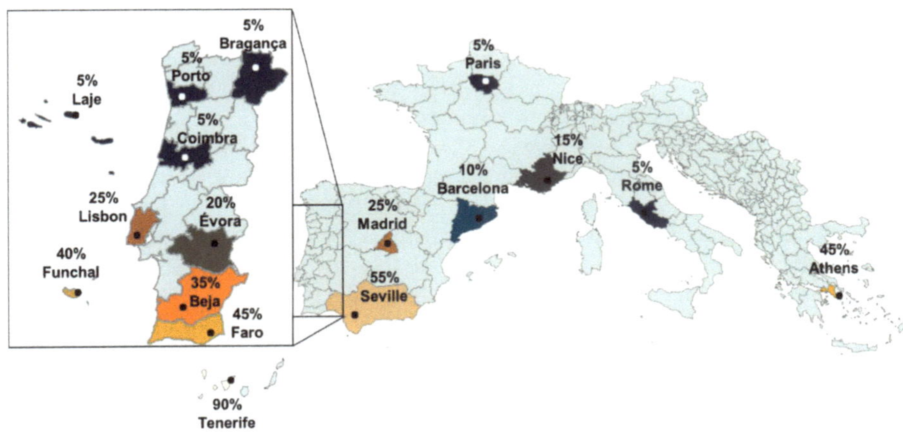

Figure 7. Optimized TSR and pseudo-color, for regions of the Mediterranean cities under study, in case B (without air conditioning) and type BD2 (with insulation).

Table 5 condenses all the optimal values of TSR, and correspondent example of pseudo-colors according to the color catalog reported by Resines Company [54], for each simulation scenario.

Tenerife is the city with the highest optimal values of TSR (85–90%) which led to greater energy savings (32%). This fact confirms the known benefits of high reflective paints when applied in hot climates with cooling needs higher than heating needs, reported elsewhere [27]. However, among the seventeen cities studied, only five (Athens, Faro, Funchal, Seville, and Tenerife) achieved optimized TSR values superior to 50%, showing that the majority of the considered cities display better overall year-round thermal behavior if hot paints (low TSR) are used.

Table 5. Summary of Optimal TSR values achieved for each scenario studied, colored with an example of suitable paint color.

City/Climate	BD1 Without Thermal Insulation		BD2 With Thermal Insulation	
	Case A	**Case B**	**Case A**	**Case B**
Madrid	25%	35%	5%	25%
Barcelona	15%	30%	10%	10%
Seville	60%	70%	65%	55%
Tenerife	85%	90%	90%	90%
Paris	5%	5%	5%	5%
Nice	20%	35%	20%	15%
Rome	20%	25%	15%	5%
Athens	55%	55%	60%	45%
Bragança	5%	20%	5%	5%
Porto	5%	25%	5%	5%
Coimbra	25%	35%	20%	5%
Lisbon	30%	40%	30%	25%
Évora	25%	40%	25%	20%
Beja	40%	45%	40%	35%
Faro	50%	55%	55%	45%
Lajes	20%	30%	25%	5%
Funchal	45%	55%	45%	40%

Dias et al. [26] performed similar simulations to study the performance of cool paints (92% TSR) on residential buildings. Their results for Porto showed an overall annual energy demand upsurge of 28%, when compared to a reference coating (50% TSR), due to the winter penalty on the building heating demand. On the other hand, the results reported here for Porto show that the annual energy demand for heating and cooling, instead of increasing, decreases c.a. 11% when optimized TSR paints are applied.

Hybrid solutions with differential winter-summer behavior thus seem necessary to grasp the most of TSR properties/properties of paints. Granadeiro et al. [21] showed the performance of thermochromic paints compared to common black (5% TSR) and common white (90% TSR) paints, using the same building for energy simulations in three different climates. By comparing their results with the energy demand reported here with optimal TSR paints for Porto (52.5 kWh·m^{-2}·y^{-1}) and Madrid (87.7 kWh·m^{-2}·y^{-1}), in non-insulated buildings (BD1), it is noticeable the lower energy demand when thermochromic paints are used: 52.4 kWh·m^{-2}·y^{-1} for Porto and 42.1 kWh·m^{-2}·y^{-1} for Madrid. Similar differences also occur if the building is thermally insulated, less 10 kWh·m^{-2}·y^{-1} and 5 kWh·m^{-2}·y^{-1} are required with thermochromics, respectively in Porto and Madrid.

For climates where cold is dominant over warmth, such as Paris (optimal 5% TSR), the impact of using thermochromic paints instead of an optimized TSR coating is not that large. Wang et al. [19] depicts ca. 7% of energy savings when a thermally responsive coating is applied in Paris, which is in line with the results of this paper that, for Paris show energy savings of 7.6% and 6.3% for cases BD1 and BD2, respectively. The use of thermochromic external paints in typically cold or hot climates would likely not justify the higher investment in such technology if an optimized TSR coating could produce almost the same benefits. Therefore, the reported TSR optimization can be considered a valuable contribution to match the best external paint coating characteristics with the building environment.

It must also be pointed out that buildings displaying an external coating with a TSR value lower than 25% become very dark and can, therefore, be considered by the public opinion not aesthetic. For the cases where the optimum value is lower than 25%, simulations results including the 25% TSR value, are shown in Appendix A, (Tables A1 and A2). To make very low TSR values compatible with the aesthetic value, the use of hot pigments is suggested [29]. The relation between the optimum TSR and the correspondent pseudo-color is merely indicative: there are numerous different colors for the same TSR value [54], and the solar reflectance also depends upon the material chosen, the presence of impurities and the surface roughness [55]. Nevertheless, the matching between the TSR values and the pseudo-colors helps to visualize the effect of increasing/decreasing the TSR in terms of lightening/darkening the external color of buildings.

For the purpose of this study, some assumptions were made to avoid unnecessary arbitrariness, and keep an overall frame sufficiently focused on clear conclusions. However, these can be interpreted, in some perspectives, as study limitations, which must be acknowledged. For example, the thermal assessment was performed based on the building's indoor temperature variations. Analyses of the operative and indoor surface temperatures could lead to a more accurate thermal comfort assessment; COP values while differing from summer to winter, were not changed when changing climates. Although we acknowledge that seasonal averaged COP values vary with the climate, presenting different COPs for each climate would require decisions such as which specific machine and variation to be considered (since some equipment has larger variations than others). Also, the thermal insulation impact is limited for the two different cases studied: the absence of thermal insulation (BD1) and 60 mm of insulation applied to the external wall and roof (BD2). Different insulation thicknesses could hypothetically lead to different optimal envelope reflectivity values, and such dependence is reported elsewhere [42]. Finally, for future developments, physical tests would be important to validate the simulation results and understand to what extent the simplifications mentioned affect the outcomes.

4. Conclusions

This article discusses the impact of using total solar reflectance (TSR) optimized paints on residential buildings, in terms of energy savings and indoor temperatures. Residential buildings with different thermal characteristics (BD1—non-insulated building and BD2—insulated building) were considered. Two different scenarios of active comfort systems were taken into account: case A—building with an HVAC system that keeps the indoor temperature between 20 °C and 25 °C; and case B—building with no HVAC system. The dynamic thermal behavior and energy demand were simulated using the ESP-r software. The cities under study were located in different Mediterranean countries: Bragança, Porto, Coimbra, Lisbon, Évora, Beja, Faro, Lajes and Funchal, in Portugal; Madrid, Barcelona, Seville, and Tenerife, in Spain; Paris and Nice, in France; Rome, in Italy; and Athens, in Greece.

The results enabled conclusions on two main aspects: The dependence of the optimal TSR values on the location/climate and the influence of the optimized paints upon the thermal performance of the buildings.

In what regards the dependence of the optimal TSR values on the location/climate, as expected, the results show that warmer climates favor higher TSR values/light colors, while cold climates require lower TSR values/dark colors. The range of optimal TSR values identified goes from 5% (e.g., Bragança and Paris) to 90% (Tenerife).

Climates with both cold winter and hot summer require intermediate values (e.g., around 30% for Madrid). The optimal TSR values are fairly similar whether or not thermal insulation is considered, as well as whether the building is considered as being in conditioned mode (case A) or full free-floating mode (case B).

In what concerns the energy savings in conditioned mode, the benefits seem to be clearer in climates that are either predominantly warm (e.g., Tenerife) or predominantly cold (e.g., Paris), with savings going up to just over 30% in the best case. For climates that have both a cold winter and a warm summer (e.g., Madrid and Rome) the savings due to TSR optimization are smaller, often lower than 5%. A similar trend was obtained when analyzing the results in terms of indoor temperature benefits in free-floating mode. Nevertheless, these outcomes require future confirmation by physical tests in a real scenario.

Overall, while these results confirm the benefits of tailored TSR optimization, they also point towards the convenience of developing paints that can perform well both in summer as in winter modes, e.g., thermochromic paints or other concepts with similar effects.

Author Contributions: Conceptualization, A.M., J.M.; methodology, V.L. and T.S.; software, T.S.; validation, M.A., V.L. and A.M.; formal analysis, T.S. and M.A.; investigation, M.A. and T.S.; data curation, T.S.; writing—original draft preparation, T.S.; writing—review and editing, M.A. and V.L.; visualization, M.A.; supervision, A.M and V.L.; funding acquisition, J.M. All authors have read and agreed to the published version of the manuscript.

Funding: This work was financially supported by: Base Funding—UIDB/00511/2020 of the Laboratory for Process Engineering, Environment, Biotechnology and Energy—LEPABE—funded by national funds through the FCT/MCTES (PIDDAC); and by SunStorage—Harvesting and storage of solar energy, with reference POCI-01-0145-FEDER-016387, funded by European Regional Development Fund (ERDF), through COMPETE 2020—Operational Programme for Competitiveness and Internationalisation (POCI), and by national funds, through FCT—Fundação para a Ciência e a Tecnologia I.P.

Acknowledgments: The authors would also like to acknowledge the contributions of Diana Dias and Vasco Granadeiro for their collaboration and also FCT and CIN S.A. for the support of Diana Dias's Ph.D. grant (SFRH/BDE/33548/2009) and to ARCP (http://www.arcp.pt) for the technical support.

Conflicts of Interest: The authors declare no conflict of interest.

List of Acronyms

BD1	Building without thermal insulation
BD2	Insulated Building
BR	Bitumen Roof
COP	Coefficient of performance
EER	Energy efficiency ratio
ETICS	External Thermal Insulation Composite Systems
GHG	Green House Gases
GR	Green Roof
HVAC	Heating, ventilation and air conditioning
IR	Infrared
PCM	Phase Chasing Materials
TSR	Total Solar Reflectance

Notation and Glossary

E_{demand}	Energy demand to keep the indoor temperature between 20 and 25 °C
f_B	Indoor temperature deviation indicator for case B
T	Indoor temperature (°C)
T_{min}	Week average daily minimum indoor temperature (°C)
T_{max}	Week average daily maximum indoor temperature (°C)
$\overline{T_M}$	Average maximum temperature of each day of the year (°C)
$\overline{T_m}$	Average minimum temperature of each day of the year (°C)
U	Heat transfer coefficient ($W{\cdot}m^{-2}{\cdot}K^{-1}$)

Appendix A

Table A1. Cases with optimal TSR value lower than 25%: Energy demand as function of TSR in conditioned mode (case A).

City	Building	TSR	TSR Value	$E_{demand}/MWh{\cdot}y^{-1}$	E_{demand} Reduction
Madrid	BD2	Reference	50%	4.82	-
		Optimal	**5%**	4.67	3.2%
		Minimum	25%	4.68	2.9%
Barcelona	BD1	Reference	50%	7.13	-
		Optimal	**15%**	6.81	4.5%
		Minimum	25%	6.84	4.1%
	BD2	Reference	50%	3.54	-
		Optimal	**10%**	3.42	3.4%
		Minimum	25%	3.43	3%

Table A1. *Cont.*

City	Building	TSR	TSR Value	E_{demand}/MWh·y^{-1}	E_{demand} Reduction
Paris	BD1	Reference	50%	12.98	-
		Optimal	**5%**	12.00	7.6%
		Minimum	25%	12.32	5.1%
	BD2	Reference	50%	6.54	-
		Optimal	**5%**	6.13	6.3%
		Minimum	25%	6.27	4.1%
Nice	BD1	Reference	50%	7.55	-
		Optimal	**20%**	7.34	2.8%
		Minimum	25%	7.34	2.8%
	BD2	Reference	50%	3.70	-
		Optimal	**20%**	3.63	1.7%
		Minimum	25%	3.63	1.7%
Rome	BD1	Reference	50%	7.58	-
		Optimal	**20%**	7.39	2.4%
		Minimum	25%	7.40	2.4%
	BD2	Reference	50%	3.71	-
		Optimal	**15%**	3.63	2.1%
		Minimum	25%	3.64	2%
Bragança	BD1	Reference	50%	10.29	-
		Optimal	**5%**	9.33	9.3%
		Minimum	25%	9.57	7.0%
	BD2	Reference	50%	5.18	-
		Optimal	**5%**	4.74	8.6%
		Minimum	25%	4.87	5.9%
Porto	BD1	Reference	50%	6.23	-
		Optimal	**5%**	5.51	11.6%
		Minimum	25%	5.63	9.6%
	BD2	Reference	50%	3.03	-
		Optimal	**5%**	2.71	10.4%
		Minimum	25%	2.78	8.1%
Coimbra	BD2	Reference	50%	2.77	-
		Optimal	**20%**	2.62	5.6%
		Minimum	25%	2.62	5.5%
Lajes	BD1	Reference	50%	3.29	-
		Optimal	**20%**	3.08	6.2%
		Minimum	25%	3.08	6.2%

Table A2. Cases with optimal TSR value lower than 25%: Indoor temperature indicators as function of TSR values in free-floating mode (case B).

City	Building	TSR	TSR Value	T_{min}/°C	T_{max}/°C	f_B	f_B Reduction
Barcelona	BD2	Reference	50%	5.6	28.4	22561	-
		Optimal	**10%**	6.4	31.1	20699	8.3%
		Minimum	25%	6.1	30	21010	6.9%
Paris	BD1	Reference	50%	−3	28.7	71881	-
		Optimal	**5%**	−2.5	33.3	66982	6.8%
		Minimum	25%	−2.7	31.1	68566	4.6%
	BD2	Reference	50%	−0.7	27.3	64454	-
		Optimal	**5%**	−0.2	28.9	59946	7.0%
		Minimum	25%	−0.4	28.1	61725	4.2%
Nice	BD2	Reference	50%	7.1	28.7	23933	-
		Optimal	**15%**	7.8	31.6	22487	6%
		Minimum	25%	7.6	30.7	22621	5.5%

Table A2. *Cont.*

City	Building	TSR	TSR Value	T_{min}/°C	T_{max}/°C	f_B	f_B Reduction
Rome	BD2	Reference	50%	6.3	28.7	24338	-
		Optimal	**5%**	7	31.4	21987	9.7%
		Minimum	25%	6.7	29.9	22547	7.4%
Bragança	BD1	Reference	50%	−0.4	32.6	53 678	-
		Optimal	**20%**	0.3	36.6	49 812	7.2%
		Minimum	25%	0.2	35.8	49 991	6.9%
	BD2	Reference	50%	1.8	29.9	46 619	-
		Optimal	**5%**	2.8	31.9	41 327	11.4%
		Minimum	25%	2.3	30.9	43 157	7.4%
Porto	BD2	Reference	50%	7.6	28.3	20539	-
		Optimal	**5%**	8.8	29.9	18354	10.6%
		Minimum	25%	8.3	29.2	18916	7.9%
Coimbra	BD2	Reference	50%	7.1	28.9	18195	-
		Optimal	**5%**	8.2	31.2	16610	8.7%
		Minimum	25%	7.7	30.1	16869	7.3%
Évora	BD2	Reference	50%	7.7	30.4	19565	-
		Optimal	**20%**	8.2	31.6	18583	5%
		Minimum	25%	8.2	31.4	18620	4.8%
Lajes	BD2	Reference	50%	12.2	27.3	7665	-
		Optimal	**5%**	12.6	28.7	6712	12.4%
		Minimum	25%	12.4	28	6940	9.5%

References

1. VijayaVenkataRaman, S.; Iniyan, S.; Goic, R. A review of climate change, mitigation and adaptation. *Renew. Sustain. Energy Rev.* **2012**, *16*, 878–897. [CrossRef]
2. Luber, G.; McGeehin, M. Climate Change and Extreme Heat Events. *Am. J. Prev. Med.* **2008**, *35*, 429–435. [CrossRef] [PubMed]
3. Stoffel, M.; Tiranti, D.; Huggel, C. Climate change impacts on mass movements—Case studies from the European Alps. *Sci. Total Environ.* **2014**, *493*, 1255–1266. [CrossRef] [PubMed]
4. De Lima, M.I.P.; Santo, F.E.; Ramos, A.M.; de Lima, J.L.M.P. Recent changes in daily precipitation and surface air temperature extremes in mainland Portugal, in the period 1941–2007. *Atmos. Res.* **2013**, *127*, 195–209. [CrossRef]
5. Ruiz-Medina, M.D.; Espejo, R.M. Integration of spatial functional interaction in the extrapolation of ocean surface temperature anomalies due to global warming. *Int. J. Appl. Earth Obs. Geoinf.* **2013**, *22*, 27–39. [CrossRef]
6. Climate Action-Climate Change Consequences. Available online: http://ec.europa.eu/clima/change/consequences/index_en.htm (accessed on 15 December 2019).
7. Figueiredo, R.; Nunes, P.; Panão, M.J.N.O.; Brito, M.C. Country residential building stock electricity demand in future climate–Portuguese case study. *Energy Build.* **2019**, *209*, 109694. [CrossRef]
8. Energy Statistics-An Overview. Available online: https://ec.europa.eu/eurostat/statistics-explained/index.php?title=Energy_statistics_-_an_overview#Final_energy_consumption (accessed on 18 December 2019).
9. Bynum, R.T. *Insulation Handbook*; Chapter 1-A Brief History of Thermal Insulation; McGraw-Hill: New York, NY, USA, 2001; p. 532.
10. Constable, G.; Somerville, B. *A Century of Innovation: Twenty Engineering Achievements that Transformed Our Lives*; Chapter-Air Conditioning and Refrigeration; Joseph Henry Press: Washington, DC, USA, 2003; p. 248.
11. Muhieldeen, M.W.; Adam, N.M.; Salman, B.H. Experimental and numerical studies of reducing cooling load of lecture hall. *Energy Build.* **2015**, *89*, 163–169. [CrossRef]
12. Cabeza, L.F.; Castell, A.; Medrano, M.; Martorell, I.; Pérez, G.; Fernández, I. Experimental study on the performance of insulation materials in Mediterranean construction. *Energy Build.* **2010**, *42*, 630–636. [CrossRef]
13. Aktacir, M.A.; Büyükalaca, O.; Yılmaz, T. A case study for influence of building thermal insulation on cooling load and air-conditioning system in the hot and humid regions. *Appl. Energy* **2010**, *87*, 599–607. [CrossRef]
14. Guillén, I.; Gómez-Lozano, V.; Fran, J.M.; López-Jiménez, P.A. Thermal behavior analysis of different multilayer façade: Numerical model versus experimental prototype. *Energy Build.* **2014**, *79*, 184–190. [CrossRef]

15. Kolaitis, D.I.; Malliotakis, E.; Kontogeorgos, D.A.; Mandilaras, I.; Katsourinis, D.I.; Founti, M.A. Comparative assessment of internal and external thermal insulation systems for energy efficient retrofitting of residential buildings. *Energy Build.* **2013**, *64*, 123–131. [CrossRef]
16. Marinosci, C.; Strachan, P.A.; Semprini, G.; Morini, G.L. Empirical validation and modelling of a naturally ventilated rainscreen façade building. *Energy Build.* **2011**, *43*, 853–863. [CrossRef]
17. De Gracia, A.; Cabeza, L.F. Phase change materials and thermal energy storage for buildings. *Energy Build.* **2015**, *103*, 414–419. [CrossRef]
18. Pisello, A.L.; Piselli, C.; Cotana, F. Thermal-physics and energy performance of an innovative green roof system: The Cool-Green Roof. *Sol. Energy* **2015**, *116*, 337–356. [CrossRef]
19. Wang, C.; Zhu, Y.; Guo, X. Thermally responsive coating on building heating and cooling energy efficiency and indoor comfort improvement. *Appl. Energy* **2019**, *253*, 113506. [CrossRef]
20. Li, H. Chapter 2-Literature Review on Cool Pavement Research. In *Pavement Materials for Heat Island Mitigation*; Li, H., Ed.; Butterworth-Heinemann: Boston, MA, USA, 2016; pp. 15–42. [CrossRef]
21. Granadeiro, V.; Almeida, M.; Souto, T.; Leal, V.; Machado, J.; Mendes, A. Thermochromic Paints on External Surfaces: Impact Assessment for a Residential Building through Thermal and Energy Simulation. *Energies* **2020**, *13*, 1912. [CrossRef]
22. Wu, Y.; Krishnan, P.; Zhang, M.-H.; Yu, L.E. Using photocatalytic coating to maintain solar reflectance and lower cooling energy consumption of buildings. *Energy Build.* **2018**, *164*, 176–186. [CrossRef]
23. Azarnejad, A.; Mahdavi, A. Building Façades' Visual Reflectance and Surface Temperatures: A Field Study. *Energy Procedia* **2015**, *78*, 1720–1725. [CrossRef]
24. Moghtadernejad, S.; Chouinard, L.E.; Mirza, M.S. Design strategies using multi-criteria decision-making tools to enhance the performance of building façades. *J. Build. Eng.* **2020**, *30*, 101274. [CrossRef]
25. Upstone, S. Measurement of Total Solar Reflectance of Paint Panels using PerkinElmer UV/Vis/NIR Spectrophotometers and UV WinLab Software. *PerkinelmerInc.* **2019**, *3*, 1–2.
26. Dias, D.; Machado, J.; Leal, V.; Mendes, A. Impact of using cool paints on energy demand and thermal comfort of a residential building. *Appl. Therm. Eng.* **2014**, *65*, 273–281. [CrossRef]
27. Becherini, F.; Lucchi, E.; Gandini, A.; Barrasa, M.C.; Troi, A.; Roberti, F.; Sachini, M.; Di Tuccio, M.C.; Arrieta, L.G.; Pockelé, L.; et al. Characterization and thermal performance evaluation of infrared reflective coatings compatible with historic buildings. *Build. Environ.* **2018**, *134*, 35–46. [CrossRef]
28. Pigment Database. Available online: http://coolcolors.lbl.gov/LBNL-Pigment-Database/database.html (accessed on 20 December 2019).
29. Levinson, R.; Berdahl, P.; Akbari, H. Solar spectral optical properties of pigments—Part II: Survey of common colorants. *Sol. Energy Mater. Sol. Cells* **2005**, *89*, 351–389. [CrossRef]
30. Levinson, R.; Berdahl, P.; Akbari, H.; Miller, W.; Joedicke, I.; Reilly, J.; Suzuki, Y.; Vondran, M. Methods of creating solar-reflective nonwhite surfaces and their application to residential roofing materials. *Sol. Energy Mater. Sol. Cells* **2007**, *91*, 304–314. [CrossRef]
31. Akbari, H. Measured energy savings from the application of reflective roofs in two small non-residential buildings. *Energy* **2003**, *28*, 953–967. [CrossRef]
32. Akbari, H.; Levinson, R.; Rainer, L. Monitoring the energy-use effects of cool roofs on California commercial buildings. *Energy Build.* **2005**, *37*, 1007–1016. [CrossRef]
33. Levinson, R.; Akbari, H.; Reilly, J.C. Cooler tile-roofed buildings with near-infrared-reflective non-white coatings. *Build. Environ.* **2007**, *42*, 2591–2605. [CrossRef]
34. Akbari, H.; Pomerantz, M.; Taha, H. Cool surfaces and shade trees to reduce energy use and improve air quality in urban areas. *Sol. Energy* **2001**, *70*, 295–310. [CrossRef]
35. Berdahl, P.; Akbari, H.; Levinson, R.; Jacobs, J.; Klink, F.; Everman, R. Three-year weathering tests on asphalt shingles: Solar reflectance. *Sol. Energy Mater. Sol. Cells* **2012**, *99*, 277–281. [CrossRef]
36. Levinson, R.; Berdahl, P.; Berhe, A.A.; Akbari, H. Effects of soiling and cleaning on the reflectance and solar heat gain of a light-colored roofing membrane. *Atmos. Environ.* **2005**, *39*, 7807–7824. [CrossRef]
37. Monitoring the Energy-Use Effects of Cool Roofs on California Commercial Buildings. Available online: http://escholarship.org/uc/item/5j49q2kg#page-2 (accessed on 15 January 2019).
38. Yuxuan, Z.; Yunyun, Z.; Jianrong, Y.; Xiaoqiang, Z. Energy saving performance of thermochromic coatings with different colors for buildings. *Energy Build.* **2020**, *215*, 109920. [CrossRef]

39. Karlessi, T.; Santamouris, M.; Apostolakis, K.; Synnefa, A.; Livada, I. Development and testing of thermochromic coatings for buildings and urban structures. *Sol. Energy* **2009**, *83*, 538–551. [CrossRef]
40. Santamouris, M.; Synnefa, A.; Karlessi, T. Using advanced cool materials in the urban built environment to mitigate heat islands and improve thermal comfort conditions. *Sol. Energy* **2011**, *85*, 3085–3102. [CrossRef]
41. Shi, X.; Abel, T.; Wang, L. Influence of two motion types on solar transmittance and daylight performance of dynamic façades. *Sol. Energy* **2020**, *201*, 561–580. [CrossRef]
42. Yuan, J.; Farnham, C.; Emura, K.; Alam, M.A. Proposal for optimum combination of reflectivity and insulation thickness of building exterior walls for annual thermal load in Japan. *Build. Environ.* **2016**, *103*, 228–237. [CrossRef]
43. Piselli, C.; Saffari, M.; de Gracia, A.; Pisello, A.L.; Cotana, F.; Cabeza, L.F. Optimization of roof solar reflectance under different climate conditions, occupancy, building configuration and energy systems. *Energy Build.* **2017**, *151*, 81–97. [CrossRef]
44. ESRU-Energy Systems Research Unit. Available online: http://www.esru.strath.ac.uk//Programs//ESP-r.htm (accessed on 8 January 2020).
45. Clarke, J.A. *Energy Simulation in Building Design*, 2nd ed.; Clarke, J.A., Ed.; Butterworth-Heinemann: Oxford, UK, 2001; pp. 64–98. [CrossRef]
46. Castell, A.; Solé, C. 11-Design of latent heat storage systems using phase change materials (PCMs). In *Advances in Thermal Energy Storage Systems*; Cabeza, L.F., Ed.; Woodhead Publishing: Cambridge, UK, 2015; pp. 285–305. [CrossRef]
47. Strachan, P. ESP-r: Summary of validation studies. In *Energy Systems Research Unit*; University of Strathclyde: Scotland, UK, 2000.
48. Leal, V.; Maldonado, E. The role of the PASLINK test cell in the modelling and integrated simulation of an innovative window. *Build. Environ.* **2008**, *43*, 217–227. [CrossRef]
49. Strachan, P.A.; Kokogiannakis, G.; Macdonald, I.A. History and development of validation with the ESP-r simulation program. *Build. Environ.* **2008**, *43*, 601–609. [CrossRef]
50. Santos, C.A.P.; Matias, L. *Coeficiente de Transmissão Térmica de Elementos da Envolvente dos Edifícios-Verão Actualizada 2006*; Laboratório Nacional de Engenharia Civil: Lisbon, Portugal, 2006.
51. Weather Atlas. Available online: https://www.weather-atlas.com/ (accessed on 16 February 2020).
52. Decree-Law No 80/2006 of April 4-Regulation on Thermal Insulation in Buildings (Regulamento das Características de Comportamento Térmico dos Edifícios). In *Official Gazette (Diário da República-I Série-A)*; Ministry of Public Works Transport and Communications: Lisbon, Portugal, 2006.
53. EnergyPlus Energy Simulation Software-Weather Data. Available online: http://apps1.eere.energy.gov/buildings/energyplus/cfm/weather_data3.cfm/region=6_europe_wmo_region_6/country=PRT/cname=Portugal (accessed on 10 December 2019).
54. Light Reflectance Values. Available online: https://www.resene.co.nz/swatches/reflectance.htm (accessed on 18 February 2020).
55. Berdahl, P.; Bretz, S.E. Preliminary survey of the solar reflectance of cool roofing materials. *Energy Build.* **1997**, *25*, 149–158. [CrossRef]

energies

MDPI

Article

Thermochromic Paints on External Surfaces: Impact Assessment for a Residential Building through Thermal and Energy Simulation

Vasco Granadeiro [1], Margarida Almeida [2], Tiago Souto [3], Vítor Leal [4], João Machado [3] and Adélio Mendes [2,*]

1 INEGI, Institute of Science and Innovation in Mechanical and Industrial, rua Dr. Roberto Frias, 4200-465 Porto, Portugal; vascogranadeiro@gmail.com
2 LEPABE, Faculdade de Engenharia, Universidade do Porto, rua Dr. Roberto Frias, 4200-465 Porto, Portugal; margarida.almeida@fe.up.pt
3 CIN—Corporação Industrial do Norte, S.A., Av. de Dom Mendo 831, 4471-909 Maia, Portugal; tiago.souto@cin.pt (T.S.); jmachado@cin.pt (J.M.)
4 DEMEC, Faculdade de Engenharia, Universidade do Porto, rua Dr. Roberto Frias, 4200-465 Porto, Portugal; vleal@fe.up.pt
* Correspondence: mendes@fe.up.pt; Tel.: +351 22 5081695; Fax: +351 22 5081449

Received: 18 March 2020; Accepted: 10 April 2020; Published: 14 April 2020

check for updates

Abstract: This work addresses the effect of using thermochromic paints in residential buildings. Two different thermochromic paint types were considered: One that changes properties through a step transition at a certain temperature, and another that changes properties in a gradual/linear manner throughout a temperature range. The studied building was a two-floor villa, virtually simulated through a digital model with and without thermal insulation, and considering thermochromic paints applied both on external walls and on the roof. The performance assessment was done through the energy use for heating and cooling (in conditioned mode), as well as in terms of the indoor temperature (in free-floating mode). Three different cities/climates were considered: Porto, Madrid, and Abu Dhabi. Results showed that energy savings up to 50.6% could be reached if the building is operated in conditioned mode. Conversely, when operated in free-floating mode, optimally selected thermochromic paints enable reductions up to 11.0 °C, during summertime, and an increase up to 2.7 °C, during wintertime. These results point out the great benefits of using optimally selected thermochromic paints for obtaining thermal comfort, and also the need to further develop stable and cost-effective thermochromic pigments for outdoor applications, as well as to test physical models in a real environment.

Keywords: thermochromic paints; residential buildings; thermal comfort; energy demand; energy efficiency; building simulation

1. Introduction

Several policy initiatives around the world are boosting efforts to decarbonize the building stock. In the European Union, 50% of the final energy consumption is used for heating and cooling, of which about 80% is used in buildings [1]. Greenhouse gas (GHG) emissions are mainly related to the consumption of fossil energy. A combination of energy efficiency measures and a renewable energy supply are therefore key to achieve the decarburization goals.

To mitigate the rise in energy demand, passive building strategies should be considered to obtain indoor thermal comfort. The American Society of Heating Refrigerating and Air-Conditioning Engineers, Inc. (ASHRAE) defines thermal comfort as "*that condition of mind that expresses satisfaction with the thermal*" [2]. Different factors, such as air temperature, humidity, air speed, physical activity, and clothing, contribute to the thermal comfort sensation. In most buildings, the heat transfer through the building envelope is the most important factor to obtain indoor thermal comfort if the building does not have mechanical heating or cooling (and thus is operated in a so-called free-floating mode) [3]

Solar reflectance is the fraction of the solar irradiance reflected by a surface [4]. There is a linear correlation between the increase of solar reflectance of a surface and the decrease of its daily peak temperature [5]. A paint may be designed to absorb the solar irradiance (thus called hot paint) and then contribute to warming up a building. Alternatively, a high solar reflectance paint (thus called cool paint) should be applied to prevent solar irradiance absorption and then to avoid overheating of the building [6]. The building surface can reject (reflect) or absorb the heat of its surroundings, leading to less or more heat retention in the building envelope through its construction materials. The reflectance of the building surface is normally characterized by the so-called total solar reflectance value (TSR). High values of TSR are associated with light colors (cool paints) while dark colors (hot paints) have a low TSR.

The color, alongside the roughness and construction material (chemical composition), is one of the main physical characteristics that affect the thermal performance of a surface [7,8]. Bansal et al. [9] studied the influence of the external surface color on the thermal performance of a building. It was concluded that a white painted building, when compared with a black one, can reduce the indoor temperature by 6 °C during winter and 8 °C in summer months, if no ventilation is considered. This difference can be decreased to 4 and 6 °C, respectively, when a rate of three air exchanges is taken into account. Cool paints are a passive thermal comfort strategy, easy to apply as passive cooling, which can be used in any type of building with a short period of investment cost recovery [5,10]. For example, the application of cool roofs is reported to produce an 18%–93% decrease in cooling loads and 11%–27% decrease in terms of the peak cooling demand in air-conditioned buildings, leading to energy savings between 2% and 44% with an average of 20%, in both residential and nonresidential buildings [11,12]. Such figures vary significantly with the local climate, direct solar gain, building envelope type, and the use or not of a heating, ventilation and air-conditioning (HVAC) system [5,8].

Despite the positive impact of using cool paints, some studies show that these paints may also have a detrimental effect on the thermal comfort of a building. Dias et al. [10] simulated the impact of increasing the TSR value of a paint, from 50% to 92%, both at the external walls and on the roof of a residential building located in three different cities of Portugal. They observed a summer maximum indoor temperature reduction of ca. 4 °C, and cooling energy savings of up to 1.9 $MWh \cdot y^{-1}$. However, these authors also observed a significant increase on the heating demand of up to 2.9 $MWh \cdot y^{-1}$ (a penalty of about 25.9%). The observed overall effect was then a significant increase in the energy consumption for obtaining indoor thermal comfort. The winter penalties of cool paints triggered the development of coatings that change reversibly their color as a thermal response to the environment [13].

Thermochromic paints contain thermochromic pigments or compounds that reversibly change color and optical properties according to the surface temperature [14,15]. The thermochromic effect is triggered by the temperature that produces: i) Changes of pH; ii) conversion of the crystalline type; iii) loss of crystalline water by heating; iv) a shift in the electron equilibrium between an electron donor and electron acceptor and; v) a ring-opening reaction of molecules by heating [16]. The temperature at which the reversible color-changing of thermochromic pigments occurs is named the transition temperature.

Thermochromic pigments or compounds can be used directly to formulate paints or can be first microencapsulated. The first reporting of thermochromic paints as coatings to a building's façade was by Ma et al. [17], in 2000. These authors added microencapsulated thermochromic dyes to a white interior and exterior facade coatings and used them to paint cubes. They showed that below the transition temperature (paint is colorful), the temperature of the cube was close to that of a cube painted with an ordinary paint, and above the transition temperature (paint is white), the temperature of the cube was 4 °C cooler. The accelerated aging tests for 400 h showed a fast color fading rate. In 2001, Ma et al. [16] repeated the reversible color-changing process more than 1000 times in an indoor environment and the pigments kept their color. Later, in 2002, the same researchers [18] studied the reversible effect, the energy-absorbing and energy-reflecting states of thermochromic building coatings, when exposed to solar radiation. Results showed that below the transition temperature, 18 °C in this case, the addition of thermochromic pigments to a white paint led to the same absorption amount of solar energy as an ordinary colored coating. On the other hand, the absorption of solar irradiance was minimized above the transition temperature since the paint turned white.

In 2009, Karlessi et al. [19] developed and tested thermochromic paints intended to be applied in buildings. These authors used thermochromic pigments, colored below 30 °C, and translucent above this temperature. During the experimental period, August–mid-September 2007 in Athens, Greece, the temperatures of the thermochromic paint coatings were lower, ranging from 23.8 to 38.4 °C, than the temperatures of color-matched cool paints, from 28.1 to 44.6 °C; and common samples, from 29.8 to 48.5 °C. In 2012, Ye et al. [20] designed and evaluated a thermochromic roof coat for energy savings. These authors used an aqueous solution of a thermosensitive polymer with a transition temperature of 32 °C; below the transition temperature, the solution was transparent and above the solution it became opaque and the paint appeared white, reflecting most of the light. It was demonstrated that this transition between light-absorbing and light-reflecting states as triggered by the temperature was spontaneous and reversible. When the concentration of the thermosensitive polymer reached 0.05 g/mL, and above 32 °C, the paint coat reached the maximum reflectance of 33%, which led to a 6 °C decrease of the surface temperature compared to when it was coated with the absence of the thermochromic polymer.

More recently, in 2019, Ascione et al. [21] investigated the ideal radiative characteristics of a thermochromic paint to obtain optimized thermal comfort in both winter and summer seasons in Naples, Italy. The simulation results showed that a thermal absorptivity (α) between 0.90 and 0.95, and a thermal emissivity (ε) between 0.65 and 0.75 would be the ideal values for a thermochromic paint applied in that climate. Such a high absorptivity is expected to produce an increase of the cooling needs during summer. However, its effect on the reduction of the heating load surpasses the small penalties on the cooling ones. Globally, a reduction of the thermal energy demand of around 21.1 $Wh{\cdot}m^{-2}$ is expected (from 138.8 $Wh{\cdot}m^{-2}$, when using a common white paint $\alpha = 0.7$ and $\varepsilon = 0.9$, to 117.7 $Wh{\cdot}m^{-2}$)

The combination of the thermochromic properties with phase change materials (PCMs) has also been studied, with the aim of reaching better passive thermal comfort. Soudian, S. and U. Berardi [22] introduced thermochromic pigments into a PCM-finish plaster to achieve a dynamic response of the facade to different surface temperatures and solar radiation. The results supported an expected variation of the emissivity according to the seasons. However, the emissivity values showed a non-linear relation between PCM melting temperatures and the different thermochromic colors.

Overall, improved thermochromic pigments are required to ensure paints that do not change over time due to aging and photo-degradation problems. Karlessi and Santamouris [23] performed tests with thermochromic coatings protected with filters that only allowed visible radiation to reach the coatings' surface. These researchers concluded that the optical properties of the thermochromic paints used were not only affected by the ultraviolet radiation but also by the 400 to 480 nm spectrum. It is crucial to develop stable and cost-effective thermochromic pigments [13,19] with proven stability not only to radiation but also to temperature and weathering.

Several thermochromic paints are already available in the market in the field of art, design, and indoor architecture [24], but no outdoor applications are available yet [13,24]. Since the cost of thermochromic materials is still high [13,19], real exposure tests to study the thermochromic paints' application in buildings are difficult to perform. The present study intends to overcome this difficulty by assessing the potential benefits of these paints and simulating a residential building considering three cities/climates: Porto in Portugal (41°09′ N 8°37′ W), Madrid in Spain (40°25′ N 3°42′ O), and Abu Dhabi in the United Arab Emirates (24°28′ N 54°22′ E).

The assessment was performed through dynamic building simulation with an open source ESP-r software [25,26]. ESP-r is an integrated energy modeling tool and has been extensively used for the simulation of buildings' performance and assessment of their energy use related to environmental control systems.

The thermal behavior and energy demand was assessed for two different thermochromic paints: i) A paint using a single thermochromic pigment that changes color (and then the TSR) at a given temperature and ii) a paint containing various pigments that allow a continuum and linear change of the color and TSR value as a function of the temperature.

The present work intends to further assess the potential of thermochromic paints for residential buildings, by trying to identify the optimal transition temperature and their impact upon the energy demand (building conditioned mode) or indoor temperature (building in free-floating mode).

2. Simulation Scenarios and Alternatives

This study assessed a single villa with two living floors, with a floor area of 90 m^2 each, and an uninhabited attic. The residential building is represented in Figures 1 and 2 and also fully described elsewhere [10]. Briefly, the ground floor has a living room, a dining room, a kitchen, a toilet, a hall, and stairs. The first floor has three bedrooms, a suite, a toilet, a bathroom, a corridor, and stairs. The attic is an open space, as mentioned before it is not inhabited.

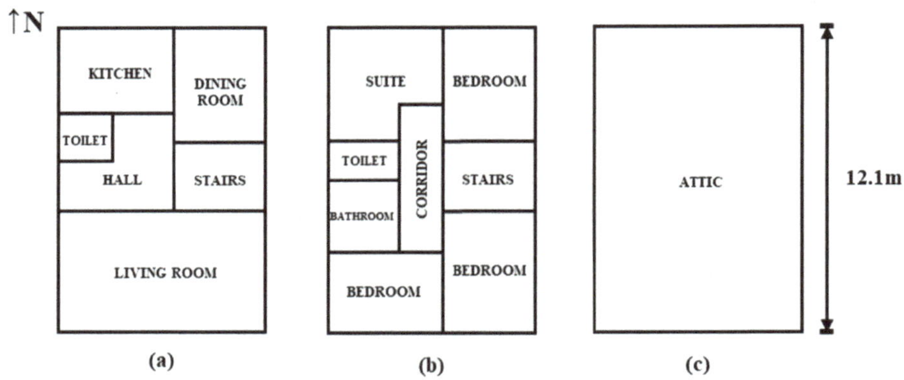

Figure 1. Plans of the building: (**a**) ground floor, (**b**) first floor, and (**c**) attic.

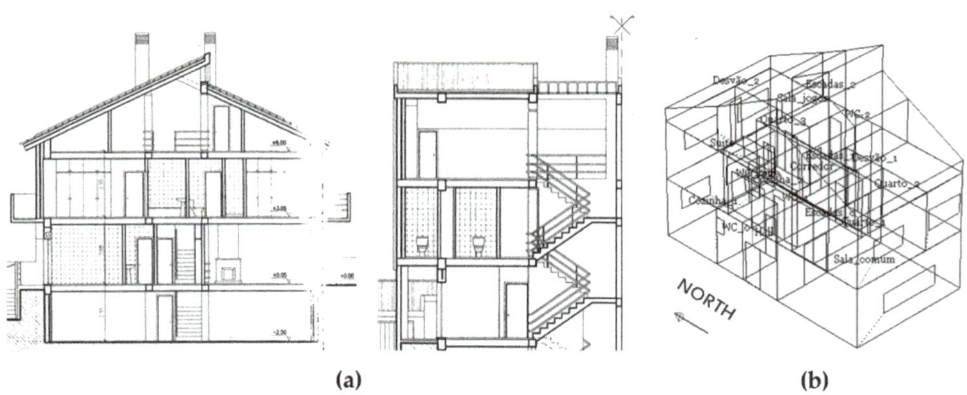

Figure 2. (**a**) Building sections, and (**b**) a 3D model of the simulated building.

All zones of the building were modelled explicitly in ESP-r. The air changes per hour and internal gains of each zone of the building were taken from the Portuguese legislation at the time of calculations [27]: A constant value of 0.6 air changes per hour in each zone. Internal gains of 4 W·m^{-2}

on the living room, dining room, kitchen, three bedrooms, and suite were considered; on the remaining zones, 1 W·m^{-2} of internal gains was assumed (except for the attic, where no internal gains were assumed). The residential building has a total area of 26.3 m^2 of windows with a single clear glass. The existence of venetian blinds covering 50% of each window was considered.

Full-year simulations were performed, with an hourly time step. Two building alternatives with different construction types were considered:

i) A building with no thermal insulation, neither on the façades nor on the roof (named building configuration BD1); and
ii) A building with 60 mm of expanded polystyrene (EPS) insulation, both on external walls and on the roof slab (named building configuration BD2). The conductivity value of this insulation material was considered to be 0.042 W·m^{-1}·K^{-1}.

The construction details of BD1 and BD2 are shown in Tables 1 and 2. The specifications of each layer of the construction (e.g., conductivity, density, specific heat, emissivity, and absorption) follow the recommendations by the Portuguese national legislation [28].

The performance of the buildings was assessed in two complementary ways:

- Through the energy needed to maintain the rooms between 20 and 25 °C during the whole year, assuming that there is an air-conditioning system (conditioned mode); and
- Through an analysis of the indoor temperatures in summer and in winter if there was no mechanical heating nor cooling available (free-floating mode).

Table 1. Construction details of the building with a single wall façade (BD1).

Construction	Layer	Thickness/mm	U/W·m^{-2}·K^{-1}
External wall	Coating Plaster Brick Plaster	0.2 30 190 10	1.3
Internal wall	Plaster Brick Plaster	10 110 10	2.1
Ground floor	Common earth Gravel Concrete Asphalt Concrete Wood floor	300 300 300 10 20 20	0.7
Window	Clear glass	8	5.6
Slab	Plaster Concrete Plaster	10 240 10	2.0
Roof slab	Clay tile Air Plaster Brick Plaster	5 3 5 110 5	1.6

Table 2. Construction details of case BD2—differences from BD1 in bold.

Construction	Layer	Thickness/mm	U/W·m^{-2}·K^{-1}
External wall	Coating	0.2	**0.5**
	Plaster	30	
	Insulation	**60**	
	Brick	190	
	Plaster	10	
Roof slab	Clay tile	5	**0.5**
	Air	3	
	Plaster	5	
	Insulation	**60**	
	Brick	110	
	Plaster	5	

All the weather data used in this article were obtained from the US Department of Energy/Energy Plus database. For the free-floating simulation, BD1 and BD2 were compared in the hottest and the coldest weeks of the year: The week's average daily maximum and minimum temperatures were determined and are presented in the following chapter.

The energy demand needed to keep the indoor temperature between 20 and 25 °C was assessed by a full-year simulation. The cooling and heating thermal energy needs were converted into the final electrical energy demand by considering as the air conditioning system a reference heat pump with an average coefficient of performance (COP) of 4 for heating and 3 for cooling (also called EER instead of COP, in some of the literature). The final heating and cooling energy demand are presented in the following chapter for both building simulation cases (BD1 and BD2).

In terms of paints, four different alternatives were considered:

- A conventional white paint with a TSR value of 90%, named WP (white paint);
- A conventional dark paint with 5% of TSR, named BP (black paint);
- A thermochromic paint with a specific transition temperature, named case ST (step transition) and;
- A thermochromic paint with a linear transition temperature, named case LT (linear transition).

Regarding the thermochromic step transition paints (ST), the thermochromic paint has a TSR of 5% below the transition temperature and a TSR of 90% above it. A parametric study on the transition temperature was performed to identify the nearly optimal value (Section 3).

A thermochromic linear transition (LT) paint was also considered. Below the linear transition temperature zone, the thermochromic paint has a TSR value of 5%, and above the linear transition temperature zone, the thermochromic paint is assumed to have a TSR value of 90%. The TSR value over the linear transition temperature zone depends linearly on the temperature and ranges between TSR values of 5% and 90%. A parametric study on the initial and final temperatures of the linear transition zone was also performed to find the optimum (Section 3).

3. Results and Discussion

The residential building cases, BD1 and BD2, were compared for each of the three reference locations. In the conditioned mode (i.e., building with heating and cooling), the results were expressed in terms of the electricity demand for heating, $E_{Heating}$, and electricity demand for cooling, $E_{Cooling}$, over one year, taking into account the COP and EER stated in Section 2. In the free-floating mode (i.e., building without active heating or cooling), the results were expressed in terms of the weekly average daily maximum temperature, T_M, and the weekly average daily minimum temperature, T_m, over the hottest and coldest week of the year, respectively.

3.1. Porto, Portugal

3.1.1. Case ST—Thermochromic Paint with a Specific Transition Temperature

Concerning the residential building configurations BD1 and BD2, a set of parametric simulations varying the value of the transition temperature were performed to determine its optimal value—the one that minimizes the energy use required to maintain the indoor temperature between 20 and 25 °C—Figure 3.

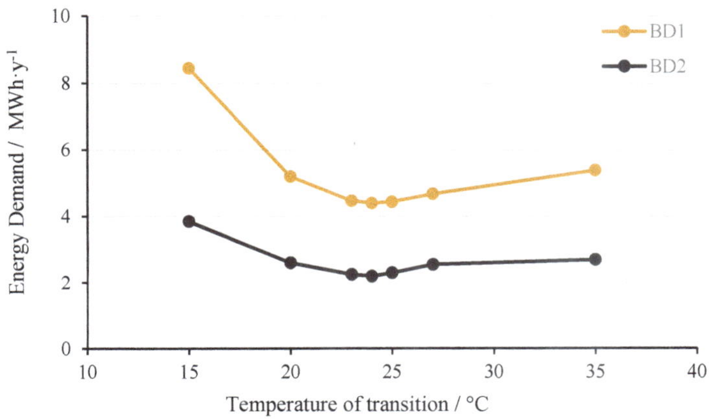

Figure 3. Annual energy demand, for building configurations BD1 and BD2, located in Porto, Portugal, considering different transition temperatures for the thermochromic paint.

As it can be seen from Figure 3, for Porto, the optimum transition temperature of the thermochromic paint is 24 °C for both BD1 and BD2. For this transition temperature, the annual energy demand is ca. 4.4 and 2.2 MWh·y^{-1}, respectively, for building configurations BD1 and BD2.

The thermal behavior of the residential buildings painted with the optimized thermochromic coat is given in Tables 3 and 4. Table 3 shows T_{Max} over the hottest week of the year and T_{min} over the coldest week of the year, in each of the three floors of each building case (BD1 and BD2). The annual energy demand, in terms of the electricity used by the reference system, for cooling, for heating, and in total are shown in Table 4 for both building configurations.

Table 3. Average maximum indoor temperature (T_{Max}) over the hottest week of the year and average minimum indoor temperature (T_{min}) over the coldest week of the year, for a building painted with an optimized thermochromic step transition paint (ST), located in Porto, Portugal.

Building Configuration	Case	T_{Max}/°C			T_{min}/°C		
		Ground Floor	First Floor	Second Floor	Ground Floor	First Floor	Second Floor
BD1	BP	30.5	32.3	35.9	8.3	8.9	7.2
	WP	24.8	27.0	24.4	6.4	6.4	4.9
	ST	26.2	27.7	26.5	8.3	8.9	7.1
	ST − BP [1]	**−4.3**	**−5.3**	**−9.4**	**0**	**0**	**−0.1**
	ST − WP [2]	**+1.4**	**+0.7**	**+2.1**	**+1.9**	**+2.5**	**+2.2**
BD2	BP	28.7	29.9	29.4	9.3	9.4	8.8
	WP	25.0	26.4	24.2	7.8	7.7	6.5
	ST	25.7	26.9	25.4	9.3	9.4	8.8
	ST − BP [1]	**−3.0**	**−3.0**	**−4.0**	**0.0**	**0.0**	**0.0**
	ST − WP [2]	**+0.7**	**+0.5**	**+1.2**	**+ 1.5**	**+1.7**	**+2.3**

[1] **ST − BP:** saving between a thermochromic step transition paint (ST) and a black paint (BP). [2] **ST − WP:** saving between a thermochromic step transition paint (ST) and a white paint (WP).

Table 4. Final energy demand for heating ($E_{Heating}$), and cooling ($E_{Cooling}$), total final energy demand (E_{Total}), and energy savings over the year, for a building painted with an optimized thermochromic step transition paint (ST), located in Porto, Portugal.

Building Configuration	Case	Energy/MWh·y^{-1}			Energy Savings/%
		$E_{Heating}$	$E_{Cooling}$	E_{Total}	
BD1	BP	4.17	1.34	5.51	
	WP	8.51	0.02	8.53	
	ST	4.33	0.09	4.42	
	ST − BP [1]	**+0.16**	**−1.25**	**−1.09**	**19.8**
	ST − WP [2]	**−4.18**	**+0.08**	**−4.11**	**48.2**
BD2	BP	2.14	0.57	2.71	
	WP	3.83	0.01	3.85	
	ST	2.20	0.04	2.23	
	ST − BP [1]	**+0.06**	**−0.53**	**−0.48**	**17.6**
	ST − WP[2]	**−1.63**	**+0.03**	**−1.61**	**41.9**

[1] **ST − BP:** saving between a thermochromic step transition paint (ST) and a black paint (BP). [2] **ST − WP:** saving between a thermochromic step transition paint (ST) and a white paint (WP).

Results with the free-floating mode show that, for Porto, a thermochromic paint with an optimized transition temperature leads to significant reductions in the weekly average maximum temperature indoors. T_{Max} decreased by 9.4 and 4.0 °C in the upper floor, and about 5 and 3 °C in the inhabited floors, respectively, for building configurations BD1 and BD2, when compared with a common black paint. During winter, T_{min} increased on average by 2 °C, when using thermochromic paints instead of common white paints. These results show a positive thermal effect in summer and winter seasons, leading to indoor temperatures closer to the thermal comfort range.

Regarding the analysis in the conditioned mode (indoors kept between 20 and 25 °C), the thermochromic paint with an optimized transition temperature originates annual energy savings up to 4.1 MWh·y^{-1} for building configuration BD1 and 1.6 MWh·y−1 for configuration BD2. These correspond to about 48.2% and 41.9% for BD1 and BD2, respectively. Since the climate in Porto leads to $E_{Heating}$ higher than $E_{Cooling}$, the impact of thermochromics' application is more expressive when compared with common white than black plaints—energy savings of around 18%.

3.1.2. Case LT—Thermochromic Paint with a Linear Transition Temperature

The use of a hypothetic thermochromic paint with a linear transition behavior was also assessed. This thermochromic paint was assumed to have a TSR value of 5% when the surface temperature of the paint is below the transition temperature range, a TSR value of 90% when the surface temperature is above this range, and a TSR value with a linear variation between 5% and 90% in the transition temperature range. Additionally, in this case, the optimal transition temperature range (defined by the low transition limit and by a high transition limit) was sought to minimize the annual energy demand required to maintain the indoor temperature between 20 and 25 °C. For an optimization process using 1 °C steps, the best result was obtained when both the low and high transition points were 24 °C, see Figure 4, which means that for this building and in the Porto climate, the best performing thermochromic paint is a step transition temperature.

Although a single transition temperature provides better energy savings, its energy demand is only 7% and 10% lower for BD1 and BD2, respectively, than the wider transition temperature range considered (18–30 °C). This slight variation allows a gradual color transition for a thermochromic paint without compromising its performance significantly.

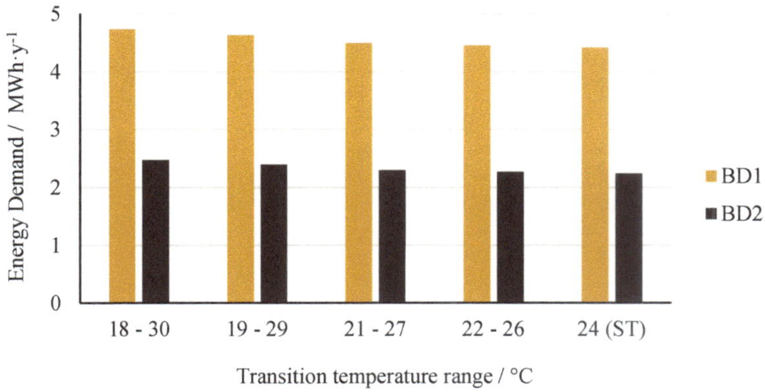

Figure 4. Annual final energy demand for building configurations BD1 and BD2 located in Porto, Portugal, considering different transition temperature ranges.

3.2. Madrid, Spain

3.2.1. Case ST—Thermochromic Paint with a Specific Transition Temperature

A similar study was performed for Madrid. Different transition temperatures were also studied, and the optimum transition temperature was selected taking into account the minimum annual energy demand of both residential building configurations, BD1 and BD2, as shown in Figure 5.

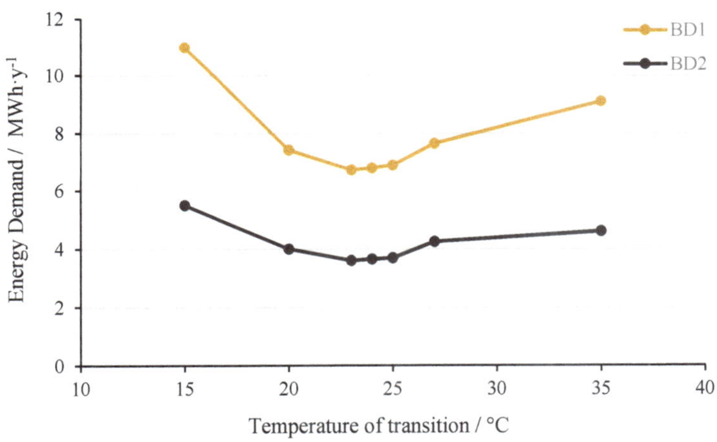

Figure 5. Annual energy demand for building configurations BD1 and BD2 located in Madrid, Spain, considering different transition temperatures for the thermochromic paint.

For Madrid, the optimum transition temperature of the thermochromic coating is 23 °C, for both building configurations BD1 and BD2. This transition temperature corresponds to annual energy demands of about 6.9 and 3.6 MWh·y^{-1} for BD1 and BD2, respectively.

The curves from Figure 5 present a similar tendency to the ones from Figure 6 (Porto case). In both climates, a transition temperature lower than the optimal value (e.g., 15 °C) would have a more detrimental impact on the energy savings than choosing higher temperatures (e.g., 35 °C). It is also visible that the changes in the transition temperature lead to smoother variations in the energy demand for insulated buildings (BD2) than buildings without insulation (BD1).

Tables 5 and 6 show the indoor thermal behavior and the annual energy demand of buildings BD1 and BD2 using a thermochromic paint with a transition temperature of 23 °C. Additionally, for Madrid, values of T_{Max} over the hottest week of the year, T_{min} over the coldest week of the year, and the annual energy demand for cooling, for heating, and in total are presented.

Table 5. Average maximum indoor temperature (T_{Max}) over the hottest week of the year and average minimum indoor temperature (T_{min}) over the coldest week of the year, for a building painted with an optimized thermochromic step transition paint (ST), located in Madrid, Spain.

Building Configuration	Case	T_{Max}/°C			T_{min}/°C		
		Ground Floor	First Floor	Second Floor	Ground Floor	First Floor	Second Floor
BD1	BP	35.5	38.4	43.1	4.2	4.1	2.2
	WP	30.4	32.6	32.1	2.6	1.7	0.2
	ST	30.4	32.6	32.1	4.2	4.1	2.2
	ST − BP [1]	**−5.1**	**−5.8**	**−11.0**	**0**	**0**	**0**
	ST − WP [2]	**0**	**0**	**0**	**+1.6**	**+2.4**	**+2.0**
BD2	BP	30.9	33.1	36.3	5.5	4.7	3.7
	WP	28.3	29.9	29.9	4.2	2.9	1.5
	ST	28.3	29.9	29.9	5.5	4.7	3.7
	ST − BP [1]	**−2.6**	**−3.2**	**−6.4**	**0**	**0**	**0**
	ST − WP [2]	**0**	**0**	**0**	**+1.3**	**+1.8**	**+2.2**

[1] **ST − BP:** saving between a thermochromic step transition paint (ST) and a black paint (BP). [2] **ST − WP:** saving between a thermochromic step transition paint (ST) and a white paint (WP).

Table 6. Final energy demand for heating ($E_{Heating}$), and cooling ($E_{Cooling}$), total final energy demand (E_{Total}), and energy savings over the year, for a building painted with an optimized thermochromic step transition paint (ST), located in Madrid, Spain.

Building Configuration	Case	Energy/MWh·y^{-1}			Energy Savings/%
		$E_{Heating}$	$E_{Cooling}$	E_{Total}	
BD1	BP	5.95	3.35	9.30	
	WP	10.74	0.52	11.26	
	ST	6.28	0.58	6.86	
	ST − BP [1]	**+0.32**	**−2.77**	**−2.45**	**26.3**
	ST − WP [2]	**−4.46**	**+0.06**	**−4.40**	**39.1**
BD2	BP	3.18	1.49	4.67	
	WP	5.15	0.32	5.47	
	ST	3.29	0.33	3.62	
	ST − BP [1]	**+0.13**	**−1.16**	**−1.05**	**22.4**
	ST − WP [2]	**−1.85**	**+0.01**	**−1.84**	**33.7**

[1] **ST − BP:** saving between a thermochromic step transition paint (ST) and a black paint (BP). [2] **ST − WP:** saving between a thermochromic step transition paint (ST) and a white paint (WP).

For Madrid, residential buildings with a thermochromic paint with a transition temperature of 23 °C could have reductions of T_{Max} of up to 11.0 °C and rises of T_{min} of up to 2.4 °C, when compared with common black and white paints, respectively. During the coldest week of the year, this thermochromic paint has the same behavior of a common black paint, while during the hottest week of the year, it has the same behavior of a common white paint. The near 6 °C decrease in T_{Max} for the first floor in case BD1 endorses the estimation predicted (6–8 °C) by Bansal et al. [9] for the surface color effect when comparing a white surface color to a black one.

Additionally, for Madrid, thermochromic paint with an optimized transition temperature always leads to annual energy savings. Maximum reductions of about 4.4 and 1.8 MWh·y^{-1} are observed for BD1 and BD2, respectively, corresponding to maximum energy savings of 39.1% and 33.7%. Ascione et al.'s [21] simulations reported energy savings of ca. 15% with an optimized thermochromic coating in Naples, Italy. That value is considerably lower than those obtained for Madrid and Porto in case BD1 and BD2. In fact, their comparison was made with a 30% TSR paint, while this study used white (90% TSR) and black paints (5% TSR) as references, which can, alongside climate variations, explain this difference.

3.2.2. Case LT—Thermochromic Paint with a Linear Transition Temperature

A hypothetic thermochromic paint with a linear transition temperature was also assessed for Madrid in Spain. Similar to Porto, a TSR value of 5% below the transition temperature range, a TSR value of 90% above this range, and a linear variation between 5% and 90% in the transition temperature range were assumed The transition temperature range was selected taking into account the minimum annual energy demand needed to keep the inside temperature of the residential buildings between 20 and 25 °C, as shown in Figure 6.

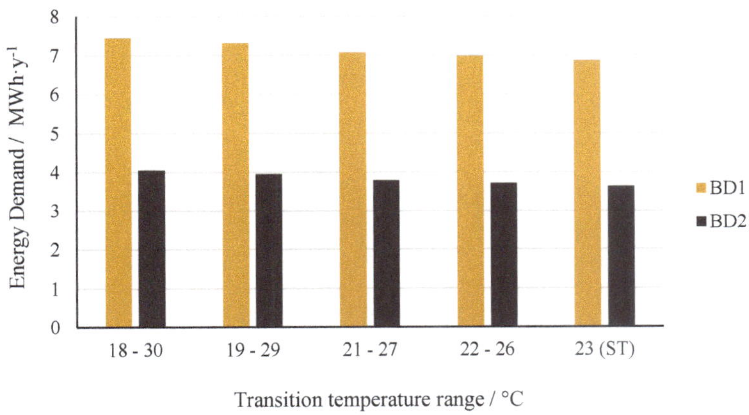

Figure 6. Annual final energy demand, for building configurations BD1 and BD2 located in Madrid, Spain, considering different transition temperature ranges.

As one can see, the same conclusion from Porto can be drawn for Madrid. Additionally, for Madrid, higher energy savings could be achieved with a thermochromic paint with the specific transition temperature of 23 °C, applied both on external walls and on the roof of residential buildings. If the larger transition temperature range is considered, 18–30 °C, the energy demand can increase of up to 8% and 12% in case BD1 and BD2, respectively. Widening the thermochromic transition seems to lead to higher penalties in insulated buildings than buildings without thermal insulation.

3.3. *Abu Dhabi, United Arab Emirates*

3.3.1. Case ST—Thermochromic Paint with a Specific Transition Temperature

Additionally, for Abu Dhabi in the United Arab Emirates, a similar study was performed and the optimum transition temperature was also selected taking into account the minimum annual energy demand of both residential building configurations, BD1 and BD2, as shown in Figure 7.

For Abu Dhabi, the optimum transition temperature of the thermochromic coating is also 23 °C, both for BD1 and BD2. This transition temperature corresponds to annual energy demands of about 6.0 and 3.4 MWh·y^{-1} for BD1 and BD2, respectively. The very small variations in energy demand for transition temperatures lower than 23 °C are a consequence of dealing with a warm and dry climate. The annual energy demand is not significantly affected by color changes in the building surface when the temperature is low because the outdoor temperature rarely achieves low values.

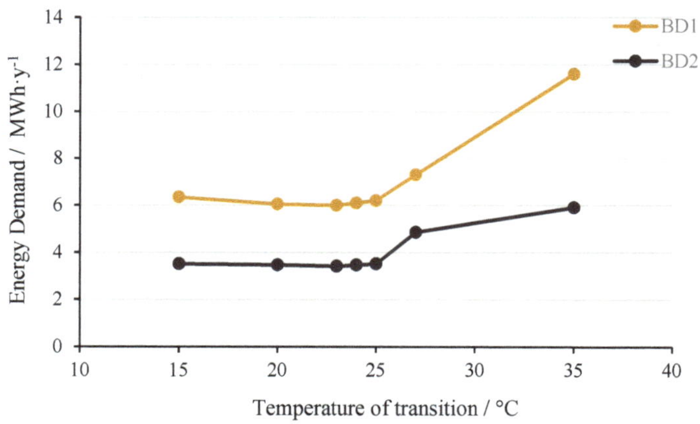

Figure 7. Annual energy demand, for building configurations BD1 and BD2 located in Abu Dhabi, United Arab Emirates, considering different transition temperatures for the thermochromic paint.

Tables 7 and 8 show the indoor thermal behavior and the annual energy demand of building configurations BD1 and BD2 using a thermochromic paint with a transition temperature of 23 °C, where values of T_{Max} over the hottest week of the year, T_{min} over the coldest week of the year, and the annual energy demand for cooling for heating and total are presented.

The residential buildings located in Abu Dhabi, United Arab Emirates, having a thermochromic paint with a transition temperature of 23 °C, could have reductions of T_{Max} of up to 7.7 °C and rises of T_{min} of up to 2.7 °C, when compared with common black and white paints, respectively.

For the three locations studied in this article, the indoor maximum temperature was reduced on average by 5 °C in Abu Dhabi, 6 °C in Porto, and 7 °C in Madrid, without considering thermal insulation. In the literature, Ma et al. [17], Karlessi et al. [19], and Ye et al. [20] reported maximum surface temperature reductions from 4 to 10 °C with the use of thermochromic paints.

Table 7. Average maximum indoor temperature (T_{Max}) over the hottest week of the year and average minimum indoor temperature (T_{min}) over the coldest week of the year, for a building painted with an optimized thermochromic step transition paint (ST), in Abu Dhabi, United Arab Emirates.

Building Configuration	Case	T_{Max} /°C			T_{min}/°C		
		Ground Floor	First Floor	Second Floor	Ground Floor	First Floor	Second Floor
BD1	BP	40.5	44.9	47.2	16.5	17.5	17.5
	WP	37.5	41.0	39.5	13.6	16.0	14.7
	ST	37.4	41.0	39.5	16.1	17.5	17.4
	ST − BP [1]	**−3.1**	**−3.9**	**−7.7**	**−0.4**	**0**	**−0.1**
	ST − WP[2]	**−0.1**	**0**	**0**	**+2.5**	**+2.5**	**+2.7**
BD2	BP	36.3	40.1	41.9	16.5	17.5	18.1
	WP	34.1	37.3	36.9	14.2	16.7	15.5
	ST	34.1	37.4	36.9	16.0	17.5	18.0
	ST − BP [1]	**−2.2**	**−2.7**	**−5.0**	**−0.5**	**0**	**−0.1**
	ST − WP [2]	**0**	**−0.1**	**0**	**−1.8**	**+0.8**	**+2.5**

[1] **ST − BP:** saving between a thermochromic step transition paint (ST) and a black paint (BP). [2] **ST − WP:** saving between a thermochromic step transition paint (ST) and a white paint (WP).

Table 8. Final energy demand for heating ($E_{Heating}$), and cooling ($E_{Cooling}$), total final energy demand (E_{Total}), and energy savings over the year, for a building painted with an optimized thermochromic step transition paint (ST), located in Abu Dhabi, United Arab Emirates.

Building Configuration	Case	Energy/MWh·y^{-1}			Energy Savings /%
		$E_{Heating}$	$E_{Cooling}$	E_{Total}	
BD1	BP	0.04	12.23	12.27	
	WP	0.46	6.11	6.57	
	ST	0.07	6.00	6.06	
	ST − BP [1]	**+0.03**	**−6.23**	**−6.21**	**50.6**
	ST − WP [2]	**−0.40**	**−0.11**	**−0.51**	**7.7**
BD2	BP	0.02	5.97	5.99	
	WP	0.17	3.38	3.54	
	ST	0.03	3.39	3.42	
	ST − BP [1]	**+0.01**	**−2.58**	**−2.56**	**42.8**
	ST − WP [2]	**−0.14**	**+0.01**	**−0.12**	**3.4**

[1] **ST − BP:** saving between a thermochromic step transition paint (ST) and a black paint (BP). [2] **ST − WP:** saving between a thermochromic step transition paint (ST) and a white paint (WP).

For Abu Dhabi, a thermochromic paint with a transition temperature of 23 °C always leads to annual energy savings. Maximum reductions of about 6.2 and 2.6 MWh·y^{-1} are observed for building configurations BD1 and BD2, respectively, corresponding to maximum energy savings of about 50.6% and 42.8%, when compared to dark coatings. However, the energy savings are less meaningful when the comparison is made with white paints—7.7% to BD1 and 3.4% to BD2. Since this climate has negligible heating needs ($\boldsymbol{E_{Heating}} \approx 0$), the use of cool paints (high TSR value) would contribute substantially to a decrease of the energy demand of the building, with winter penalties of around 3%–8%. A similar response would be expected from a typically cold climate, with negligible cooling needs ($\boldsymbol{E_{Cooling}} \approx 0$). A thermochromic paint application would have a very small positive impact when compared to a dark coating. Nonetheless, the overall energy performance is expected to be always better with thermochromic paints, with a more or less notable effect, because the paint is expected to respond thermally to its environment, independently of the climate.

3.3.2. Case LT—Thermochromic Paint with a Linear Transition Temperature

The use of a hypothetic thermochromic paint with a linear transition temperature was also assessed for Abu Dhabi in the United Arab Emirates. A thermochromic paint with a TSR value of 5% below the transition temperature range, a TSR value of 90% above this range, and a linear variation between 5% and 90% in the transition temperature range were assumed. The transition temperature range was selected taking into account the minimum annual energy demand needed to keep the indoor temperature between 20 and 25 °C, as shown in Figure 8.

As observed in Figure 8, only the transition temperature range of 22 to 24 °C has, for case BD1, a slightly lower annual energy demand compared with case ST (23 °C). This small difference is not sufficient to consider the use of a thermochromic paint with a linear transitional temperature instead of a thermochromic paint with a step transition one beneficial.

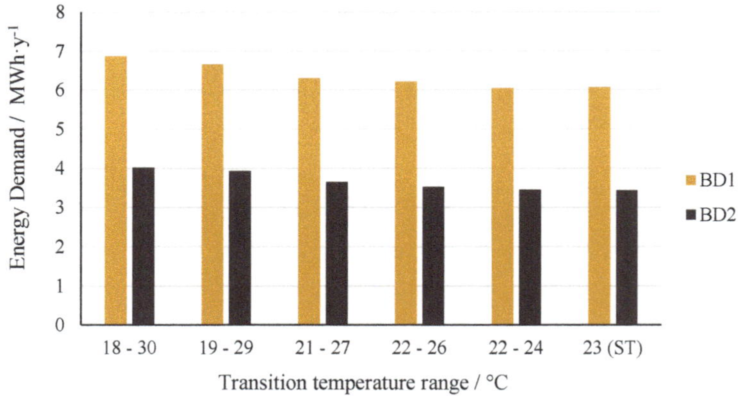

Figure 8. Annual final energy demand for building configurations BD1 and BD2, located in Abu Dhabi, the United Arab Emirates, considering different transition temperature ranges.

4. Conclusions

The thermal behavior of a residential building was simulated, assessing the impact of using thermochromic paints with a step transition or with linear transition, applied both on external walls and on the roof. This simulation study was performed using the open source software ESP-r, for building configurations with and without thermal insulation, and virtually located in three different cities/climates: Porto in Portugal, Madrid in Spain, and Abu Dhabi in the United Arab Emirates.

For the three different locations, the optimized step transition temperatures were determined; the results indicated that these temperatures are 24 °C for Porto and 23 °C for Madrid and Abu Dhabi. Considering the thermochromic paint with a linear transition, it was found that the optimal transition range was very narrow and around the optimal transition temperature of the step transition case.

Regarding the assessment in free-floating mode, i.e., without any active mechanical heating or cooling, the highest indoor temperature reduction in summer was achieved in Madrid, from 43.1 to 32.1 °C, in the upper floor of the detached house. The highest indoor temperature increment during the winter was observed for Abu Dhabi, from 14.7 to 17.4 °C, also in the upper floor. In general, the use of thermochromic paints allowed maximum indoor temperature reductions from 2.2 to 11 °C, when compared to dark coatings. Additionally, minimum indoor temperature increments from 0.8 to 2.7 °C were achieved, when compared to white paints.

In terms of energy use, when operating the building in the conditioned mode to keep the indoor temperature between 20 and 25 °C, thermochromic paints led to energy savings of up to 48%, 39%, and 8%, comparably to common white paints, and up to 20%, 26%, and 50.6% when compared to common black paints, for Porto, Madrid, and Abu Dhabi, respectively.

Regarding the effect of thermal insulation, it decreased ca. 50% of the building energy demand. The impact of thermochromic paints in terms of energy savings is 2%–8% higher in buildings without thermal insulation than insulated buildings.

It was also possible to conclude that the use of thermochromic paints has a more evident impact in temperate climates. The seasonal temperature variation must be significant to justify the need for a dynamic response of the surface by changing its optical properties and color.

Overall, these virtual simulation results confirmed that thermochromic paints can deliver benefits both in summer mode as in winter, unlike cool paints, which provide summer benefits but have a penalty in winter. Nevertheless, these outcomes require future confirmation by physical tests in a real scenario. Additionally, the chemical composition of the paint/coating is essential for understanding the final performance of these materials. Further developments on the chemical formulation of such paints to ensure longevity and cost reductions should thus be encouraged.

Glossary/Nomenclature/Abbreviations

List of Acronyms	
BD1	Building configuration without thermal insulation
BD2	Thermal insulated building configuration
BP	Black paint
COP	Coefficient of performance
EER	Energy efficiency ratio
EPS	Expanded Polystyrene
GHG	Greenhouse gases
HVAC	Heating, ventilation and air-conditioning
LT	Linear transition
ST	Step transition
TSR	Total solar reflectance
WP	White paint
Notation and Glossary	
E_{Heating}	final energy demand for heating (MWh·y^{-1})
E_{Cooling}	final energy demand for cooling (MWh·y^{-1})
E_{Total}	total final energy demand for air conditioning (MWh·y^{-1})
T_{Max}	week average daily maximum indoor temperature, (°C)
T_{min}	week average daily minimum indoor temperature (°C)
α	absorptivity
ε	emissivity

Author Contributions: Conceptualization, A.M., J.M.; methodology, V.G.; software, V.G.; validation, M.A., V.L. and A.M.; formal analysis, T.S. and M.A.; investigation, M.A and T.S.; data curation, V.G.; writing—original draft preparation, T.S. and V.G.; writing—review and editing, M.A. and V.L.; visualization, M.A.; supervision, A.M and V.L.; funding acquisition, J.M. All authors have read and agreed to the published version of the manuscript.

Funding: This work was financially supported by: Base Funding—UIDB/00511/2020 of the Laboratory for Process Engineering, Environment, Biotechnology and Energy—LEPABE—funded by national funds through the FCT/MCTES (PIDDAC).

Acknowledgments: The authors would also like to acknowledge the contributions of Diana Dias for their collaboration and also FCT and CIN S.A. for the support of Diana Dias's Ph.D. grant (SFRH/BDE/33548/2009) and to ARCP (http://www.arcp.pt) for the technical support.

Conflicts of Interest: The authors declare no conflict of interest.

References

1. Council of Europe. Directive (EU) 2018/844 of the European Parliament and of the Council of 30 May 2018 amending Directive 2010/31/EU on the Energy Performance of Buildings and Directive 2012/27/EU on Energy Efficiency (Text with EEA Relevance). Available online: http://data.europa.eu/eli/dir/2018/844/oj (accessed on 20 January 2020).
2. ASHRAE—American Society of Heating Refrigerating and Air-Conditioning Engineers. *ANSI/ASHRAE Standard 55-2013—Thermal Environmental Conditions for Human Occupancy*; American Society of Heating Refrigerating and Air-Conditioning Engineers: Atlanta, GA, USA, 2013.
3. Frend, R. *Determination of Thermal Insulation Properties of Buildings and Structures' External Envelope Using Infrared Thermal Imaging*; University of Huddersfield: Huddersfield, UK, 2011.
4. Li, H. Chapter 2—Literature Review on Cool Pavement Research. In *Pavement Materials for Heat Island Mitigation*; Li, H., Ed.; Butterworth-Heinemann: Boston, MA, USA, 2016; pp. 15–42. [CrossRef]
5. Pisello, A.L. State of the art on the development of cool coatings for buildings and cities. *Sol. Energy* **2017**, *144*, 660–680. [CrossRef]
6. Upstone, S. *Measurement of Total Solar Reflectance of Paint Panels Using PerkinElmer UV/Vis/NIR Spectrophotometers and UV WinLab Software*; PerkinElmer, Inc.: Waltham, MA, USA, 2019; p. 3.
7. Doulos, L.; Santamouris, M.; Livada, I. Passive cooling of outdoor urban spaces. The role of materials. *Sol. Energy* **2004**, *77*, 231–249. [CrossRef]

8. Premier, A. Façade cool coatings: An experiment on colour and surface quality. *Intell. Build. Int.* **2019**, 1–18. [CrossRef]
9. Bansal, N.K.; Garg, S.N.; Kothari, S. Effect of exterior surface colour on the thermal performance of buildings. *Build. Environ.* **1992**, *27*, 31–37. [CrossRef]
10. Dias, D.; Machado, J.; Leal, V.; Mendes, A. Impact of using cool paints on energy demand and thermal comfort of a residential building. *Appl. Therm. Eng.* **2014**, *65*, 273–281. [CrossRef]
11. Akbari, H.; Konopacki, S. Energy effects of heat-island reduction strategies in Toronto, Canada. *Energy* **2004**, *29*, 191–210. [CrossRef]
12. Synnefa, A.; Santamouris, M.; Apostolakis, K. On the development, optical properties and thermal performance of cool colored coatings for the urban environment. *Sol. Energy* **2007**, *81*, 488–497. [CrossRef]
13. Santamouris, M.; Synnefa, A.; Karlessi, T. Using advanced cool materials in the urban built environment to mitigate heat islands and improve thermal comfort conditions. *Sol. Energy* **2011**, *85*, 3085–3102. [CrossRef]
14. Asdrubali, F.; Desideri, U. Chapter 6—Building Envelope. In *Handbook of Energy Efficiency in Buildings*; Butterworth-Heinemann: Amsterdam, The Netherlands, 2019; pp. 295–439. [CrossRef]
15. Seeboth, A.; Lötzsch, D. *Thermochromic and Thermotropic Materials*; Pan Stanford: Singapore, 2014.
16. Ma, Y.; Zhu, B.; Wu, K. Preparation and solar reflectance spectra of chameleon-type building coatings. *Sol. Energy* **2001**, *70*, 417–422. [CrossRef]
17. Ma, Y.; Zhu, B.; Wu, K. Preparation of reversible thermochromic building coatings and their properties. *J. Coat. Technol.* **2000**, *72*, 67–71. [CrossRef]
18. Ma, Y.; Zhang, X.; Zhu, B.; Wu, K. Research on reversible effects and mechanism between the energy-absorbing and energy-reflecting states of chameleon-type building coatings. *Sol. Energy* **2002**, *72*, 511–520. [CrossRef]
19. Karlessi, T.; Santamouris, M.; Apostolakis, K.; Synnefa, A.; Livada, I. Development and testing of thermochromic coatings for buildings and urban structures. *Sol. Energy* **2009**, *83*, 538–551. [CrossRef]
20. Ye, X.; Luo, Y.; Gao, X.; Zhu, S. Design and evaluation of a thermochromic roof system for energy saving based on poly(N-isopropylacrylamide) aqueous solution. *Energy Build.* **2012**, *48*, 175–179. [CrossRef]
21. Ascione, F.; Bianco, N.; Iovane, T.; Mauro, G.M.; Napolitano, D.F. Development of an analytical model to investigate the effects of the extraflux versus the sky and the ground and optimization of the radiative characteristics of a thermochromic paint for a typical Italian location. *AIP Conf. Proc.* **2019**, *2191*, 020011. [CrossRef]
22. Soudian, S.; Berardi, U. Cementitious plasters for façade finishing with phase change materials and thermochromic pigments. *IOP Conf. Ser. Mater. Sci. Eng.* **2019**, *609*, 062023. [CrossRef]
23. Karlessi, T.; Santamouris, M. Improving the performance of thermochromic coatings with the use of UV and optical filters tested under accelerated aging conditions. *Int. J. Low-Carbon Technol.* **2013**. [CrossRef]
24. Ritter, A. *Smart Materials in Architecture, Interior Architecture and Design*; Birkhäuser Basel: Basel, Switzerland, 2007.
25. ESRU—Energy Systems Research Unit. Available online: http://www.esru.strath.ac.uk//Programs//ESP-r.htm (accessed on 16 January 2020).
26. Clarke, J.A. *Energy Simulation in Building Design*, 2nd; Clarke, J.A., Ed.; Butterworth-Heinemann: Oxford, UK, 2001; pp. 64–98. [CrossRef]
27. *Decree-Law No 80/2006 of April 4—Regulation on Thermal Insulation in Buildings (Regulamento das Características de Comportamento Térmico dos Edifícios)*; In Diário da República: Lisboa, Portugal, 2006.
28. Santos, C.A.P.; Matias, L. *Coeficiente de Transmissão Térmica de Elementos da Envolvente dos Edifícios—Verão Actualizada 2006*; Laboratório Nacional de Engenharia Civil: Lisbon, Portugal, 2006.

energies

MDPI

Article

Thermal and Energy Performance Assessment of the Prefab Electric Ondol System for Floor Heating in a Residential Building

Wongeun Lee [1], Taesub Lim [1,*] and Daeung Danny Kim [2,*]

[1] Department of Architectural Engineering, Seoil University, Seoul 02192, Korea; wklee34@seoil.ac.kr
[2] Department of Architectural Engineering, King Fahd University of Petroleum and Minerals (KFUPM), Dhahran 31261, Saudi Arabia
* Correspondence: francis9@seoil.ac.kr (T.L.); dkim@kfupm.edu.sa (D.D.K.)

Received: 1 September 2020; Accepted: 9 October 2020; Published: 2 November 2020

Abstract: In South Korea, radiant floor heating has been used from old housing to the recently constructed residential buildings, which is called "Ondol". The Ondol system is generally a water-based system and it uses hot water as a heat medium provided by boilers fueled by natural gas. With great effort to reduce greenhouse gas emissions, electric Ondol panels have been increasingly applied to the recent residential buildings for floor heating. While the prefab electric Ondol panels were developed with the demand for dry construction method, the information about the prefab electric Ondol system is not sufficient. For the present study, the thermal performance of the prefab electric Ondol panels was investigated through field measurement. In addition, the heating energy and economic performance of the electric panel were compared with the conventional Ondol system. As a result, a significant surface temperature difference was observed. Moreover, the heating cost for the prefab electric Ondol system was more expensive than the conventional system, even though a heat loss was observed by the operation of the conventional system.

Keywords: prefab electric Ondol system; thermal performance; heating energy; heating cost; residential building

1. Introduction

As one of the main contributors to building energy consumption in South Korea, a residential building sector has accounted for more than 65% of total buildings in 2017 [1]. According to the report provided by the Korea Energy Economics, about 10% of the total energy was used for residential buildings [2]. Specifically, natural gas and electricity accounted for 46% and 28% of household end-use energy consumption in 2015, respectively, which were the most-used energy sources for residential buildings [2]. In addition, more than half of the annual energy consumption for residential buildings was used for space heating and cooling, and others were used for water heating, lighting, and miscellaneous equipment [3,4].

While the energy consumption for cooling has been recently increased, a significant amount of energy still has used for heating in residential buildings [5,6]. Accordingly, there are several types of heating methods available including central gas heating, district heating, and individual gas heating. As an individual gas boiler has become available with the development of infrastructures since 2000, the individual gas heating method has been dominantly used for residential buildings [7]. Traditionally, most residential buildings in South Korea have preferred to use radiant floor heating. According to the sample data of housing units in the study of Park et al., all housing units have equipped a hydronic radiant floor heating system [8]. For a radiant heating method, it is imperative to use fossil fuels for a domestic boiler, district, or central heating in South Korea in that the greenhouse gas emissions are still increased. Therefore, it is necessary to find alternative heat sources for radiant heating in residential buildings for the goal of reducing greenhouse gas emissions [9].

The use of radiant heating enables to provide more thermal comfort to occupants in houses as well as an opportunity for the energy-saving more than the conventional air heating systems [10,11]. Many studies have performed investigations for the performance of the radiant heating method. According to the study of Lin et al., the thermal performance of a water-based radiant heating system was compared with a convective heating system in residential buildings. Even though there was little difference between the two systems, the convective heating system might cause local discomfort based on the occupants' surveys [12]. Sun et al. also investigated the thermal performance of a radiant heating system using a heat pipe [13]. In addition, the capillary tube was employed for the radiant floor heating system [14]. While water has been generally used for radiant heating systems as a heat source, radiant heating systems using electric cables have been recently adapted. Lodi et al. analyzed the efficiency of electric radiant heating systems regarding thermal comfort in old buildings [15]. As another type of electric panel, thermoelectric heating panels are used consisting of a radiant plate and several thermoelectric modules [16]. In the case of the study of Fang et al., phase change materials as a thermal storage medium were used for the electric radiant floor heating [17].

For the present study, the energy and economic performance of the electric radiant floor heating system in a residential building were assessed through the measurement and energy simulation. According to the Act on the Promotion of the Development, Use, and Diffusion of New and Renewable Energy, Korea Ministry of Trade, renewable energy systems should be designed for newly constructed residential buildings in South Korea [18]. Therefore, the electricity consumption for electric radiant floor heating systems can be offset by renewable energy systems such as electricity generated by solar PV panels. Moreover, this electric-based radiant heating system can contribute to reducing greenhouse gas emissions. However, there were a few studies for the investigation of the performance of the electric Ondol panels. Thus, this study will evaluate the thermal performance and the economic impact of the electric Ondol systems and discuss the results obtained through the measurement and the simulation. Moreover, the outcomes of the present study will be used to develop more energy and thermally efficient electric Ondol panels.

2. Prefab Electric Ondol System

Traditionally, Ondol has been used for an underfloor heating system in residential buildings, which meant a warm stone [19,20]. From the early nineteenth century to the recent construction, the Ondol system has been used in residential buildings as a representative floor heating method because of the advantage of the radiant heating [19,21,22]. By using hot water as a heating medium, hot water provided by a boiler system circulates pipes embedded in the concrete slab in that the circulation of hot water within the Ondol system can provide radiant heating. Comparing with the conventional air heating method, the Ondol system is more environmentally sustainable and cheaper regarding the life cycle cost [23–25].

In general, the Ondol system was constructed by the wet construction method on site. Nowadays, the modular construction method with prefab Ondol panels is increasingly adapted in residential buildings to enhance the construction quality and shorten the construction period [26]. The use of prefab Ondol panels can make buildings lighter than those made by the wet construction method as well as provide an opportunity to reduce construction waste [27]. While most prefab Ondol panels have used hot water from boilers fueled by natural gas or oil, another type of Ondol panel system using electricity was proposed as shown in Figure 1. According to the study of Jeong, about 10% of total energy-saving was achieved by using the prefab electric Ondol system [28]. However, there are a few studies about the performance of the prefab electric Ondol panels. To assess the performance of the prefab electric Ondol panels more accurately, the present study compared the performance of the prefab electric Ondol panels with the conventional Ondol systems in residential buildings.

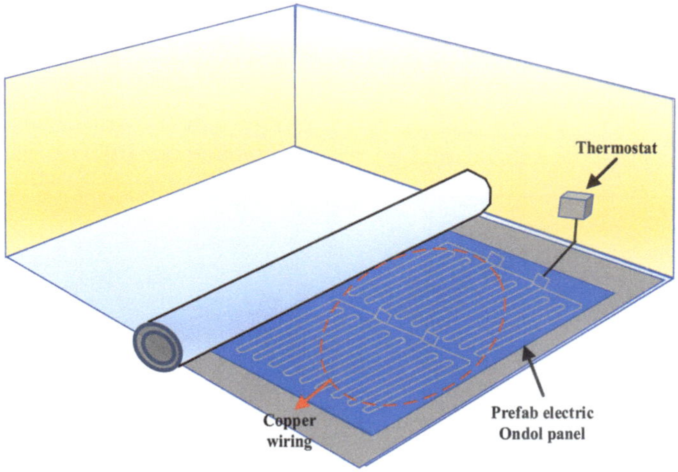

Figure 1. Prefab electric Ondol panel.

3. Methodology

To assess the thermal performance of the prefab electric Ondol panels, thermal performance measurement was conducted in a residential building. In addition, energy performance, especially heating was analyzed by using energy simulation.

3.1. Building Description

For the present study, an apartment building located in Seoul in South Korea was selected. The apartment building has 56 units with 16 floors. On the 8th floor of the apartment building, the unit was selected with a total area of 125 m^2. The measurement was conducted in the living room of the unit facing the southeast. The specification of the building envelops of the unit was presented in Table 1.

Table 1. The specification of the envelope systems of the reference building.

U-value of walls	0.62 W/m^{2}K
U-value of ceiling	0.69 W/m^{2}K
U-value of window systems	2.8 W/m^{2}K
Shading coefficient	0.6
Air infiltration	2.1 cm^2/m^2
Internal heat gain	3 occupants Lighting: 5.4 W/m^2 Equipment: 7.0 W/m^2
Design temperature	26 °C for cooling and 20 °C for heating

3.2. Thermal Measurement

As shown in Figure 2, 9 K-type thermocouples were located on the floor with 50 cm intervals to figure out the surface temperature of the floor heated by using the prefab electric Ondol panels. To monitor the indoor air temperature, a thermocouple was located at 1.5 m from the floor. In addition, a thermocouple was also located outdoor. The measurement was conducted from April 27th to May 4th in 2018 and the temperature data were recorded with 10 min intervals by the datalogger. Moreover, a surface temperature on the floor was visualized by using a thermal imaging camera. The energy consumption was also monitored to compare with the energy simulation. The equipment used for the measurement was presented in Table 2.

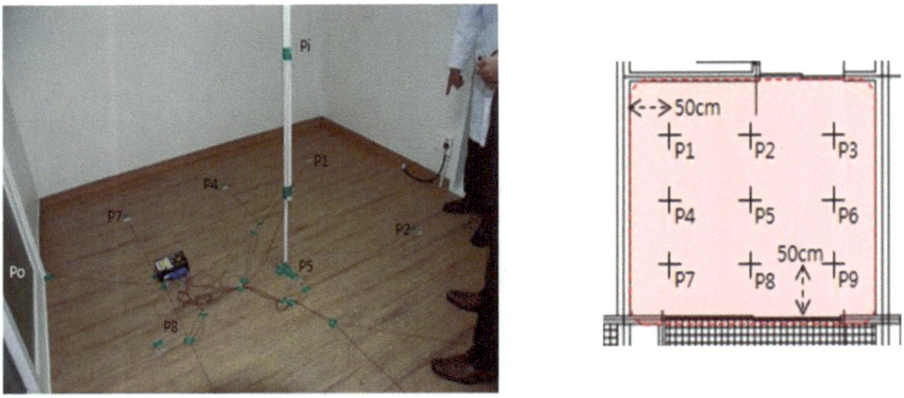

(a)

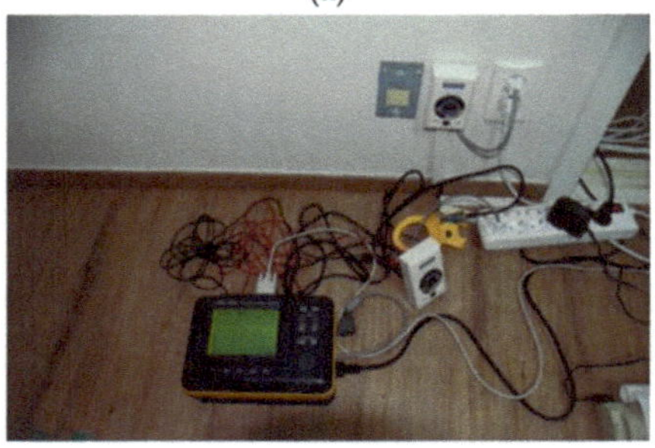

(b)

Figure 2. Thermal measurement: (**a**) The locations of thermocouples, (**b**) monitoring electricity consumption.

Table 2. Specifications of the equipment.

Power meter (3166 Clamp on Power HiTESTER, US)	- Voltage: 150 V to 600 V - Frequency range: 45 Hz to 66 Hz - Accuracy: AC Voltage: ±0.2% reading. ±0.1% full scale AC Current: ±0.2% reading. ±0.1% full scale
Datalogger (1250 series Remote Squirrel meter/logger, UK)	- Operation range: Temperature −40 °C to +85 °C - Accuracy: Temperature: ±0.3 °C - Response time: <0.7 s (start-up 3 s)
Thermal imaging camera (testo 865, Germany)	- IR resolution of 160 x 120 pixels - Measuring range: −20 °C to 280 °C - Accuracy of ±2 °C - Emissivity: 0.01 to 1

3.3. Energy Simulation

To assess the energy performance of the prefab electric Ondol panels, the energy simulation was performed by using eQuest 3.61, which is the comprehensive energy simulation tool to evaluate design parameters of energy conservation measures as well as provide detailed information of building energy use [29]. Using this software, the reference apartment building was modeled as shown in Figure 3.

In addition, the energy simulation was performed by using the specification of the reference building in Table 1. The residential building was operated for 24 hours and the HVAC system ran between 5 pm to 7 am. Moreover, the design temperatures for heating and cooling were set at 20 °C and 26 °C, respectively. For the conventional Ondol system, a central water-to-water heat pump system was modeled for providing hot water, while the data obtained from the measurement were used for the prefab electric ondol model. The annual energy consumption of the unit by using the conventional Ondol system fueled by natural gas was compared with that operated by the prefab electric Ondol panels. For the simulation, the TMY weather data of Seoul in South Korea was utilized. The mean air temperature in Seoul was about 13 °C. The lowest and highest air temperatures were −12.6 °C and 35.4 °C, respectively. The mean wind speed was 2.2 m/s.

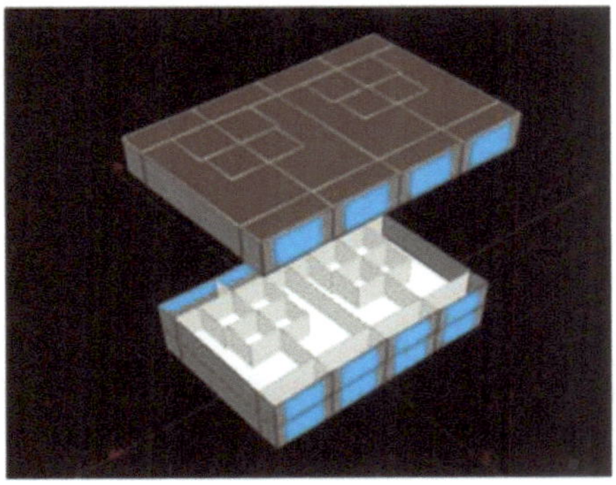

Figure 3. Computational model of the apartment building.

By using the coefficient of variation of the root mean squared error (CV(RMSE)) provided by ASHRAE Guideline 14, the monthly energy consumption of the reference residential building with the conventional Ondol system was compared with the energy simulation [30]. The models will be declared to be calibrated if they produce CV(RMSE)s within ±15% with monthly energy data.

$$RMSE = \sqrt{\frac{\sum_{i=1}^{n} (M_i - S_i)^2}{n}} \quad (1)$$

$$C_V(RMSE) = \frac{RMSE}{M_{avg}} \times 100 \quad (2)$$

where M_i is the energy consumption of the residential building, while S_i is the monthly energy consumption by energy simulation. n is the period and M_{avg} is the average for the energy consumption of the residential building.

4. Result

4.1. Thermal Performance of the Prefab Electric Ondol Panels

Figure 4 shows the air temperature distributions of the surface on the floor and the indoor and outdoor on May 2nd and 3rd. From 0 to 19:30 h on May 2nd, the heating was off and the indoor air temperature and the surface temperature on the floor were maintained at about 19.5 °C because of the thermal capacity of the building envelopes. From 19:30 h on May 2nd to 24 h on May 3rd, heating was provided by using the prefab electric Ondol panels. While the setpoint temperature for the Ondol panels was set at 60 °C at full-load, the maximum surface temperature on the floor was maintained at about 50 °C. In addition, the minimum surface temperature on the floor was about 28 °C. Even though the surface temperature on the floor was increased about 7 °C by the operation of the Ondol panels, the increase in the indoor air temperature was only about 4 °C with/without the Ondol panels. Thus, the indoor air temperature was about 24 °C. The operation of the Ondol panels took 1.4 kWh of electricity. Moreover, the surface temperature on the floor was visualized by using the thermal imaging camera (Figure 5). As shown in Figure 4, the maximum and minimum surface temperature difference was more than 20 °C and the surface temperature was higher than the indoor air temperature. It can be seen that water from conventional Ondol systems can transfer radiant and convective heat into the indoor fully because of higher heat capacity of water than that of air, while the electric Ondol panels only heated the floor directly. This caused the lower indoor air temperature than the surface temperature on the floor.

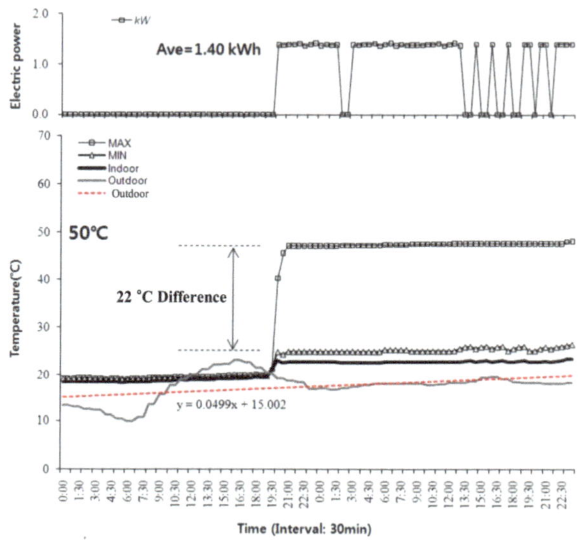

Figure 4. Temperature distribution and the electricity consumption.

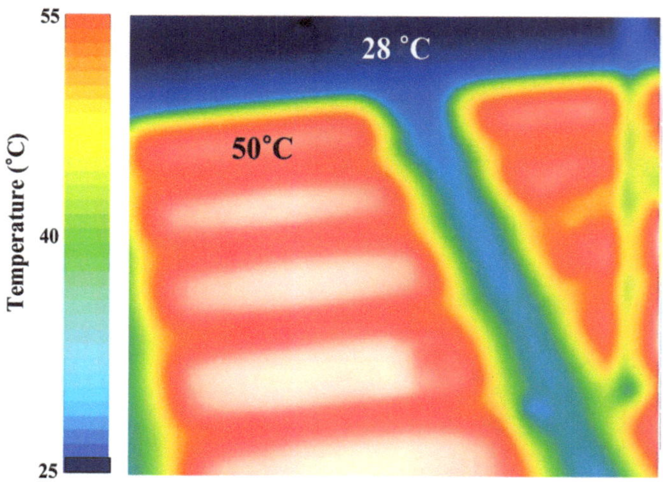

Figure 5. Temperature distribution by the thermal imaging camera.

4.2. The Analysis of the Energy Consumption

4.2.1. The Comparison between Energy Simulation and the Monthly Energy Consumption

To confirm the validity of the energy model, the monthly energy consumption of the selected unit of the apartment building was compared with the energy consumption prediction. The energy simulation modeled the conventional Ondol system fueled by natural gas. As can be shown in Table 3, the total energy consumption of the selected unit and the energy simulation were 7862 kWh and 8611 kWh, respectively. In addition, the root mean squared errors (CV(RMSEs) ranged from 0.18 to 7.77. Even though there was much difference between the data and the prediction in December, all the values were within the acceptable range, the predicted results by the simulation met the requirement by ASHRAE Guideline 14 [30].

Table 3. The monthly energy consumption comparison.

Month	Energy Consumption (kWh)			CV(RMSE) (%)
	The Selected Unit of the Apartment Building	Energy Simulation	Difference	
January	769.3	860.0	−90.7	4.00
February	664.5	779.0	−114.5	5.04
March	640.2	745.0	−104.8	4.62
April	576.0	580.0	−4.0	0.18
May	589.2	512.0	77.2	3.40
June	607.7	648.0	−40.3	1.78
July	748.4	799.0	−50.6	2.23
August	792.5	823.0	−30.5	1.34
September	612.2	786.0	−173.8	7.66
October	564.4	644.0	−79.6	3.51
November	578.5	680.0	−101.5	4.47
December	719.3	755.0	−35.7	1.57

4.2.2. The Heating Cost Analysis

Considering the heating energy consumption, the annual heating cost for the conventional Ondol system and the prefab electric Ondol panel system was analyzed for 8 months (October to May) (Figure 6). Tables 4 and 5 show the heating energy consumption and the cost of these two heating methods. The total heating cost for the conventional Ondol system and the prefab electric Ondol system were $140.5 (US dollar) and $313.2 (US dollar), respectively. As a result, the prefab electric Ondol system requires about 53% of the additional heating cost.

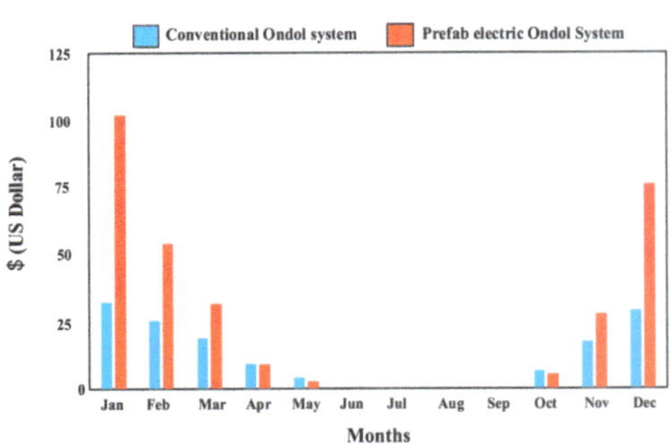

Figure 6. The heating cost comparison between the conventional Ondol system and the prefab electric Ondol system.

Table 4. The conventional Ondol system.

Month		Jan	Feb	Mar	Apr	May	Jun	July	Aug	Sep	Oct	Nov	Dec	Total
Heating energy consumption (kWh)		624.7	488	374.4	175.8	66.3	0	0	0	0	119	348.6	559.4	2756
Heating cost	Demand charge ($US dollar)	1.13	1.13	1.13	1.13	1.13	0	0	0	0	1.13	1.13	1.13	-
	Energy charge ($US dollar)	32	25	18	8.6	3.2	0	0	0	0	5.8	17	28.6	-
Monthly total heating cost ($US dollar)		33	19.4	19.4	9.7	4.4	0	0	0	0	6.9	18	29.7	$140.5/year

Table 5. The prefab electric Ondol panel system.

Month		Jan	Feb	Mar	Apr	May	Jun	July	Aug	Sep	Oct	Nov	Dec	Total
Heating energy consumption (kWh)		571	446	342	161	61	0	0	0	0	109	319	511	2520
Heating cost	Demand charge ($US dollar)	8.2	4.5	2.4	0.6	0.03	0	0	0	0	0.6	2.4	8.2	-
	Energy charge ($US dollar)	95	50.3	30	9	2.7	0	0	0	0	5.1	26	68.4	-
Monthly total heating cost ($, US dollar)		103	55	32	9.5	2.7	0	0	0	0	5.6	28.4	77	$313.2/year

5. Discussion

The present study assessed the thermal and energy performance of the prefab electric Ondol panel comparing with the conventional Ondol system. Considering the result of the thermal measurement, a notable surface temperature difference was observed when the prefab electric Ondol system was applied. In addition, the increase in the surface temperature on the floor was about 30 °C, while about a 4 °C increase in the indoor air temperature was observed. This can cause thermal stratification, as well as occupants, can be dissatisfied thermally. Finally, it may require a significant amount of energy for thermal comfort. Considering the balance point temperature and the heat gain/loss by building envelopes. Thus, it is also necessary to conduct thermal measurement in a chamber by varying climate conditions.

As shown in Figure 7, the use of the prefab electric Ondol panels cost 50% more than that of the conventional Ondol system, while the heating energy consumption of the prefab electric Ondol system was about 8.6% lower than that of the conventional Ondol system. This was caused by the progressive electricity billing during the winter period (December to February) in South Korea, while the price of natural gas was the flat rate. Therefore, it requires to find out the criteria for assessing the performance of the prefab electric Ondol panel system more accurately.

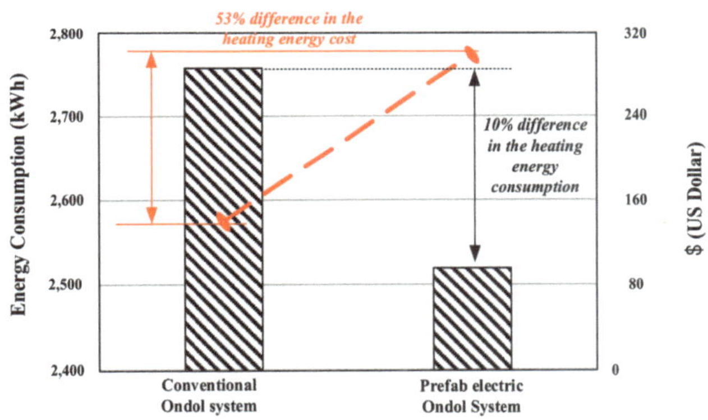

Figure 7. The heating energy and cost comparison.

Another important design consideration for the use of the electric Ondol panel system is electromagnetic waves. According to the National Radio Research Agency in South Korea, the permissible exposure limit is 833 mG, which is quite bigger than the permissible limit of the United States (2 mG) and Swiss (10 mG) [31,32]. Even though the electromagnetic wave of the prefab electric Ondol panel system ranges generally 0.4–0.5 mG, it requires further investigation for the electromagnetic waves from this electric Ondol panels and re-assessment with the permissible limit of other countries.

6. Conclusions

With an increasing demand for the dry construction method, the use of prefab Ondol panels has simplified building construction methods and shortened construction periods, where Ondol floor systems have been dominant. In addition, the electric Ondol panel system has become attractive due to its simplicity and electricity provided by the solar PV panels. The newly constructed residential buildings in South Korea should equip renewable energy systems. Even though the use of the prefab electric Ondol panel system has been rapidly increased, there were a few studies about the performance of this panel system.

For the present study, the thermal performance of the prefab electric Ondol system was investigated through the on-site measurement. In a unit of an apartment building, the surface temperature on the floor and the indoor and outdoor air temperatures were measured. As a result, a significant surface temperature difference was observed. In addition, the use of the prefab electric Ondol system rarely influenced indoor air temperature. The increase in the indoor air temperatures was quite lower than the increase in surface temperature on the floor with/without the use of the electric panels. This was also visualized by using a thermal imaging camera. Moreover, the heating cost for these two systems were evaluated by using energy simulation. The heating energy consumption in the unit with the prefab electric Ondol panel was about 10% lower than that by the conventional Ondol system fueled by natural gas. However, more than 50% of the heating cost was required for the electric Ondol system than the conventional one. It was caused by different energy sources.

Regarding the views of thermal and heating cost performance, the advantage for the use of the prefab electric Ondol panel system is not clear, even though this system has several advantages in terms of construction. For further study, it is necessary to investigate electromagnetic waves from the electric Ondol panels for occupants' safety in buildings. In addition, it requires to conduct measurements with various types of Ondol panels during the winter. To figure out the surface temperature distribution, it is also necessary to employ computational fluid dynamics simulation. Moreover, the life cycle cost analysis is required for the complete investigation of the economic impact of the electric Ondol system.

Author Contributions: Data curation, W.L. and T.L.; formal analysis, W.L. and T.L.; investigation, W.L. and T.L.; software, D.D.K.; validation, D.D.K.; writing—original draft, D.D.K. All authors have read and agreed to the published version of the manuscript.

Funding: The present research has been conducted by the Research Grant of Seoil University in 2020.

Conflicts of Interest: The authors declare no conflict of interest.

References

1. The Annual Energy Consumption in 2017. Korea Energy Agency. Available online: http://www.Energy.Or.Kr/renew_eng/main/main.Aspx (accessed on 15 August 2018).
2. A Report for the Energy Consumption in 2017. Korea Energy Economics Institute. Available online: http://www.Keei.Re.Kr/main.Nsf/index.Html (accessed on 15 August 2018).
3. Song, D.; Choi, Y.-J. Effect of building regulation on energy consumption in residential buildings in korea. *Renew. Sustain. Energy Rev.* **2012**, *16*, 1074–1081. [CrossRef]
4. Choi, I.Y.; Cho, S.H.; Kim, J.T. Energy consumption characteristics of high-rise apartment buildings according to building shape and mixed-use development. *Energy Build.* **2012**, *46*, 123–131. [CrossRef]
5. Jeong, Y.-S.; Lee, S.-E.; Huh, J.-H. Estimation of CO_2 emission of apartment buildings due to major construction materials in the republic of korea. *Energy Build.* **2012**, *49*, 437–442. [CrossRef]
6. Kim, M.J.; Cho, M.E.; Kim, J.T. Energy use of households in apartment complexes with different service life. *Energy Build.* **2013**, *66*, 591–598. [CrossRef]
7. Jang, H.; Jones, L.; Kang, J. Prioritisation of old apartment buildings for energy-efficient refurbishment based on the effects of building features on energy consumption in south korea. *Energy Build.* **2015**, *96*, 319–328. [CrossRef]
8. Park, J.S.; Lee, S.J.; Kim, K.H.; Kwon, K.W.; Jeong, J.-W. Estimating thermal performance and energy saving potential of residential buildings using utility bills. *Energy Build.* **2016**, *110*, 23–30. [CrossRef]
9. Sim, J.; Sim, J. The effect of new carbon emission reduction targets on an apartment building in south korea. *Energy Build.* **2016**, *127*, 637–647. [CrossRef]
10. Laouadi, A. Development of a radiant heating and cooling model for building energy simulation software. *Build. Environ.* **2004**, *39*, 421–431. [CrossRef]
11. Karakoyun, Y.; Acikgoz, O.; Yumurtacı, Z.; Dalkilic, A.S. An experimental investigation on heat transfer characteristics arising over an underfloor cooling system exposed to different radiant heating loads through walls. *Appl. Therm. Eng.* **2020**, *164*, 114517. [CrossRef]
12. Lin, B.; Wang, Z.; Sun, H.; Zhu, Y.; Ouyang, Q. Evaluation and comparison of thermal comfort of convective and radiant heating terminals in office buildings. *Build. Environ.* **2016**, *106*, 91–102. [CrossRef]

13. Sun, H.; Wu, Y.; Lin, B.; Duan, M.; Lin, Z.; Li, H. Experimental investigation on the thermal performance of a novel radiant heating and cooling terminal integrated with a flat heat pipe. *Energy Build.* **2020**, *208*, 109646. [CrossRef]
14. Cho, J.; Park, B.; Lim, T. Experimental and numerical study on the application of low-temperature radiant floor heating system with capillary tube: Thermal performance analysis. *Appl. Therm. Eng.* **2019**, *163*, 114360. [CrossRef]
15. Lodi, C.; Magli, S.; Contini, F.M.; Muscio, A.; Tartarini, P. Improvement of thermal comfort and energy efficiency in historical and monumental buildings by means of localized heating based on non-invasive electric radiant panels. *Appl. Therm. Eng.* **2017**, *126*, 276–289. [CrossRef]
16. Shen, L.; Tu, Z.; Hu, Q.; Tao, C.; Chen, H. The optimization design and parametric study of thermoelectric radiant cooling and heating panel. *Appl. Therm. Eng.* **2017**, *112*, 688–697. [CrossRef]
17. Fang, Y.; Ding, Y.; Tang, Y.; Liang, X.; Jin, C.; Wang, S.; Gao, X.; Zhang, Z. Thermal properties enhancement and application of a novel sodium acetate trihydrate-formamide/expanded graphite shape-stabilized composite phase change material for electric radiant floor heating. *Appl. Therm. Eng.* **2019**, *150*, 1177–1185. [CrossRef]
18. Berger, J.; Gasparin, S.; Dutykh, D.; Mendes, N. Accurate numerical simulation of moisture front in porous material. *Build. Environ.* **2017**, *118*, 211–224. [CrossRef]
19. Yeo, M.-S.; Yang, I.-H.; Kim, K.-W. Historical changes and recent energy saving potential of residential heating in korea. *Energy Build.* **2003**, *35*, 715–727. [CrossRef]
20. Kim, D.-K. The natural environment control system of korean traditional architecture: Comparison with korean contemporary architecture. *Build. Environ.* **2006**, *41*, 1905–1912. [CrossRef]
21. Lee, J.; McCuskey Shepley, M.; Choi, J. Exploring the localization process of low energy residential buildings: A case study of korean passive houses. *J. Build. Eng.* **2020**, *30*, 101290. [CrossRef]
22. Song, G.-S. Buttock responses to contact with finishing materials over the ondol floor heating system in korea. *Energy Build.* **2005**, *37*, 65–75. [CrossRef]
23. Chu, W.-S.; Kim, M.-S.; Lee, K.-T.; Bhandari, B.; Lee, G.-Y.; Yoon, H.-S.; Kim, H.-S.; Park, J.-I.; Bilegt, E.; Lee, J.-Y.; et al. Design and performance evaluation of korean traditional heating system—Ondol: Case study of nepal. *Energy Build.* **2017**, *138*, 406–414. [CrossRef]
24. Chung, M.; Park, C.; Lee, S.; Park, H.-C.; Im, Y.-H.; Chang, Y. A decision support assessment of cogeneration plant for a community energy system in korea. *Energy Policy* **2012**, *47*, 365–383. [CrossRef]
25. Bae, C.; Chun, C. Research on seasonal indoor thermal environment and residents' control behavior of cooling and heating systems in korea. *Build. Environ.* **2009**, *44*, 2300–2307. [CrossRef]
26. Lee, W.-h.; Kim, K.-w.; Lim, S.-h. Improvement of floor impact sound on modular housing for sustainable building. *Renew. Sustain. Energy Rev.* **2014**, *29*, 263–275. [CrossRef]
27. Kim, Y.K.; Lee, T.W. A study on the heating characteristics of radiant floor panel using heat pipes with the double wick. *Soc. Air-Cond. Refrig. Eng. Korea* **2012**, *24*, 183–189.
28. Yong-ho, J. The performance evaluation of electric ondol system with the ptc thermistor function. *Korean Inst. Archit. Sustain. Environ. Build. Syst.* **2011**, *5*, 153–160.
29. Equest. Available online: http://www.Doe2.Com/equest/ (accessed on 15 August 2018).
30. American Society of Heating, Refrigerating and Air Conditioning Engineers. Ashrae Guideline 14-2002, Measurement of Energy and Demand Savings—Measurement of Energy, Demand and Water Savings American Society of Heating, Refrigerating and Air Conditioning Engineers, Inc.: Atlanta, CA, USA, 2002.
31. National Radio Research Agency, South Korea. Available online: https://rra.Go.Kr/ko/index.Do (accessed on 1 November 2018).
32. Radio Frequency Safety, Federal Communications Commission. Available online: https://www.Fcc.Gov/ (accessed on 1 November 2018).

energies

MDPI

Article

Heating Performance Analysis of an Air-to-Water Heat Pump Using Underground Air for Greenhouse Farming

Taesub Lim [1], Yong-Kyu Baik [1,*] and Daeung Danny Kim [2,*]

1 Department of Architectural Engineering, Seoil University, Seoul 02192, Korea; francis9@seoil.ac.kr
2 Architectural Engineering Department, King Fahd University of Petroleum and Minerals (KFUPM), Dhahran 31261, Saudi Arabia
* Correspondence: ykbaik29@seoil.ac.kr (Y.-K.B.); dkim@kfupm.edu.sa (D.D.K.)

Received: 29 June 2020; Accepted: 27 July 2020; Published: 28 July 2020

check for updates

Abstract: As one of the main businesses in Jeju-do in South Korea, specialized local products are grown in greenhouses. For greenhouse farming, it is preferable to use geothermal heat pump systems for energy conservation because of the stable temperature of the ground. In the same manner, heat pumps using underground air is recommended for greenhouse farming since underground air can easily be obtained from porous volcanic rocks in Jeju-do. However, direct usage of the underground air is not feasible for planting in the greenhouse or livestock care because the underground air is relatively humid and its temperature is low. For the present study, the heating performance of an air-to-water heat pump which used underground air as a heat source for greenhouse farming during the winter was assessed through measurements. In addition, the economic impact of the air-to-water heat pump (AWHP) was compared with a conventional air heater. According to the results, an AWHP can save more than 70% of the total heating costs compared with a conventional air heater. In sum, the utilization of the air-to-water heat pump using underground air can have a positive impact on reducing energy consumption as well as provide direct economic benefits.

Keywords: air-to-water heat pump; greenhouse; heating load; underground air

1. Introduction

While facing significant global warming, a lot of concerns have focused on energy use by buildings, which accounts for more than half of the total energy consumption [1,2]. Thus, a reduction in building energy consumption can lead to a reduction in CO_2 emissions. According to previous studies, energy savings can be achieved by applying appropriate thermal insulation to building envelopes and the use of advanced mechanical systems [3–7]. As building occupants spend more time in an indoor environment, the importance of the selection of proper mechanical systems has become a main concern for building stakeholders regarding thermal comfort and energy conservation [8].

Among current technologies for providing both heating and cooling, heat pumps have generally applied for residential and commercial buildings by implementing demand response measures which enable one to shift building energy loads [9–11]. These systems can maintain indoor thermal comfort as well as provide a potential for energy savings in buildings by extracting energy from the air, water, and soil [12–15]. Even though heat pumps often have excessive system complexity, they have increasingly been used to building applications [14,16]. Among the broad range of heat pump applications, heat pump technology can also be utilized in the area of agriculture.

According to a review article of Daghigh et al., heat pumps can be used to dry agricultural and forest products even under low-temperature conditions [17]. In addition, Kosan et al. have analyzed the efficiency of photovoltaic-thermal assisted heat pump drying systems [18]. Kumar et al. have used a closed-loop heat pump system for drying herbal leaves [19]. Besides, several studies have proposed drying methods based on heat pump technology for preserving agricultural products [20–24]. Among other applications, heat pump technology has been utilized to control the thermal performance in greenhouses. According to a study by Yang et al., the temperature in a greenhouse could be controlled by using an earth-air heat exchanger system [25]. Moreover, Ozgener and Hepbasli have studied the performance of a solar-assisted geothermal heat pump system for greenhouse heating under the climatic conditions in Turkey, where the night temperature drops to 14 °C while the day temperature is about 31 °C [26–28]. Ozgener has also proposed the use of a hybrid solar-assisted geothermal heat pump technology with a small wind turbine system for greenhouse heating [29].

For agricultural applications in South Korea, heat pumps have also applied to greenhouse space heating. In the case of Jeju-do in South Korea, where about 20% of the people work in agriculture cultivating local specialties like tangerines and mandarins. In the winter, these local specialties and others such as livestock and flowers are generally raised or planted in greenhouses where heating is provided by air-source heat pumps. Since the ground temperature is stable at a depth of 3 m–50 m, it is preferable to use the ground as a heat source for both heating and cooling (i.e., geothermal heat pumps) [25,30]. Like geothermal heat pumps, air source heat pump systems using underground air are recommended for greenhouse farming. Since Jeju-do is a dormant volcano in which the geographical features consist of volcanic and sedimentary rocks, basanites, and so on, it can form underground air and water tables between rocks and sediment [31]. Thus, it is easy to obtain underground air from porous volcanic rocks, in which the underground air temperature ranges around 15 °C–18 °C. However, the use of this underground air for planting or livestock care is not feasible because the underground air is relatively humid and its temperature is low. Most studies have investigated the performance of the air-source heat pumps and there have been few studies about air-to-water heat pumps using heat transferred from underground air. For the present study, the heating performance of an air-to-water heat pump (AWHP) system intended for use in greenhouse farming was assessed. To analyze the performance of the AWHP, the heating capacity and COP were analyzed through measurements. In addition, the economic impact of the heating performance of the AWHP for greenhouse farming was compared with a conventional air heater.

2. Heat Transfer from Underground Air to Heat Pumps

Figure 1 shows the schematic diagram of the AWHP. Temperature sensors were located at the underground air intake of the air/water direct contact heat exchanger, the compressor and evaporator, the fan-coil unit, and the heat storage tank. In addition, water flow meters were located at each unit of the AWHP. For the specification, the air/water direct contact heat exchanger is a cylinder of 1 m diameter with a height of 2.5 m. From a borehole, underground air is drawn by the turbo blower to the air/water direct contact heat exchanger through a cylinder of 0.3 m diameter. Similar to the process of cooling towers, the airflow at the lower part of the heat exchanger is mixed with water sprayed from the top. In addition, this heat exchanger utilizes plastic splash type fill to enhance the contact between air and water. During the process, the heat is transferred from the humid underground air at 16 °C to water. Therefore, as a heat source, the water flow is sent to the AWHP to provide further greenhouse heating. In the AWHP, heat transferred from the air/water direct contact heat exchanger is delivered to the evaporator. Passing the compressor, water is heated at the condenser. The control system uses a Programmable Logic Controller (PLC).

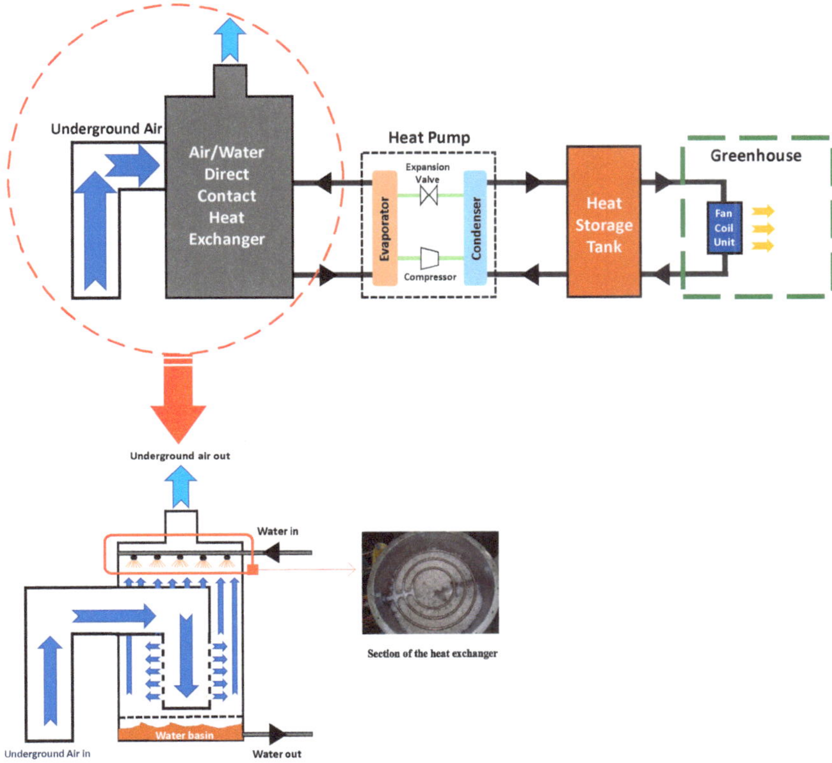

Figure 1. A schematic diagram of the AWHP.

3. Method

3.1. The AWHP at the Greenhouse

To analyze the heating performance of the AWHP, the AWHP was installed at the greenhouse for flower production located in Jeju-do in South Korea. The size of the greenhouse was 330 m^2. The specifications of the AWHP are presented in Table 1.

3.2. COP of the AWHP

For the control of the AWHP, a blower, an air/water direct contact heat exchanger, water pumps were operated to heat or cool the water in the heat storage tank to reach the setpoint temperature. In the case that the setpoint temperature of the AWHP does not reach the heating or cooling set temperatures in the greenhouse, water pumps of the heat storage tank and the fan of the fan-coil unit were operated. To measure the Coefficient of Performance (COP) for the AWHP in a heating application, T-type thermocouples and ultrasonic flowmeters (PT868, Panametrics, Boston, MA, USA) were located at the condenser, the evaporator, and the heat storage tank. In addition, the power usage of the AWHP was measured by using a power meter (CW240, Yokogawa, Tokyo, Japan). The specifications of the equipment are presented in Table 2.

Table 1. Specifications of the AWHP.

Component			Specification
Air/water direct contact heat exchanger	Material		Stainless steel
	Diameter (mm)		1000
	Height (mm)		2500
	Heat transfer fluid		Normal water
Heat pump	Compressor	Type	High-temperature scroll type
		Capacity	35 kW (10 RT)
		Voltage	380 V (3 Phase)
	Condenser/Evaporator		Flat type heat exchanger
	Refrigerant		R22
Heat storage tank	Diameter (mm)		2000
	Height (mm)		2000
	Heat storage fluid		Normal water
Turbo blower	Capacity		7.5 kW
	Airflow rate		102 m^3/min

Table 2. Specifications of the equipment.

PT868	- Non-Intrusive Liquid Flowmeter - Velocity range: −12.2 m/s to 12.2 m/s - Accuracy: 2% - Fluid temperature: −40 °C to 150 °C
CW240	- Voltage: 150 V to 1000 V - Frequency range: 45 Hz to 65 Hz - Accuracy: ±0.2% reading - Display interval: Approx. 0.5 s

Before calculating the COP, the heating capacity was calculated using Equation (1) below:

$$Q_h = \rho_w \cdot V_w \cdot c_w \cdot (T_{w,o} - T_{w,i}) \tag{1}$$

where Q_h is the heating capacity of the AWHP (kcal/h, 1 kcal/h = 0.0012 kW). In addition, ρ_w, V_w, and c_w are water density (kg/m^3), water flow (m^3/s), and the specific heat capacity of water (kJ/kg °C), respectively. $T_{w,o}$ and $T_{w,i}$ are the water temperatures at the outlet and inlet of the condenser (°C). Using Equation (2) [32], the COP was calculated for open-loop and closed-loop systems. In an open-loop system, heated water from the condenser is dumped directly to the outside, while water at the heat storage tank circulates and is reheated at the condenser and returns to the heat storage tank for a closed-loop system:

$$COP_h = \frac{Q_h}{P_{HP}} \tag{2}$$

where COP_h is the COP of the AWHP for the heating application. Moreover, P_{HP} is the power usage of the AWHP (kW). To compare the heating energy consumption and temperature distributions in the greenhouse, a conventional heat pump system was also installed in the same size of the greenhouse.

4. Results

4.1. The Temperature Profiles of the System During the Condenser Operation

It is important to figure out the relationship between the temperature changes of each component of the AWHP and the outdoor temperature conditions. The measurement results are shown in Figure 2. For the open-loop system, the heating performance was observed for 7 min until the water was drawn off, while it took 3 h to heat 6 tons of water from 18 °C to 50 °C for the closed-loop system on 24 March 2015.

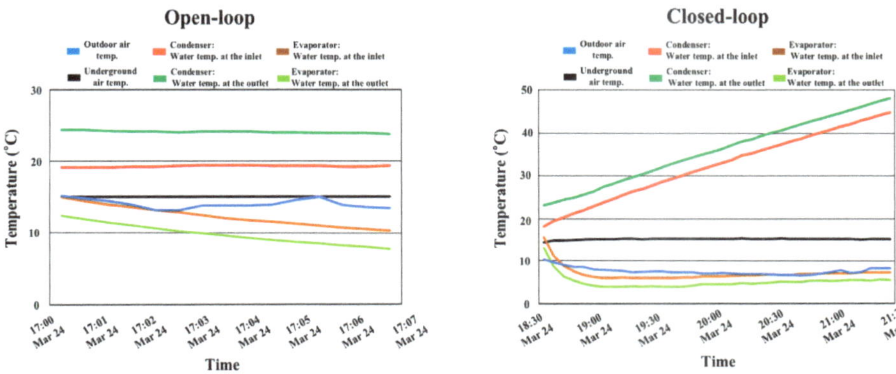

Figure 2. Temperature profiles of the system for open-loop and closed-loop application.

When measuring the water temperature of the open-loop system, the outdoor air temperature was about 13.8 °C, while the underground air temperature at the inlet of the air/water direct contact heat exchanger was maintained at 15 °C. Because of the short observation, there were a little temperature changes at both inlet and outlet of the condenser. In the case of the evaporator, about a 5 °C decrease in water temperature was observed at both inlet and outlet due to the rapid water runoff. For the closed-loop system, the outdoor air temperature was slightly decreased from 10 °C to 8 °C, while the underground air temperature was also maintained at 15 °C during the measurement. In the case of the evaporator, both water temperatures at the inlet and outlet were dropped from about 15 °C to 5 °C after 10 min from the beginning of the measurement. It was caused by the low temperature in the condenser at the early operation, which required a large amount of heat transfer. After the temperature in the condenser was increased, the amount of heat transfer was decreased. Therefore, the temperatures at the inlet and outlet of the evaporator were slightly increased to 8 °C after the temperature was slightly increased in the condenser. Simultaneously, the underground air temperature at the outlet of the air/water direct contact heat exchanger was also increased.

4.2. Heating Capacity for the Heating Application

Figure 3 shows the heating capacity for the open-loop and closed-loop applications. The observed heating capacity for the open-loop and closed-loop systems was ranged from 35,000 to 41,000 kcal/h and 27,000 to 40,000 kcal/h, respectively. In addition, the heat extracted by the air/water direct contact heat exchanger for the open-loop system was about 28,000 kcal/h. For the closed-loop system, the extracted heat was decreased from 30,000 kcal/h to 18,000 kcal/h.

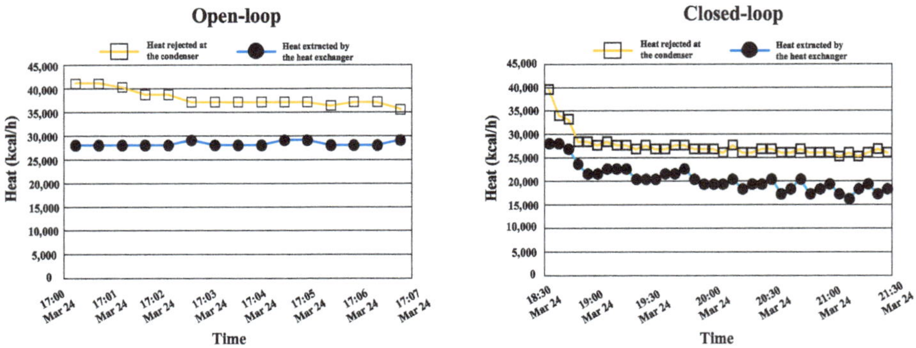

Figure 3. The heating capacity for the open-loop and closed-loop systems.

Based on the values of the heating capacity and electricity consumption, the COP values were calculated using Equation (2). As can be seen in Figure 4, the COP values were ranged from 4.3 to 5.5 for the open-loop system, while the COP value was decreased from 5.0 to 2.5 for the closed-loop system as the inlet water temperature of the condenser was increased to 50 °C. For energy consumption, the electricity usage for the open-loop system was decreased, while it was increased for the closed-loop system. It can be seen that the continuous operation of the AWHP for increasing the water temperature in the heat storage tank caused an increase in electricity consumption.

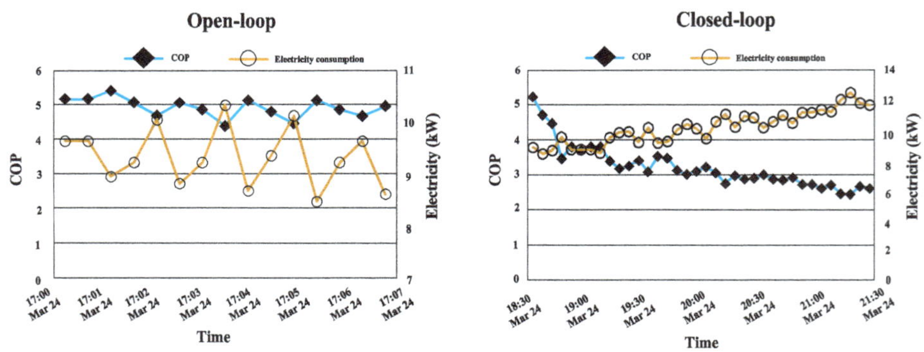

Figure 4. The COP and energy consumption for the open-loop and closed-loop systems.

Figure 5 shows the comparison of air temperature in the greenhouse operated by the AWHP and a conventional air heater. When the outdoor air temperature varied from 7 °C to 12 °C on 19 April 2015, the air temperature in the greenhouse heated by the conventional air heater was ranged from 19 °C to 26 °C, while the greenhouse heated by the AWHP showed a small air temperature fluctuation which ranged from 20 °C to 22 °C. Based on the results, the AWHP can provide more stable heating than the conventional air heater.

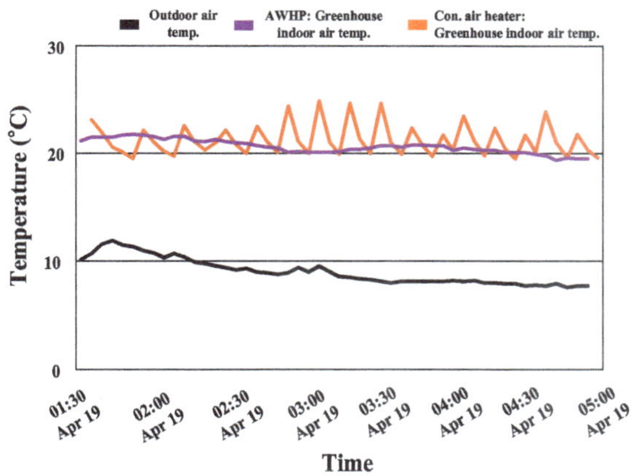

Figure 5. Comparison of heating performance between the AWHP and a conventional air heater.

4.3. Economic Assessment of the AWHP

To analyze the economic impact of the AWHP, the heating cost was compared with that of a conventional air heater (Table 3). Since each system used different energy sources, the total heating cost for these two systems was compared. From February to June in 2016, the AWHP consumed 4064 kWh/m^2 of electricity at a cost of ₩45.5/kWh (South Korea Won, i.e., $0.04 (US Dollar)). In the case of the conventional air heater, about 889.8 liters of diesel/m^2 were consumed at ₩840 per liter (South Korea Won, i.e., $0.70 (US Dollar)). As a result, the AWHP only consumed about 25% of the total heating cost of the conventional air heater.

Table 3. Heating cost comparison for the AWHP and a conventional air heater.

System	Energy Consumption	Total Heating Cost		Comparison (%)
		South Korea Won	US Dollar	
AWHP	4064 kWh/m^2 (Electricity)	₩184,912	$154	24.7
Conventional air heater	889.8 liters of diesel/m^2	₩747,432	$621	100

5. Discussion and Conclusions

As one of the main businesses in Jeje-do in South Korea, agriculture is supported by the South Korean Government. Heat pump systems using underground air are recommended for greenhouse farming, however, the direct use of underground air may harm greenhouse products because of the high humidity and low temperature of the underground air. The present study assessed the heating performance of an AWHP for greenhouses in Jeju-do through measurements. Specifically, the AWHP used the heat extracted from water in which the heat was transferred from underground air through an air/water direct contact heat exchanger.

During the measurement period, the outdoor air temperature was in the range of 7 °C to 12 °C, while the underground air temperature was maintained at 15 °C. When the water temperature of the heat storage tank ranged 40 °C–45 °C, the COP was 2.1 to 2.7 and the heating capacity was 34.9 kW–44.2 kW. Comparing the heating cost with the conventional air heater from February to June, the AWHP consumes about 25% of the total heating cost.

When delivering underground air at a rate of 102 m^3/min, the underground air temperature dropped from 15 °C to 10 °C and 32.6 kW of thermal energy was obtained. This was slightly lower than that generated by conventional heat pumps since the heat energy was transferred twice through the air/water direct contact heat exchanger and the evaporator. In addition, the COP for the closed-loop system decreased from 5.0 to 2.5. Thus, it can be seen that an increase in water temperature in the heat storage tank during the early AWHP operation reduced the amount of heat absorbed by the condenser and evaporator. The heat reduction ultimately caused the AWHP to operate continuously leading to an increase in the electricity consumption of the AWHP.

Considering the outcome of the present study, the AWHP system is more energy and cost-efficient than a conventional mechanical system. Moreover, the utilization of the AWHP for greenhouse farming can reduce greenhouse gas emissions. For further study, it is necessary to find a feasible solution to increase the COP for the closed-loop application. In addition, energy-advanced or more sophisticated controlled conventional air heaters could be used for more accurate comparison with the AWHP. Furthermore, life cycle cost analysis will be conducted considering factors such as capital investments.

Author Contributions: T.L. and Y.-K.B. designed and performed the experiments; D.D.K. wrote the manuscript and analyzed the data. All authors have read and agreed to the published version of the manuscript.

Funding: This research received no external funding.

Conflicts of Interest: The authors declare no conflict of interest.

References

1. Pérez-Lombard, L.; Ortiz, J.; Pout, C. A review on buildings energy consumption information. *Energy Build.* **2008**, *40*, 394–398. [CrossRef]
2. Gonzato, S.; Chimento, J.; O'Dwyer, E.; Bustos-Turu, G.; Acha, S.; Shah, N. Hierarchical price coordination of heat pumps in a building network controlled using model predictive control. *Energy Build.* **2019**, *202*, 109421. [CrossRef]
3. Jim, C.Y. Air-conditioning energy consumption due to green roofs with different building thermal insulation. *Appl. Energy* **2014**, *128*, 49–59. [CrossRef]
4. Stavrakakis, G.M.; Androutsopoulos, A.V.; Vyörykkä, J. Experimental and numerical assessment of cool-roof impact on thermal and energy performance of a school building in Greece. *Energy Build.* **2016**, *130*, 64–84. [CrossRef]
5. Zagorskas, J.; Zavadskas, E.K.; Turskis, Z.; Burinskienė, M.; Blumberga, A.; Blumberga, D. Thermal insulation alternatives of historic brick buildings in Baltic sea region. *Energy Build.* **2014**, *78*, 35–42. [CrossRef]
6. Jiang, A.; O'Meara, A. Accommodating thermal features of commercial building systems to mitigate energy consumption in Florida due to global climate change. *Energy Build.* **2018**, *179*, 86–98. [CrossRef]
7. Hong, G.; Kim, D.D. Airtightness of electrical, mechanical and architectural components in South Korean apartment buildings using the fan pressurization and tracer gas method. *Build. Environ.* **2018**, *132*, 21–29. [CrossRef]
8. Martinez, A.; Díaz de Garayo, S.; Aranguren, P.; Astrain, D. Assessing the reliability of current simulation of thermoelectric heat pumps for nearly zero energy buildings: Expected deviations and general guidelines. *Energy Convers. Manag.* **2019**, *198*, 111834. [CrossRef]
9. D'Ettorre, F.; De Rosa, M.; Conti, P.; Testi, D.; Finn, D. Mapping the energy flexibility potential of single buildings equipped with optimally-controlled heat pump, gas boilers and thermal storage. *Sustain. Cities Soc.* **2019**, *50*, 101689. [CrossRef]
10. Clauß, J.; Georges, L. Model complexity of heat pump systems to investigate the building energy flexibility and guidelines for model implementation. *Appl. Energy* **2019**, *255*, 113847. [CrossRef]
11. Neirotti, F.; Noussan, M.; Simonetti, M. Towards the electrification of buildings heating—real heat pumps electricity mixes based on high resolution operational profiles. *Energy* **2020**, *195*, 116974. [CrossRef]

12. Li, S.; Gong, G.; Peng, J. Dynamic coupling method between air-source heat pumps and buildings in China's hot-summer/cold-winter zone. *Appl. Energy* **2019**, *254*, 113664. [CrossRef]
13. Lozano Miralles, J.A.; López García, R.; Palomar Carnicero, J.M.; Martínez, F.J.R. Comparative study of heat pump system and biomass boiler system to a tertiary building using the life cycle assessment (LCA). *Renew. Energy* **2020**, *152*, 1439–1450. [CrossRef]
14. Potočnik, P.; Vidrih, B.; Kitanovski, A.; Govekar, E. Analysis and optimization of thermal comfort in residential buildings by means of a weather-controlled air-to-water heat pump. *Build. Environ.* **2018**, *140*, 68–79. [CrossRef]
15. Yu, Q.d. Applied research on water loop heat pump system based on a novel mechanism of energy conversion. *Appl. Therm. Eng.* **2019**, *153*, 575–582. [CrossRef]
16. Le, K.X.; Huang, M.J.; Shah, N.N.; Wilson, C.; Artain, P.M.; Byrne, R.; Hewitt, N.J. Techno-economic assessment of cascade air-to-water heat pump retrofitted into residential buildings using experimentally validated simulations. *Appl. Energy* **2019**, *250*, 633–652. [CrossRef]
17. Daghigh, R.; Ruslan, M.H.; Sulaiman, M.Y.; Sopian, K. Review of solar assisted heat pump drying systems for agricultural and marine products. *Renew. Sustain. Energy Rev.* **2010**, *14*, 2564–2579. [CrossRef]
18. Koşan, M.; Demirtaş, M.; Aktaş, M.; Dişli, E. Performance analyses of sustainable PV/T assisted heat pump drying system. *Sol. Energy* **2020**, *199*, 657–672. [CrossRef]
19. Ashok Kumar, M.; Kumaresan, G.; Rajakarunakaran, S. Experimental study of moisture removal rate in Moringa leaves under vacuum pressure in closed-loop heat pump dryer. *Mater. Today Proc.* **2020**. [CrossRef]
20. Singh, A.; Sarkar, J.; Sahoo, R.R. Experimental energy, exergy, economic and exergoeconomic analyses of batch-type solar-assisted heat pump dryer. *Renew. Energy* **2020**, *156*, 1107–1116. [CrossRef]
21. Hasan Ismaeel, H.; Yumrutaş, R. Investigation of a solar assisted heat pump wheat drying system with underground thermal energy storage tank. *Sol. Energy* **2020**, *199*, 538–551. [CrossRef]
22. Kuan, M.; Shakir, Y.; Mohanraj, M.; Belyayev, Y.; Jayaraj, S.; Kaltayev, A. Numerical simulation of a heat pump assisted solar dryer for continental climates. *Renew. Energy* **2019**, *143*, 214–225. [CrossRef]
23. Fadhel, M.I.; Sopian, K.; Daud, W.R.W.; Alghoul, M.A. Review on advanced of solar assisted chemical heat pump dryer for agriculture produce. *Renew. Sustain. Energy Rev.* **2011**, *15*, 1152–1168. [CrossRef]
24. Tunckal, C.; Doymaz, İ. Performance analysis and mathematical modelling of banana slices in a heat pump drying system. *Renew. Energy* **2020**, *150*, 918–923. [CrossRef]
25. Yang, L.-H.; Huang, B.-H.; Hsu, C.-Y.; Chen, S.-L. Performance analysis of an earth–air heat exchanger integrated into an agricultural irrigation system for a greenhouse environmental temperature-control system. *Energy Build.* **2019**, *202*, 109381. [CrossRef]
26. Ozgener, O.; Hepbasli, A. Performance analysis of a solar-assisted ground-source heat pump system for greenhouse heating: An experimental study. *Build. Environ.* **2005**, *40*, 1040–1050. [CrossRef]
27. Ozgener, O.; Hepbasli, A. Experimental performance analysis of a solar assisted ground-source heat pump greenhouse heating system. *Energy Build.* **2005**, *37*, 101–110. [CrossRef]
28. Ozgener, O.; Hepbasli, A. Exergoeconomic analysis of a solar assisted ground-source heat pump greenhouse heating system. *Appl. Therm. Eng.* **2005**, *25*, 1459–1471. [CrossRef]
29. Ozgener, O. Use of solar assisted geothermal heat pump and small wind turbine systems for heating agricultural and residential buildings. *Energy* **2010**, *35*, 262–268. [CrossRef]
30. Ozgener, O.; Hepbasli, A. Modeling and performance evaluation of ground source (geothermal) heat pump systems. *Energy Build.* **2007**, *39*, 66–75. [CrossRef]
31. Jeju Special Self-Governing Province. Available online: https://www.Jeju.Go.Kr/ (accessed on 28 June 2020).
32. Dizaji, H.S.; Jafarmadar, S.; Khalilarya, S. Novel experiments on cop improvement of thermoelectric air coolers. *Energy Convers. Manag.* **2019**, *187*, 328–338. [CrossRef]

energies

MDPI

Article

PoDIT: Portable Device for Indoor Temperature Stabilization: Concept and Theoretical Performance Assessment

Vítor Leal * and Raul Teixeira

Department of Mechanical Engineering, Faculty of Engineering of the University of Porto, Rua Dr. Roberto Frias, 4200-465 Porto, Portugal; raul.santosteixeira@gmail.com
* Correspondence: vleal@fe.up.pt; Tel.: +351-22-508-1762

Received: 28 September 2020; Accepted: 6 November 2020; Published: 16 November 2020

check for updates

Abstract: This work introduces the concept of a new Portable Device for Indoor Temperature Stabilization (PoDIT), to be considered as a low-cost, quick and easy to implement remediation strategy when, for social, economic, or technical reasons, the improvement of the building envelope and/or the adoption of air conditioning are not possible. The main goal is to attenuate the maximum indoor temperature during summer and/or heat waves. The system, which is modular, consists of a certain mass of encaged phase change material (PCM) that stays indoors during the daytime and is transported to the outdoors (e.g., a balcony) during the night to discharge the heat accumulated during the daytime. Both natural convection and forced convection variants were considered. The results showed that, in the configurations and for the reference room and weather considered, the adopting 4 modules of the device can lead to reductions in the maximum room air temperature close to 3 °C, with natural convection. Adopting a fan to impose forced convection at the surfaces of the device can lead to temperature attenuations in excess of 4 °C, as it ensures full solid–liquid commuting and therefore optimal use of the PCM thermal storage capability.

Keywords: passive cooling; low energy cooling; heat waves; climate adaptation; social housing; nearly zero energy buildings

1. Introduction

There is growing evidence that, even in developed countries, many households do not have comfortable and/or healthy indoor environmental conditions, especially in what regards to indoor temperature. The evidence comes from direct measurements [1,2], but also indirectly from an observed correlation between extreme cold and hot spots and the mortality rates [3–5].

For example, in Figure 1, it can be observed that the daily mortality in Portugal was actually considerably higher in some days of 2013 and 2018 than in March and April of 2020, during the first peak of the COVID-19 pandemic in Europe [6].

Theoretically, a possible solution for the problem would to deploy air conditioning to all most residential buildings, or at least to those with lower energy-efficient envelopes. However, this would present several problems:

(i) Because of how electric systems are operated, in the short term the growth in electricity demand caused by AC systems would cause an increase in the operation of thermal power plants. This would increase the greenhouse gas emissions. In the medium or long term, the electricity could come from additional renewable sources, but there is competition for the use of electricity from other more urgent purposes (e.g., mobility). Likely, it would have additional costs also;

(ii) When not properly selected and installed, AC systems often lead to discomfort and health issues themselves [7,8];

(iii) Many households cannot afford to pay for the cost of installing the AC systems and/or the electricity that they would consume.

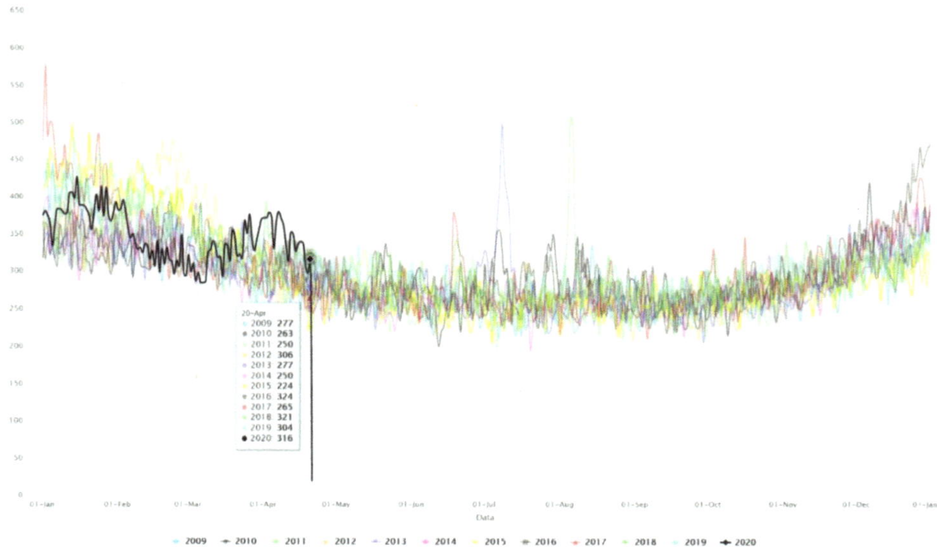

Figure 1. Daily mortality in Portugal from 2009 to April 2020 [6].

In this framework, it is of outmost importance to devise ways of improving the indoor temperatures, especially in summer, without requiring the installation of AC systems in every residential building.

A possibility to achieve this goal is to use the so-called "thermal flush" technique [9,10] in a way that can be easily adopted in existing buildings. This technique has traditionally made use of the storage of heat in the thermal mass of the building (thermal inertia), and then promote night ventilation to discharge the stored heat into the outdoor air. While still a valid concept, there are many cases where its effect in practice is limited by factors such as:

(i) Impossibility to ventilate during the night, for safety or noise reasons;

(ii) Low effectiveness of the ventilation, due to the location of the building and/or absence of wind and/or existence of obstacles surrounding the building;

(iii) Low thermal mass in the building—a trend in many countries due to the adoption of cheaper and/or modular materials e.g., in the interior walls [11,12].

Trying to overcome the limitations above, and to make a more effective use of the thermal flush concept, this paper presents and analyses the predicted performance of an innovative phase-change portable device that is intended to be indoors during the day, and moved outside (e.g., to a terrace, veranda or window hanger) during the night.

The idea of using phase change materials (PCMs) to assist cooling of buildings has been introduced since at least the early 2000s. The applications have been reviewed in recent papers such as those by Ostermann et al. [13], Pomianowski et al. [14], Iten et al. [15], Souayfane et al. [16], and Faraj et al. [17].

Faraj et al. [17] divide the applications between the Passive and the Active technologies. Passive ones include uses in Walls, and Trombe walls, glazing windows, floors and tiles, shutters and blinds, ceilings and roofs, sola façades, and solar chimneys Active ones include integration into evaporative and radiative cooling systems, ventilated façades, ice storage, thermally activated building structures (TABS), AC systems, and ventilated Trombe walls. No systems similar to the one proposed here were found.

Garcia [18] stresses that one of the main weaknesses of PCMs employed in passive cooling systems is that, though the peak cooling load is delayed, it still ends up discharged indoors. He then proposed an innovative concept for a dynamic building envelope with PCM, which however is likely to be difficult to be adopted at existing buildings without significant retrofit work—one of the intended key advantages of the system now proposed.

2. Device Description and Specifications

2.1. Concept and Shape

The device consists of a certain mass of PCM (phase change material) embodied in a container, with a shape that promotes an easy heat exchange with the surrounding environment (air and surfaces).

Besides ensuring mechanical robustness, the portability specification implies two main technical requirements: both weight and dimensions must both be low enough ensure that it can be carried frequently between the indoors and the outdoors (and vice-versa) by any person who is healthy but not necessarily athletic.

For this proof-of-concept study, a mass limit value of around 20 kg per device was considered. Given the mass specifications, the selection of PCM device shell (box) material led to the choice of a lightweight but resistant material with a high thermal conductivity: an aluminum sheet with a thickness of 2 mm was selected as the box material.

Regarding dimensions, the maximization of surface area was used as a criterion, thus adopting a geometry of thin quadrangular plates. The dimensions adopted in the PCM envelope box took into account the maximization of the surface area, but also the volumetric expansion verified in the phase change materials considered in this study. Thin plates were adopted to improve the spread of heat inside the devices.

Figure 2 presents a schematic representation with the overall dimensions of the PCM device. The shape may nevertheless be the object of optimization at a later stage of product development. It is also possible to consider the deployment of several independent modules into room.

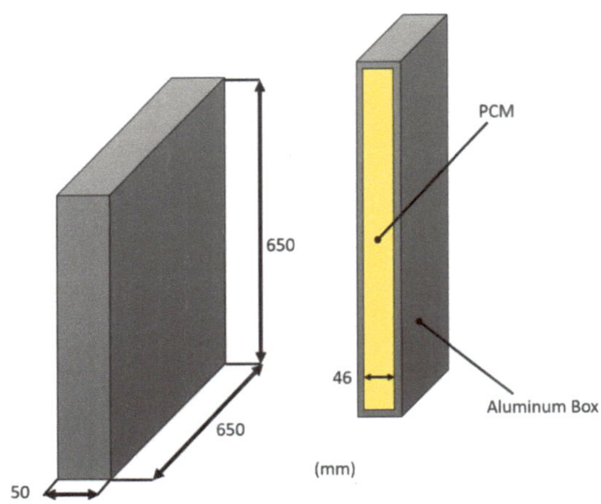

Figure 2. Schematic representation of the device for this proof-of-concept study (1 module).

2.2. PCM Characteristics

Suitable PCM were searched among those that are commercially available as technically developed/mature products. The key criteria was the phase change temperature. For an initial scenario under "normal summer weather", a PCM with phase change at 24 °C was selected [19], which therefore implies that the device will work to "try" to limit the indoor temperature to 24 °C, as is somewhat in line with conventional HVAC comfort requirements. An alternative PCM with a transition temperature of 29 °C was also considered [19], for sensitivity assessment (Table 1).

For the scenario under heat wave conditions, it proved that the 24 °C transition temperature was too low (among other things this temperature is too low for an efficient discharge outdoors during the night), and the C19H40 Nonadecane parafine, which has a transition temperature of 32 °C, was considered instead [20]. This type of material is widely used in buildings thermal applications and has advantageous properties in the context of the proposed application: congruent melting, good nucleating properties, chemically stable, non-toxic, non-corrosive, and metal casing compatibility [21]. The 32 °C transition temperature is high when compared with the conventional HVAC setpoints around 24–25 °C, but it is nevertheless representative of the more extreme conditions that may happen during heat waves. In these cases, the device will work to "try" to limit the indoor temperature to 32 °C, which despite being high in terms of comfort avoids the most dangerous temperatures regarding health. The PCM with a transition temperature of 29 °C was also considered, for sensitivity assessment.

A significant limitation of the phase change materials is their low thermal conductivity, which slows their ability to store and release thermal energy in time periods compatible with the requirements, thus decreasing the overall system performance.

A technique to enhance the thermal conductivity of phase change materials was proposed by X. Py [22] where the PCM is embedded inside a porous graphite matrix. The main advantage of this new composite material (PCM and graphite) is the increase in heat conductivity without much reduction in energy storage and the decrease of volume change in paraffins [23].

A parametric study showed that a thermal conductivity greater than 5 $Wm^{-1}K^{-1}$ did not significantly decrease PCM liquefaction or solidification time, representing the best compromise between increased thermal conductivity and decreased thermal storage capacity [24]. This thermal conductivity can be achieved with a graphite volumetric fraction of only 3%.

Table 1 summarizes the key physical characteristics of the PCM materials, already considering the effect of doping with graphite, calculated according to the correlation by Py et al. [22]. Table 2 shows the resulting key physical characteristics of the device, considering both the casing and the PCM.

Table 1. Key physical characteristics of the three composite materials considered, computed according to the method proposed by Py et al. [22].

PCM—Graphite Composite		Paraffine $C_{19}H_{40}$	PureTemp 24	PureTemp 29
Graphite volume fraction, φ	(—)	3%	3%	3%
Volumetric mass of the matrix	($kg\ m^{-3}$)	64.6	64.6	64.6
Phase change	(°C)	32	24	29
Fusion latent heat Δ	($kJ\ kg^{-1}$)	204	171	187
Specific heat in solid phase	($J\ kg^{-1}\ K^{-1}$)	1713	2696	1693
Specific heat in liquid phase	($J\ kg^{-1}\ K^{-1}$)	2081	2873	1851
Volumetric mass of the composite in liquid phase	($kg\ m^{-3}$)	811.5	822.2	890.2
Volumetric mass of the composite in solid phase		950.4	987.3	977.2
Heat conductivity of the composite in liquid phase	($Wm^{-1}\ K^{-1}$)	5	5	5
Heat conductivity of the composite in solid phase	($Wm^{-1}\ K^{-1}$)	5	5	5

Table 2. Key physical characteristics of the device, considering the casing and the phase change material (PCM) composite.

Height	0.65	m
Length	0.65	m
Width	0.05	m
Box Thickness	0.002	m
Heat Transfer Area	0.845	m^2
Aluminum Box Mass	5.2	kg
PCM Composite Mass	15.6	kg
Total Mass (Box + PCM)	20.8	kg
Box Material	Aluminum Plate 1050A H24 [25]	
Volumetric Mass Density	2700	kg m^{-3}
Specific Heat	900	J kg^{-1} K^{-1}
Thermal Conductivity	229	Wm^{-1} K^{-1}

3. Test Room and Study Scenarios

In order to assess the expectable order of magnitude of the effect of the device in the indoor temperature, a reference room realistically representative of a room in a real building was considered. Figure 3 shows a schematic representation of the reference room with the respective dimensions. The reference compartment is a small room with 14 m^2 of floor area and 2.7 m in height, with two exterior walls facing south and east, both with a 2 m^2 window, a horizontal roof terrace, two interior walls and interior floor. There is also a 2 m^2 door in the interior wall (D). It is considered that the adjacent rooms have similar thermal behavior. These characteristics, especially those related to the solar orientation and aperture, imply that the room is prone to some overheating—therefore among those that are the primary target of the system being developed.

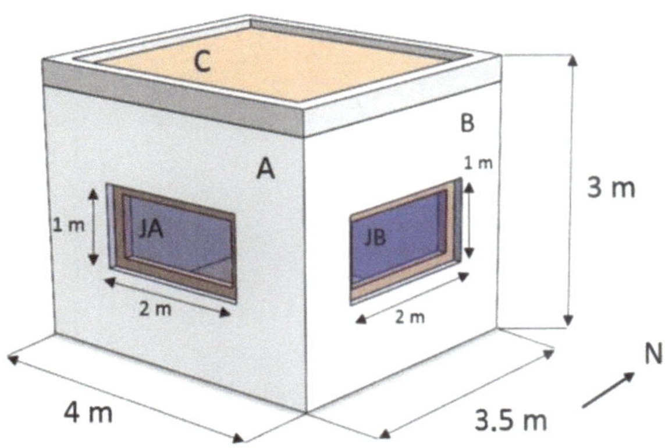

Figure 3. Schematic representation of the reference room for assessment of the impact upon indoor temperature.

Regarding the building materials, two variants were considered: one using high-density materials such as concrete and bricks (high thermal inertia variant), and another using low density materials such as plasterboard (low thermal inertia variant). The high thermal inertia solution was based in typical Portuguese single-family buildings, with ceramic bricks and expanded polystyrene EPS insulation on the outside. The low thermal inertia solution uses plasterboard at the interior walls, and the EPS isolation is located on the inside of the structuring elements. Table 3 shows the details the constructions, while Table 4 shows the properties of the materials.

Table 3. Room constructions considered in the High and Low thermal inertia variants.

	Hight Thermal Inertia		Low Thermal Inertia	
Element	Material	Thickness (m)	Material	Thickness (m)
Exterior Wall A, B	Interior Plaster Ceramic Brick EPS Insulation Exterior Plaster	0.02 0.15 0.03 0.02	Interior Plaster EPS Insulation Ceramic Brick Exterior Plaster	0.02 0.03 0.15 0.02
Interior Wall C, D	Plaster Ceramic Brick Plaster	0.02 0.11 0.02	Plasterboard Air Gap Plasterboard	0.018 0.12 0.018
Roof E	Exterior Protection (Inert) EPS Insulation Shape Layer Light Slab Interior Plaster	0.05 0.03 0.05 0.15 0.02	Exterior Protection (Inert) Shape Layer Light Slab EPS Insulation Interior Plaster	0.03 0.04 0.15 0.03 0.02
Floor F	Wood floor Screed Light Slab Plaster	0.01 0.04 0.15 0.02	Wood floor EPS Insulation Screed Light Slab Plaster	0.01 0.03 0.04 0.15 0.02

Table 4. Materials properties considered in the room constructions.

Material	Thermal Conductivity κ ($Wm^{-1} K^{-1}$)	Volumetric Density, ρ ($kg\ m^{-3}$)	Specific Heat, c_P ($J\ kg^{-1} K^{-1}$)
Plaster	1.3	1900	837
Holed Ceramic Brick 15	0.38	800	936
Holed Ceramic Brick 11	0.41	900	936
Lightweight Slab (with form layer)	0.85	1320	965
Cavernous Concrete	0.25	700	1000
Inert	2	2000	500
Screed	1.3	1800	880
Plasterboard	0.25	750	1000
Wood Floor	0.17	750	2400

4. Heat Transfer Modelling

The modeling approach followed the classic analogy of thermal systems with electrical Resistance-Capacitance (RC) circuits applied to buildings [26]. This approach has been used with very good results as long as the discretization of the system is adequate [27–29].

Figure 4 shows the RC circuit representing the heat transfer inside the device and between the device and the room air (*Ti*).

Besides the device itself, it is important to model the room and the interaction between the device and the surrounding environment (room during the day, and the outdoor environment during the night). Figure 5 shows the model representation for this integrated room + device system. The model was implemented by own coding in the MATLAB software (MATLAB. 9.7. The MathWorks Inc.; Natick, MA, USA, 2018.)

For this work two weather conditions were considered: Typical Summer and Heat Wave. For the Typical Summer weather profile, hourly climatic data from the database of Portugal's National Energy and Geology Laboratory (LNEG) was used. The reference location of Torre de Moncorvo, inner Portugal, was selected as representative of a location prone to overheating–though not the warmest within the country. It is characterized by a summer/cooling season lasting 2.7 months, typical maximum temperatures above 37 °C, and high incidence of solar radiation.

This weather profile adopted is show in Figure 6, with a maximum temperature of 37.8 °C, and a minimum of 15.9 °C. A sequence of similar days was considered.

The Heat Wave weather profile represents a hypothetical extreme scenario during heat waves and characterized by very high maximum temperatures and tropical nights, where the minimum temperature does not drop below 20 °C. This weather profile was based in historical record temperatures observed in Portugal and was developed considering a maximum temperature of 45 °C and a minimum of 25 °C. The hourly profile was built according to a procedure for generating 24 h temperature data sequences recommended by ASHRAE, adopting the normalized daily temperature profile as fractions of the daily temperature range [29]. The solar radiation data for the Heat Wave weather was kept the same as for the normal summer weather, which was already quite high as can be seen in Figure 6.

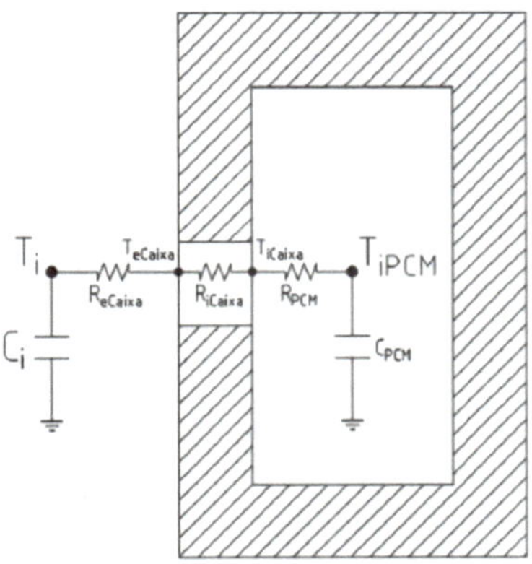

Figure 4. RC circuit representing the heat transfer inside the device and between the device and the room air. *Ti* represents the room air temperature; TiPCM is the PCM temperature [24].

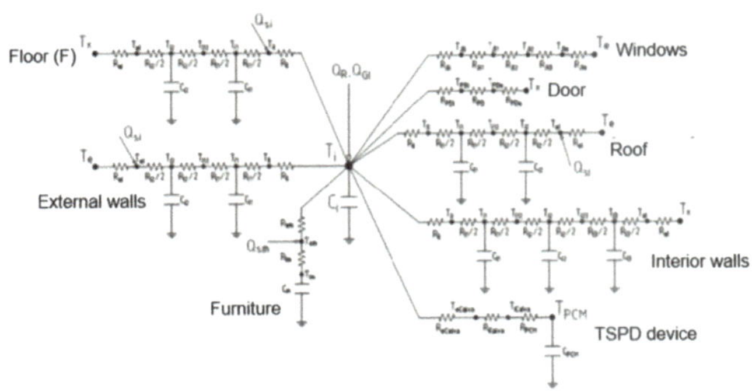

Figure 5. RC model of the room + Portable Device for Indoor Temperature Stabilization (PoDIT) device [24]. Q_{si} represent solar incidence on the floor and external walls, Q_{GI} the internal gains and Q_R the heat gains or losses due to ventilation/air renovation.

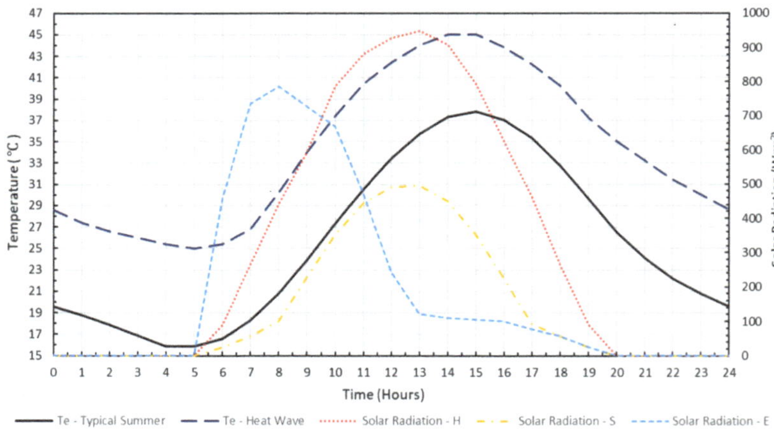

Figure 6. Temperature and solar radiation (Horizontal, South, and East) hourly profiles considered in the typical summer and in the heat wave conditions/scenarios.

5. Results

5.1. Results for Normal Summer Conditions and Natural Convection

The thermal behavior of the device and its impact upon the indoor air temperature was initially characterized for the normal summer condition, and considering natural convection at the surfaces of the device (both when placed indoors, during the day, and when placed outdoors during the night).

Figures 7 and 8 show the indoor air temperature of the room, for the low and high room inertia scenarios, considering none, one or four Portable Device for Indoor Temperature Stabilization (PoDIT) modules. The peak temperature attenuation observed in the low inertia room is about 0.8 °C with one PoDIT module (1 plate), and about 2.7 °C if four modules (4 plates) are used. For the high inertia room, the observed attenuation is about 0.5 °C with one PoDIT module (1 plate), and about 2 °C with four modules (4 plates).

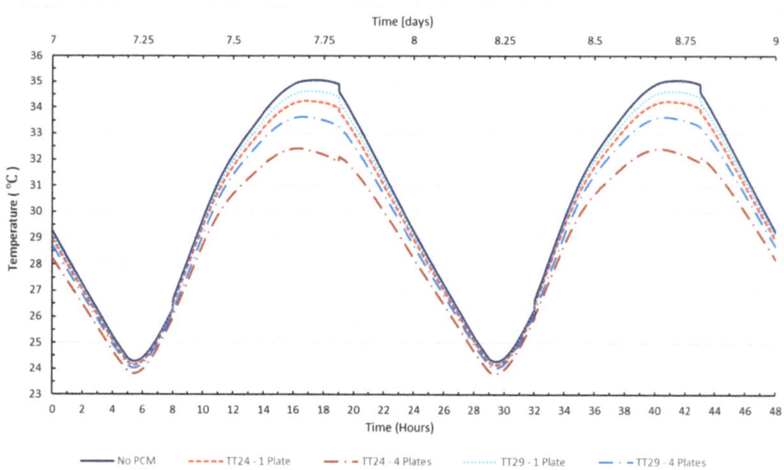

Figure 7. Indoor temperatures for the room with low thermal inertia, for the cases of no PCM, PCM with transition/melting temperature of 24 °C, and transition/melting temperature at 29 °C.

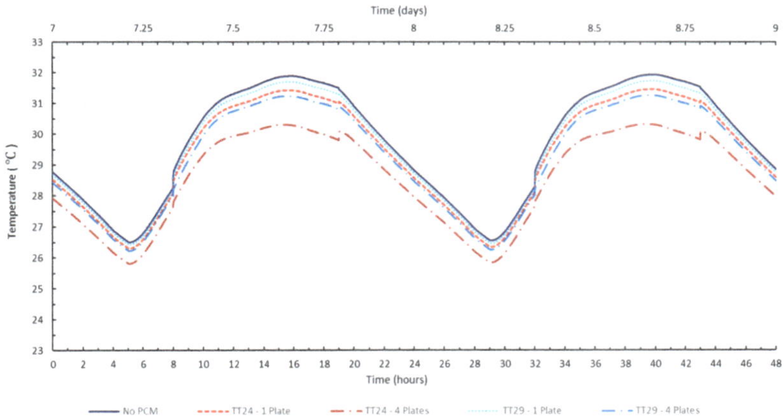

Figure 8. Indoor temperatures for the room with high thermal inertia, for the cases of no PCM, PCM with transition/melting temperature of 24 °C, and transition/melting temperature at 29 °C.

Figure 9 shows the temperature of the PCM and its liquid fraction in the same period of Figure 7, for low thermal inertia case. It can be seen that the PCM never becomes fully liquid. Thus, it does not make use of its maximum thermal storage capability, which hinders its performance/contribution to indoor temperature attenuation.

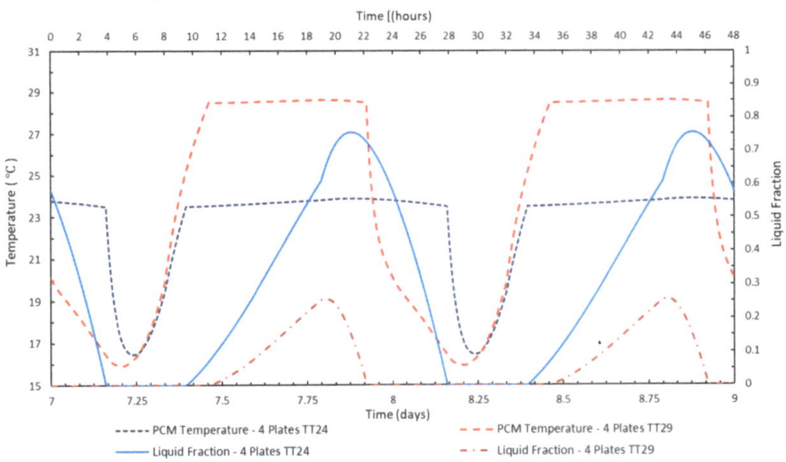

Figure 9. PCM temperature and liquid fraction for the room in low inertia configuration, during normal summer conditions, with natural convection.

5.2. Results for Normal Summer Conditions and Forced Convection

In order to overcome the shortcoming illustrated in Figure 9 (PCM never becoming fully liquid), it was decided to consider a design variant with a fan blowing air at 5 m/s through the external surfaces of the device, triggering forced convection and thus intensifying the heat transfer to and from the PCM. Figure 10 shows the results of the PCM and its liquid fraction with these new conditions. They clearly show that in this circumstance the PCM transition temperature of 24 °C does become fully liquid, thus making full use of its thermal storage capability. The PCM with a transition temperature of 29 °C improves considerably compared to Figure 9, but still does not become fully liquid.

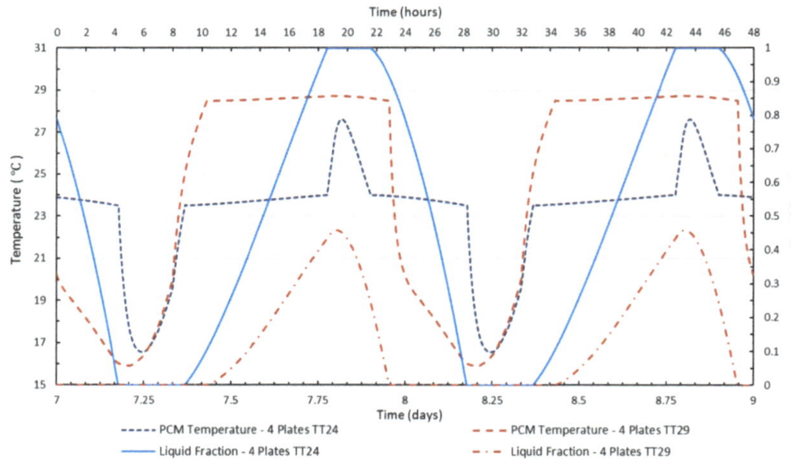

Figure 10. PCM temperature and liquid fraction for the room in low inertia configuration, during normal summer conditions, with forced convection.

Figures 11 and 12 show the indoor temperatures for 2 days in different building inertia and device alternatives, for the case with forced convection at the place surfaces. With one PoDIT module (1 plate), the observed peak temperature attenuation is about 1.1 °C for the low inertia room, and about 0.9 °C for the high inertia room. If four modules are used, then the peak temperature attenuation becomes slightly above 4 °C in the low thermal inertia building and close to 3 °C in the high thermal inertia room. Table 5 summarizes the impact of the device in the indoor temperature, showing reductions vs the "no device case" for the maximum room air temperature and for the room air temperature at 2 AM.

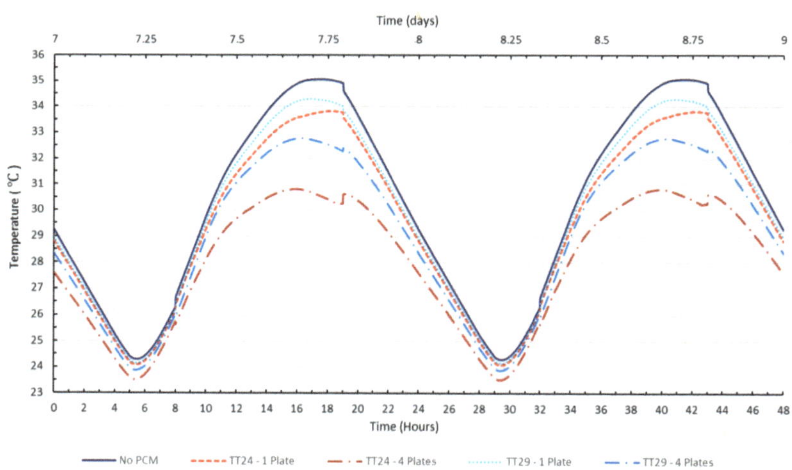

Figure 11. Indoor temperatures for the low inertia room and normal climate, with forced convection at the place surfaces.

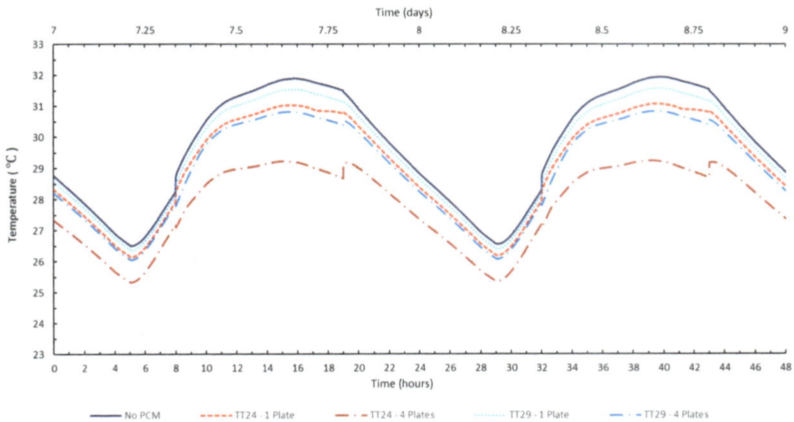

Figure 12. Indoor temperatures for the high inertia room and normal climate, with forced convection at the place surfaces.

Table 5. Impact of the PoDIT system upon indoor air temperature of the reference room, for several configurations and conditions, considering the reference normal weather.

		Natural Convection				Forced Convection			
		TT24		TT29		TT24		TT29	
		T_{Max} (°C)	ΔT (°C)	T_{Max} (°C)	ΔT (°C)	T_{Max} (°C)	ΔT (°C)	T_{Max} (°C)	ΔT (°C)
Low Inertia room	No PCM	35.1	–	35.1	–	35.1	–	35.1	–
	1 PCM Plate	34.3	0.8	34.6	0.4	33.8	1.2	34.3	0.8
	4 PCM Plates	32.4	2.6	33.7	1.4	30.8	4.3	32.8	2.3
High Inertia room	No PCM	31.9	–	31.9	–	31.9	–	31.9	–
	1 PCM Plate	31.5	0.5	31.8	0.2	31.1	0.9	31.6	0.4
	4 PCM Plates	30.3	1.6	31.3	0.7	29.2	2.7	30.8	1.1

5.3. Heat Wave Conditions

Figures 13 and 14 show the indoor temperatures during the heat wave conditions (as defined in Figure 6/Section 4), for the cases of natural or forced convection upon the system device respectively. Table 6 summarizes the impact of the PoDIT device in terms of reduction on the maximum indoor temperature. It can be seen that, for the reference room studied, the device with the TT32 PCM enables reductions in the maximum indoor temperature between 1.8 and 2.9 °C for the high thermal inertia room, and between 2.9 and 4.6 °C for the low inertia room. The PCM with a transition temperature of 29 °C yields lower temperature attenuations.

These results show that, at least in the harsh conditions considered for this study, the PoDIT device cannot be seen as a general substitute of air conditioning—the indoor temperatures are too high under any of the scenarios. It can nevertheless provide a valuable contribution for the attenuation of discomfort and health consequences.

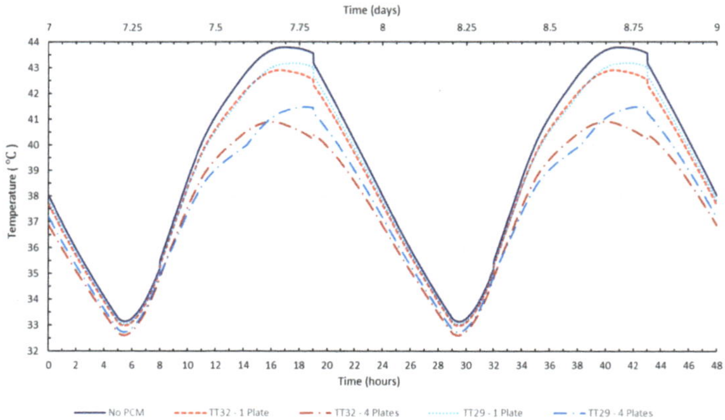

Figure 13. Indoor temperatures, for the low inertia case with natural convection at the place surfaces, under heat wave conditions.

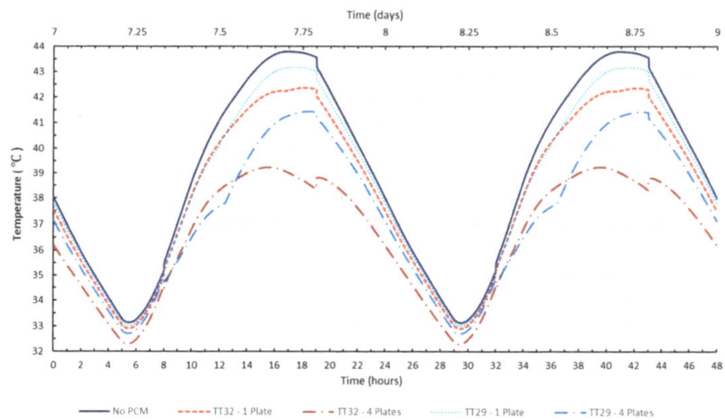

Figure 14. Indoor temperatures, for the low inertia case with forced convection at the place surfaces.

Table 6. Impact of the PoDIT system upon indoor air temperature of the reference room, for several configurations and conditions, considering the weather under heat wave conditions.

		Natural Convection				Forced Convection			
		TT24		TT29		TT24		TT29	
		T_{Max} (°C)	ΔT (°C)	T_{Max} (°C)	ΔT (°C)	T_{Max} (°C)	ΔT (°C)	T_{Max} (°C)	ΔT (°C)
Low Inertia room	No PCM	43.8	–	43.8	–	43.8	–	43.8	–
	1 PCM Plate	42.9	0.9	43.2	0.6	42.4	1.4	43.2	0.6
	4 PCM Plates	40.9	2.9	41.5	2.3	39.2	4.6	41.5	2.3
High Inertia room	No PCM	40.6	–	40.6	–	40.6	–	40.6	–
	1 PCM Plate	40.1	0.5	40.2	0.4	39.7	0.9	40.2	0.4
	4 PCM Plates	38.9	1.8	39.0	1.7	37.7	2.9	39.0	1.6

6. Conclusions

This work introduces the concept of a new Portable Device for Indoor Temperature Stabilization (PoDIT), to be considered as a low-cost, quick and easy to implement remediation strategy when, for social, economic, or technical reasons, the improvement of the building envelope and/or the adoption of air conditioning are not possible.

The system, which is modular, consists of a certain mass of engaged PCM that stays indoors during the daytime and is transported to the outdoors (e.g., a balcony) during the night to discharge the heat accumulated during the daytime. Both natural convection and forced convection (requiring a fan, in this case) variants were considered.

The results showed that, in the configurations and for the reference room and weather considered, adopting 4 modules of the device can lead to reductions in the maximum room air temperature close to 3 °C with natural convection, and more than 4 °C if a fan imposing forced convection is adopted. Adopting a fan proved important to ensure full solid–liquid commuting and therefore optimal use of the PCM thermal storage capability.

For the room studied, which has high solar aperture, and in the inner Iberian peninsula climate considered, the reduction achieved alone is not sufficient to ensure thermal comfort, as indoor air temperatures in excess of 30 °C were still obtained. Nevertheless, under stress conditions as is the case of heat waves, any attenuation of 3 or 4 °C may be important to avoid critical consequences in terms of health. Furthermore, the reference room considered has very high solar aperture.

Among the limitations of the system are the need for the existence for a terrace or balcony close to the room so that it can be placed outdoors during the night and the occupation of internal space when placed indoors.

As for future work, there is room for studying alternative/more complex system configurations, both with forced and with natural convection options, that could eventually lead to improved performance. Experimental measurements may assist a definitive validation of the models too and/or assist exploring alternative configurations. It could also address the performance of such system in other combinations of room and climate and room characteristics—e.g., to see how it operates in central and northern European cases, for example, where it may have the potential to avoid the usage of air conditioning. It would also be of interest to assess its potential use in some small commercial stores that have no air-conditioning, as well as to assess potential contribution to retain solar gains in winter. A more complete assessment of impact of the impact upon comfort, considering radiant and operative temperatures, may also be of interest at some point.

Author Contributions: Conceptualization: V.L.; methodology, V.L and R.T.; modelling: R.T. and V.L.; programming and simulations: R.T.; writing—original draft preparation, V.L. and R.T. All authors have read and agreed to the published version of the manuscript.

Funding: This research received no external funding.

Acknowledgments: The authors thank the valuable contribution of Helena Corvacho, from the Department of Civil Engineering at FEUP, who kindly commented options for making the reference room realistic from a constructions perspective.

Conflicts of Interest: The authors declare no conflict of interest.

List of Acronyms and Abbreviations

ASHRAE	American society of heating refrigerating and air conditioning
EPS	Expanded Polystyrene
GHG	Greenhouse effect gases
HVAC	–Heating, Ventilation and Air Conditioning
PCM	Phase Change Material
PoDIT	Portable Device for Indoor Temperature stabilization

References

1. Ostermana, E.; Tyagib, V.V.; Butalaa, V.; Rahimb, N.A.; Stritih, U. Review of PCM based cooling technologies for buildings. *Energy Build.* **2012**, *49*, 37–49. [CrossRef]
2. Pomianowskia, M.; Heiselberga, P.; Zhangb, Y. Review of thermal energy storage technologies based on PCM application in buildings. *Energy Build.* **2013**, *67*, 56–69. [CrossRef]
3. Iten, M.; Liu, S.; Shukla, A. A review on the air-PCM-TES application for free cooling and heating in the buildings. *Renew. Sustain. Energy Rev.* **2016**, *61*, 175–186. [CrossRef]
4. Souayfanea, F.; Fardouna, F.; Biwoleb, P.-H. Phase change materials (PCM) for cooling applications in buildings: A review. *Energy Build.* **2016**, *129*, 396–431. [CrossRef]
5. Faraj, K.; Khaled, M.; Faraj, J.; Hachem, F.; Castelain, C. Phase change material thermal energy storage systems for cooling applications in buildings: A review. *Renew. Sustain. Energy Rev.* **2020**, *119*, 109579. [CrossRef]
6. de Gracia, A. Dynamic building envelope with PCM for cooling purposes—Proof of concept. *Appl. Energy* **2019**, *235*, 1245–1253. [CrossRef]
7. Santamouris, M.; Alevizos, S.M.; Aslanoglou, L.; Mantzios, D.; Milonas, P.; Sarelli, I.; Karatasou, S.; Cartalis, K.; Paravantis, J.A. Freezing the poor—Indoor environmental quality in low and very low income households during the winter period in Athens. *Energy Build.* **2014**, *70*, 61–70. [CrossRef]
8. Magalhães, S.M.C.; Leal, V.M.S.; Horta, I.M. Predicting and characterizing indoor temperatures in residential buildings: Results from a monitoring campaign in Northern Portugal. *Energy Build.* **2016**, *119*, 293–308. [CrossRef]
9. Royé, D.; Codesido, R.; Tobías, A.; Taracido, M. Heat wave intensity and daily mortality in four of the largest cities of Spain. *Environ. Res.* **2020**, *182*, 109027. [CrossRef]
10. Nori-Sarma, A.; Anderson, G.B.; Rajiva, A.; ShahAzhar, G.; Gupta, P.; Pednekar, M.S.; Son, J.; Peng, R.D.; Bell, M.L. The impact of heat waves on mortality in Northwest India. *Environ. Res.* **2019**, *176*, 108546. [CrossRef]
11. Ma, W.; Zeng, W.; Zhou, M.; Wang, L.; Rutherford, S.; Lin, H.; Liu, T.; Zhang, Y.; Xiao, J.; Zhang, Y.; et al. The short-term effect of heat waves on mortality and its modifiers in China: An analysis from 66 communities. *Environ. Int.* **2015**, *75*, 103–109. [CrossRef] [PubMed]
12. DGS—Portuguese Directorate General for Health. eVM Online Monitoring of the Mortality in Portugal. Available online: https://evm.min-saude.pt (accessed on 30 April 2020).
13. Rupp, R.F.; Kim, J.; de Dear, R.; Ghisi, E. Associations of occupant demographics, thermal history and obesity variables with their thermal comfort in air-conditioned and mixed-mode ventilation office buildings. *Build. Environ.* **2018**, *135*, 1–9. [CrossRef]
14. Gupta, S.; Khare, M.; Goyal, R. Sick building syndrome—A case study in a multistory centrally air-conditioned building in the Delhi City. *Build. Environ.* **2007**, *42*, 2797–2809. [CrossRef]
15. Bhamare, D.K.; Rathod, M.K.; Banerjee, J. Passive cooling techniques for building and their applicability in different climatic zones—The state of art. *Energy Build.* **2019**, *198*, 467–490. [CrossRef]
16. Solgi, E.; Kari, B.M.; Fayaz, R.; Taheri, H. The impact of phase change materials assisted night purge ventilation on the indoor thermal conditions of office buildings in hot-arid climates. *Energy Build.* **2017**, *150*, 488–497. [CrossRef]
17. Paulo, M.D.; Veiga, M.d.R.; de Brito, J. Gypsum coatings in ancient buildings. *Constr. Build. Mater.* **2007**, *21*, 126–131. [CrossRef]
18. Sabapathy, K.A.; Gedupudi, S. On the influence of concrete-straw-plaster envelope thermal mass on the cooling and heating loads for different climatic zones of India. *J. Clean. Prod.* **2020**, *276*, 123117. [CrossRef]
19. Puretemp Website. Available online: http://www.puretemp.com/stories/puretemp-technical-data-sheets (accessed on 14 October 2020).
20. Sharma, A.; Tyagi, V.V.; Chen, C.R.; Buddhi, D. Review on thermal energy storage with phase change materials and applications. *Renew. Sustain. Energy Rev.* **2009**, *13*, 318–345. [CrossRef]
21. Alva, G.; Lin, Y.; Fang, G. An overview of thermal energy storage systems. *Energy* **2018**, *144*, 341–378. [CrossRef]
22. Py, X.; Olives, R.; Mauran, S. Paraffin/porous-graphite-matrix composite as a high and constant power thermal storage material. *Int. J. Heat Mass Transf.* **2001**, *44*, 2727–2737. [CrossRef]

23. Marín, J.M.; Zalba, B.; Cabeza, L.F.; Mehling, H. Improvement of a thermal energy storage using plates with paraffin-graphite composite. *Int. J. Heat Mass Transf.* **2005**, *48*, 2561–2570. [CrossRef]
24. dos Santos Teixeira, R.M. Avaliação e Pré-Dimesionamento de um Dispositivo de Mudança de Fase para Mitigação de Ondas de calor. Master's Thesis, University of Porto, Porto, Portugal, 2019.
25. Coelho, P. *Tabelas de Termodinâmica*, 4th ed.; Lidel: Lisbon, Portugal, 2017. (In Portuguese)
26. Sonderegger, R.C. Dynamic Models of House Heating Based on Equivalent Thermal Parameters. Ph.D. Thesis, Princeton University, Princeton, NJ, USA, 1978.
27. Bacher, P.; Madsen, H. Identifying suitable models for the heat dynamics of buildings. *Energy Build.* **2011**, *43*, 1511–1522. [CrossRef]
28. Bagheri, A. Development of Simplified RC Models with Physically Deducible Parameters Using the Time-Constant Concept for the Calculation of the Heating Load Demand in Single Detached Dwellings and Districts. Ph.D. Thesis, University of Mons, Mons, Belgium, 2019.
29. America Society of Heating, Refrigeration and Air Conditioing Engineers. ASHRAE Fundamentals (2007), 2009 ASHRAE Handbook—Fundamentals (SI Edition). Atlanta, USA. Available online: https://app.knovel.com/web/toc.v/cid:kpASHRAE37/viewerType:toc/ (accessed on 12 November 2020).

Publisher's Note: MDPI stays neutral with regard to jurisdictional claims in published maps and institutional affiliations.

energies

MDPI

Review

A Review of the Measures and Instruments to Promote Efficiency and Renewable Energy in Domestic Water Heating

Dania Ortiz [1,*], Vítor Leal [2] and Isabel Azevedo [3]

1 Faculty of Engineering, University of Porto (FEUP), 4200-465 Porto, Portugal
2 Department of Mechanical Engineering, Faculty of Engineering, University of Porto (FEUP), 4200-465 Porto, Portugal; vleal@fe.up.pt
3 Institute of Science and Innovation in Mechanical and Industrial Engineering (INEGI), 4200-465 Porto, Portugal; iazevedo@inegi.up.pt
* Correspondence: up201608014@fe.up.pt

Received: 7 September 2020; Accepted: 8 October 2020; Published: 15 October 2020

check for updates

Abstract: This paper identifies and characterizes the technical measures and policy instruments that can be used to promote energy efficiency and the use of renewable sources for domestic hot water (DHW). DHW presents a considerable potential for abatement of greenhouse gas emissions around the world. Measures were characterized in terms of level of transformation, impact and scope, among others. Policy instruments were characterized in terms of target groups, competences required for implementation, major challenges and nature of the instruments. A matrix showing the applicability of policy instruments per technical measure was derived, enabling policy makersto better choose articulated measures and policy instruments for their policy packs.

Keywords: domestic hot water; water heating; energy efficiency; climate change mitigation; climate policy; policy making; policy instruments

1. Introduction

The residential sector constitutes one of the greatest sources of untapped potential for energy efficiency improvements and reduction of GHG emissions. The CO_2 emissions reduction potential for the building stock in 2020 is estimated at 21–54% in developed economies, 26–46% in economies in transition and 18–41% in developing countries [1,2].

Water heating is one of the most basic energy services. It is the second largest energy use segment in the residential sector, accounting for 4% to 10% of the total energy used in developed economies [3]. Heating water for domestic activities comprises approximately 25% of households' energy use in Australia [4], 20% in Canada [5], 14% in the European Union [6], and 13.2% in US [7]. The importance of guaranteeing access to hot water and the weight of this service in the overall energy use in the residential sector make the achievement of more efficient and sustainable production and use of domestic hot water a global policy objective.

The socio-economic benefits of energy efficiency improvements associated with DHW are widely recognized, including the reduction of energy costs and the increase of wellbeing. Over the last decades, several governments have aimed for a better management of hot water use and promoted the use of renewable sources and/or more energy efficient systems.

Currently, the main challenge is to fulfill the sector needs with the most sustainable alternatives. Despite the large untapped potential for reduction of GHG emissions associated with DHW and the demonstrated political will to achieve it, there is still a lack of scaling-up of both technological and non-technological opportunities for improvement [1]. Among the existing works, a thorough review on possible technical measures that may lead to the reduction of GHG emissions associated with DHW (through increase energy efficiency and/or the use of renewable energy sources) and the policy instruments that are more adequate for the respective implementation is still lacking and this gap may be hampering policy makers from taking informed decisions. Indeed, the identification of the technical measures to be promoted and the choice of policy instruments with which the later are promoted are key for a successful policy-making process (Figure 1).

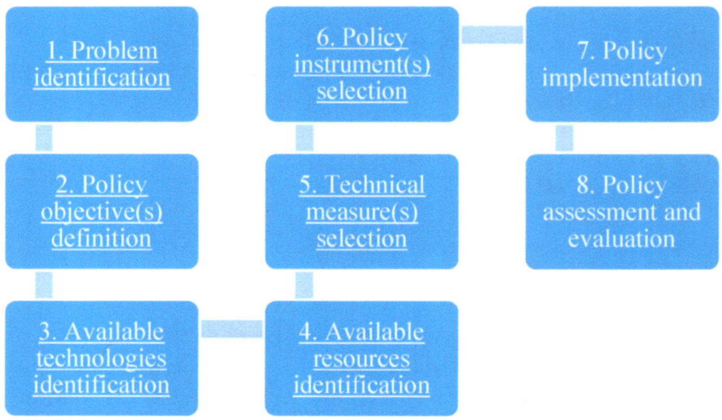

Figure 1. Water Heating System (WHS) Policy Process.

This paper aims to support WHS policy decision makers, with a focus on steps 5 and 6 as the steps where technical measures and policy instruments are to be selected, considering the WHS policy process described in Figure 1. The identification of their characteristics will allow the decision maker to better suit their selection criteria (preferences, resources and objectives) with the available options.

Regarding the organization of the paper, Section 2 provides an overview of the current and projected use of DHW around the world, including the associated energy use and GHG emissions. Section 3 describes the methodology followed to identify and characterize both the technical measures and the policy instruments. Section 4 presents the results of the review and proposes a classification scheme for the technical measures and for the policy instruments. Section 5 provides a discussion around the findings of the review, including on the relationship between technical measures and the policy instruments that are used for their implementation. Finally, Section 6 presents the conclusions.

2. Domestic Hot Water Situation Across the World

Water heating energy use varies across countries and regions, but, in general, it is associated with a high share of energy use. More than 50% of the residential energy use corresponds to thermal use, with a considerable contribution from water heating [2]. For the year 2000, the energy use associated with water heating in the residential sector corresponded to 480 Mtoe and reached 580 Mtoe in 2017 [8]. The increase in energy use can be associated with the increased use of hot water, which is not sufficiently counteracted by energy efficiency improvements. Moreover, future projections point to a significant increase of the associated energy use in the following years, especially in developing countries. Indeed, the growth in hot water use may imply a significant increase in the associated energy use or, on the other hand, reducing the hot water use would result in direct energy savings with their associated CO_2 reductions. For example, reducing the use of hot water in approximately 20% in the United States could lead to a reduction of 41 million MWh of electricity consumption and 240 billion cubic feet in the use of natural gas, corresponding to a reduction of about 38.3 million tons of CO_2 emissions [9].

Nowadays, the dominant technologies in the production of domestic hot water in buildings are either fossil fuels based or low-efficiency electrical technologies, with the exception of some developing economies where the traditional use of solid biomass is predominant [8].

Given the low-efficiency carbon-intensive technologies and the projected increase in the use of hot water, DWH comprises a large potential for energy efficiency improvements, and the reduction of GHG emissions [7]. The average level of emissions associated with the production of hot water varies significantly among the different DHW technologies, ranging from 0.02 $mtCO_{2\ eq.}$/Yr with a heat pump to 0.06 $mtCO_{2\ eq.}$/Yr with an electric water heater and to 0.87 $mtCO_{2\ eq.}$/Yr with a gas water heater [10].

Currently, the promotion of both energy efficiency and renewable energy use for water heating within the residential sector is already one of the priorities in energy policy at different governance levels. The set of solutions that can be promoted is large, given the wide variety of technologies available. The potential for reducing the CO_2 emissions associated with the production of hot water with the use of natural gas is estimated to be around 25.8 million tons of CO_2, while the adoption of electrical SWH technologies could lead to a reduction from 20.9 to 45.0 million tons [11].

3. Materials and Methods

In this paper, the identification and characterization of both technical measures and policy instruments used to promote energy efficiency and renewable energy use in DHW is based on an extensive literature review, identifying the WHS-related policies across the world.

To ensure the coherence of the review process, it was first necessary to define these concepts: "technical measures" and "policy instruments". Herein, technical measures refer to the actions that imply an actual change in the energy system and are implemented with the goal of achieving the established policy objectives. Policy instruments are tools of governance to influence targeted individuals or a group's behavior in order to achieve strategic public objectives [12]. Policy instruments can be considered as mechanisms that promote or support the implementation of the technical measures. Policies are normally executed through instruments or mechanisms that enable the implementation of the technical measures. Most of the instruments have been gathered from the policies where the technical measures were identified.

For the review, two databases on energy efficiency policies were taken as the starting point:

- The energy efficiency policies and programs database published by the International Energy Agency and retrieved from IEA (https://www.iea.org/policies) [13]. The list of policies was obtained by selecting the following filters: "heating and cooling" as sector and "Renewable Energy and Energy Efficiency" as topic.
- The solar thermal incentive database organized by and retrieved from Solarthermalworld (https://www.solarthermalworld.org/incentive) [14].

The selection of the two databases was based on the fact that these comprise a collection of several DHW policies, gathered from different national and international studies. They gather detailed information concerning several DHW policies, including location, timeframe, responsible authority, etc. More than 20 WHS policies across the world were retrieved from this review, including their correspondent technical measures and policy instruments.

To guarantee a more comprehensive review, an additional search was performed in the open browser google, under the expression "water heating efficiency technical measures". The first 50 sources displayed under this search were reviewed. This number was considered as a reasonable compromise between comprehensiveness and feasibility. The search lead to the identification of additional measures and consequently to the identification of additional policy instruments.

Finally, additional measures and instruments not found in the reviewed documents (e.g., prizes and awards, fines and tax increase, technology phase-out) were added to the list, supported by specific search based on personal experience and experts' opinions.

The review resulted in the identification of individual technical measures and policy instruments previously adopted in the promotion of energy efficiency and renewable energy use for domestic hot water. The identification of the combination of technical measures the policy instrument(s) used for their implementation was also performed. The results were organized in a table and are presented in Appendix A.

4. Results

4.1. Technical Measures

4.1.1. Identification of Technical Measures

The review of the energy efficiency policies led to the identification of a wide variety of technical measures, which were organized by type. Most of the reviewed policies promote a single technical measure, with only a few opting for the combination of several measures.

Measure 1: Adjusting the scale of the water heating system

1.1. Replacing collective system with individual WHS

Independent water heating systems (as opposed to collective) provide individuals with the possibility of controlling the devices' maintenance and operation, allowing them to make technical improvements when required. Moreover, individual metering allows for a detailed monitoring of the consumption pattern of each user, leading to a better identification of inefficiencies (technical or behavioral).

1.2. Installation of a collective WHS (building or district systems) in substitution of individual systems

Moving from individual to collective WHS is also feasible and may have advantages, depending on the context. The most adequate solution may vary depending on the characteristics of each building, but also criterion like the geographical location and the energy sources. A collective system could refer to a WHS which provides several dwellings in the same building or even several buildings. District heating is an example of a collective heating system (WHS included). The choice for a collective system may allow for the transition to more efficient and less carbon-intensive technologies that can only be effective in large-scale system. This could include the incorporation of renewable energy or excess heat.

Measure 2: Replacement of water heating technologies for more efficient or less-carbon intensive ones

Technology replacement is one of the most straightforward measures, as the replacement will normally imply the installation of a "better" technology (more efficient or cleaner technology). Differences in fuels, together with technical characteristics can be seen among the wide spectrum of options, varying from renewable energy systems, to electrical input devices and hybrid systems, existing in various configurations. The replacement trends vary from country to country, probably due to the context specificities as the benefits of the different technologies depend on the local conditions. The most common technology replacements identified through the revision of the different WHS energy efficiency policies are presented in Table 1. In general, electric and gas-based devices are replaced by cleaner source-powered technologies, as solar-powered devices.

Measure 3: Installation of solar thermal technologies (to complement WHS)

This measure corresponds to the installation of solar thermal panels to be combined with the already existing water heating device. This is an alternative to the replacement of the whole system, comprising the addition of the solar thermal system to the WHS already installed, reducing the use of energy by the later.

Table 1. Most common technology replacements.

		New Technology				
		Natural gas water heater	Natural gas central boiler	Solar water heater	Solar radiant heater	Renewable hot water (from DHC) *
Initial Technology	Fossil fuel water heater	X		X		X
	Electric tankless water heater	X				
	Electric storage water heater	X		X		
	Fossil fuel central heater		X		X	X
	Electric heat pump					X

* Renewable water heater includes district heating and cooling networks (DHC) and small-scale combined heat and power (CHP) units. It does not include solar water heaters (Source: own elaboration).

Measure 4: Insulation of boilers

Adding new or improving existing boilers' insulation can lead to energy savings. The insulation of already installed boilers can reduce the heat losses, and it may be a cheaper option than replacing the whole device.

Measure 5: Insulation of heating pipes

The piping system that is used in the distribution of the hot water by the different usage points can also be insulated, leading to a decrease in the heat losses of the whole WHS. The impact of this measure may depend on several factors, ranging from the outdoor temperature to the material of the pipes and the length of the distribution system.

Measure 6: Installation of (or replacement for) efficient water appliances

Water heating systems are combined with other appliances to provide end-use service, as could be the showerheads or the bathroom and kitchen taps. Thus, the installation of efficient (water saving or energy saving) devices may lead to a better use of the hot water. For instance, the installation of devices that reduce the water flow may lead to significant reductions of the total water use and, consequently, of the associated energy use.

Measure 7: Regular maintenance of Water Heating Systems

The performance of regular inspections may improve the daily operation of the hot water system and accelerate the identification and correction of malfunctions, and even the replacement of the devices (when necessary). The benefits of performing regular maintenance of the heating devices are numerous, including: to prevent system irregularities and cope with them opportunely, to guarantee that the energy performance does not get degraded faster than it should and to increase lifetime of the equipment.

Measure 8: Installation of smart systems

Nowadays, smart systems are available to improve the management of WHS at different scales (both individual and collective systems). These systems can include wireless monitoring and data collection systems. They could improve the monitoring of the heaters' operation status, allow for automatic temperature control and to program schedules and other timing functionalities (tele control). In addition, monitoring functionalities can also comprise performance and malfunctions identification. Furthermore, some of the smart systems currently available allow for the automation of the WHS to turn on and off according to the individuals' preferences, as the hourly energy price or the availability of renewable energy sources (solar and wind).

Measure 9: Reduction in the use of hot water

Aiming to lower the water and energy consumption, actions like taking shorter showers or cleaning (dishes or clothes) with cold water can directly decrease DHW demand. The technical measures retrieved from the literature are shown in Table 2 and briefly described in the previous paragraphs.

Table 2. List of technical measures, type and source of identification.

Source	Policy Description			Technical Measure(s) Promoted	Type of Technical Measure(s)
	Name	Country/Region	Implementation Timeframe		
(Saskenergy, 2019) [15]	Energy Star Loan Program	Canada	2015	Device change (from fuel to gas)	Replacement of water heating technologies for more efficient ones Installation of solar thermal technologies
(FortisBC, 2019) [16]	British Columbia Energy Star Water Heater Program + Furnace and Boiler Replacement Program	Canada	2014		
(Energywise, 2019) [17]	Warm Up New Zealand: Heat Smart	New Zealand	2009	Device change (from oil boiler; Gas or propane furnace/boiler to more efficient ones)	
(Energywise, 2019) [17]	Insulation Programmes 2009–2018	New Zealand	2009		
(Domácnostiam, 2019) [18]	Slovakian Green Homes incentive programme	Slovakia	2019	Installation of renewable-based systems (SWH, photovoltaics, heat pumps, small wind-mills)	
(Enova, 2016) [19]	"Enovatilskuddet" (Enova grants)	Norway	2015	Installation of solar water heaters, pellet boilers and heat pumps	
(BAFA, 2019) [20]	Market Rebate Programme for Renewable Energy	Germany	2012	Installation of SWH systems, heat pumps, pellets tanks, woodchip equipment	
(Regulator for Energy and Water, 2019) [21]	Solar water heaters	Malta	2010	Installation of solar water heaters and heat pumps	
(Sklad, 2019) [22]	Eco Fund, Slovenian Environmental Public Fund	Slovenia	2009	Installation of Solar water heating systems and other renewable technologies	
(Bydlení, 2019) [23]	Nova Zelena Usporam	Czech Republic	2013	Installation Solar thermal systems, biomass boilers and heat pumps	
(Klima- und Energiefonds, 2019) [24]	Demoprojekte Solarhaus	Austria	2017		

Table 2. *Cont.*

<table>
<tr><th rowspan="2">Source</th><th colspan="3">Policy Description</th><th rowspan="2">Technical Measure(s) Promoted</th><th rowspan="2">Type of Technical Measure(s)</th></tr>
<tr><th>Name</th><th>Country/Region</th><th>Implementation Timeframe</th></tr>
<tr><td>(Facility, 2014) [25]</td><td>Western Balkan Sustainable Energy Financing Facility II</td><td>The Western Balkan countries</td><td>2013</td><td rowspan="8">Solar thermal systems addition</td><td rowspan="9">Installation of solar thermal technologies (to complement WHS)</td></tr>
<tr><td>(Ministerio de Industria, 2014) [26]</td><td>Tax Rebate</td><td>Chile</td><td>2016</td></tr>
<tr><td>(ANME, 2019) [27]</td><td>Prosol II</td><td>Tunisia</td><td>2005</td></tr>
<tr><td>(Conservation, 2019) [28]</td><td>The National Financing Mechanism for Solar Water Heaters</td><td>Lebanon</td><td>2010</td></tr>
<tr><td>(Ministerio de Industria, 2014) [26]</td><td>Solar Plan</td><td>Uruguay</td><td>2012</td></tr>
<tr><td>(Economy Ministery, 2019) [29]</td><td>Programme for partial subsidising of purchased and installed solar water heaters for households</td><td>Macedonia</td><td>2017</td></tr>
<tr><td>(Residential Energy Efficiency Credit Line, 2019) [30]</td><td>Residential Energy Efficiency Credit Line 3</td><td>Bulgaria</td><td>2016</td></tr>
<tr><td>(California Solar Initiative, 2019) [31]</td><td>California Solar Initiative (CSI)-Thermal Program</td><td>California</td><td>2017</td></tr>
<tr><td>(Agencia Andaluza de la Energía, 2019) [32]</td><td>Andalusia: Sustainable Construction Incentive Programme</td><td>Andalusia</td><td>2017</td><td>Installation of solar WHS (for new construction buildings)</td></tr>
<tr><td>(Arctic Energy Alliance, 2019) [33]</td><td>Northwest Territories Energy Efficiency Incentive Program</td><td>Canada</td><td>2014</td><td>Electric hot water cylinder wraps installation</td><td>Insulation of boilers</td></tr>
<tr><td>(Aranda et al. 2017) [34]</td><td>Analysis of energy efficiency measures and retrofitting solutions for social housing buildings in Spain as a way to mitigate energy poverty</td><td>Spain</td><td>2017</td><td>Installation of efficient taps and showerheads</td><td>Installation of (or replacement by) efficient water appliances</td></tr>
<tr><td>(Savickas & Bielskus Directive, 2015) [35]</td><td>Technical measures to decrease heat energy consumption of final customer in multi-apartment buildings according to Energy Efficiency</td><td>Lithuania</td><td>2015</td><td>Independent heat substation for heat and hot water preparation; Hot water system balancing; Hot water metering for each final customer of the building; Smart intelligent wireless monitoring and data collection system</td><td>Individual/collective optimization of the water heating system. Installation of smart systems</td></tr>
<tr><td>(Poortinga and Steg, 2002) [36]</td><td>Viable behavioural and technological energy-saving measures</td><td>The Netherlands</td><td>2002</td><td>Rinsing dishes with cold water; shorter showers; Insulation of heating pipes</td><td>Reduction in the use of hot water. Insulation of heating pipes</td></tr>
</table>

4.1.2. Characterization of Technical Measures

In order to facilitate the assessment of the technical measures identified in the review process, these were characterized according to the following features: level of transformation; level of energy savings; scope (single vs. multi-apartment solution) and need for technical/specialized intervention (organized in Table 3).

Table 3. Characterization of the technical measures.

Technical Measure	Level of Transformation	Level of Energy Savings	Scope	Requirement of Trained Technician
1. Adjusting the scale of the water heating system.	Transformative	High Impact	Multi-apartment /Single apartment	Yes
2. Replacement of water heating technologies by more efficient or less-carbon intensive ones	Transformative	High Impact	Multi-apartment /Single apartment	No
3. Installation of solar thermal technologies (to complement WHS)	Transformative	High Impact	Multi-apartment /Single apartment	Yes
4. Insulation of boilers	Marginal	Low Impact	Multi-apartment /Single apartment	No
5. Heating pipes insulation	Marginal	Low Impact	Multi-apartment /Single apartment	No
6. Installation of (or replacement by) efficient water appliances	Marginal	Low Impact	Multi-apartment /Single apartment	Yes
7. Regular maintenance of water heating systems	Marginal	Low Impact	Multi-apartment /Single apartment	Yes
8. Installation of smart systems	Marginal	Low Impact	Multi-apartment /Single apartment	Yes
9. Reduction in the use of hot water	Marginal	Low Impact	Single apartment	No

- Level of transformation refers to the degree of change in the WHS due to the technical measure implementation. "Marginal" refers to small changes in the WHS, while "transformative" refers to significant interventions or changes. For example, the installation of more efficient appliances, as showerheads, can be considered a marginal transformation, while the installation of a solar thermal system to complement already existing WHS is a transformative measure. The analysis of this feature is important to understand if the measure can be implemented without a significant burden for the users, and without significant changes in the WHS system.
- Level of energy savings measures the energy efficiency improvements and it is classified as high or low, referring to the degree of energy-saving potential behind each measure, implying a change or considerable reduction in the fuel used, e.g., changing the shower head for more efficient ones saves up to 30% (low level of energy savings), while replacing technologies for efficient technologies can reach up to 90% (high level of energy savings) [37].
- Scope assesses if the measure can be applied to a single apartment and/or in a multi-apartment. This is important as it allows to identify the adequacy of the measures according to the dwelling typology.
- Need for technical/specialized intervention refers to how the measure implementation requires technical assistance or can be implemented directly by the final user. Its relevance relies on understanding if the measure requires specific workforce or skills for it to be implemented.

4.2. Policy Instruments

4.2.1. Identification of Policy Instruments

The policy instruments retrieved from the literature were organized by type. The different types of policy instruments identified are briefly described in the following paragraphs and also presented in Table 4.

Policy Instrument 1: Minimum Energy Performance Standard (MEPS)

As a well developed energy efficiency standard, MEPS establishes the minimum level of performance that a device must have to be sold in the market. With the aim of defining the required performance levels, standby losses are generally considered, for which testing procedures including standing loss tests are carried out [38]. MEPS normally lead consecutively to selling and installing products that are at least as efficient as defined in the standards. The use of energy standards is widely spread, whereas international agencies have participated through the creation of international references. The definition of standards can also happen by geographical areas, as has happened in North America (Canada, USA and Mexico) and in Australia and New Zealand [39]. The first standard creation for WHS in the EU was established in 1998 for gas boilers [40]. International standards have served as tools for national governments to base some regulations and achieve objectives [41]. Depending on the objectives, standards can be mandatory or voluntary.

Policy Instrument 2: Mandatory Standards

They are set as the obligatory requirements to be fulfilled by certain technologies. The obligation behind the mandatory standards provide support to policy makers to fasten the outcomes behind the supported policies.

2a: Voluntary Standards

Out coming as best practice from industry guidelines, voluntary standards can address a diversity of products. Standards may benefit final consumers by raising the quality or any other product characteristic. The application of voluntary standards is not exclusive to the industrial sector.

2b: Technology phase-out

Prohibit the installation of certain technologies represents another measure to cope with the energy-inefficient technologies, leading to ban the production and/or distribution of certain devices or systems with certain characteristics [42].

Policy Instrument 3: Rating System/Building Index

Rating systems as evaluation tools, can serve to create a benchmark among buildings, providing an alternative for evaluating and comparing criteria (e.g., efficiency, use, costs). Different scales can be used through the ratings; the National Energy Performance rating system in the US uses a 1 to 100 scale, whereas the higher the score, the better energy performer the evaluated facility is [43]. One of the purposes of rating systems is to provide practical information to stakeholders and decision makers about the efficiency and inefficiency of buildings [44]. Building index is an organized form of building evaluation, whereas a score is given to every evaluated facility. They both can serve as a basis to building certification. More specifically, building certification is one way to certificate the energy performance. Buildings are required to guarantee the energy characteristics defined in the regulation. A framing option to certificate a building in some countries is the Nearly Zero Energy Buildings (NZEB). This approach helps to reduce the building footprints (carbon emissions), improving the energy efficiency. The scope of terms and categories under the concept of NZEBs embraces a wide variety. Under the scope of a NZEBs high energy performance buildings are included, but can also refer to the production of the energy consumed in the building by renewable sources produced on-site or nearby [45].

Policy Instrument 4: Labelling

Energy labelling, as the way to categorize products under certain characteristics, provide information regarding the product technical data (e.g., efficiency). By labelling, the energy efficiency requirements in the manufacture of the product is promoted, leading to impulse market take-up of certain products or technologies.

Policy Instrument 5: Certification

Certificating certain type(s) of equipment (boilers and WHS) may include the compliance of MEPS, minimum device age or any other characteristic validated under the regulation or market criteria. For WHS, some examples include minimum efficiency levels. It is closely related to Labelling.

Policy Instrument 6: Information, education and awareness campaigns

Education and awareness campaigns provide information to consumers, service providers and/or any other targeted stakeholders. Information aimed to be communicated to the targeted audience may comprise the potential for energy savings, available WHS devices and appliances. It is frequent, through this instrument, to inform the final consumer on the benefits and drawbacks associated with the different WHS technologies available in the market and advise them on how to improve their own performance. The provision of more specific information through training or workshops is another option, usually targeting a more specific audience as technicians and other service providers. In general, this type of instrument is relatively easy to implement and does not require significant investments [46].

Policy Instrument 7: Grants and Subsidies

Subsidies, as a financial support, aim to reduce the capital cost of a technology, enabling the market growth of the technology. These instruments have widely been used across the countries, and most of the times, the incentives become unnecessary, after the market has been developed [47].

Policy Instrument 8: Tax credits/tax incentives

This type of instruments corresponds to a positive financial incentive, meaning that incentivizes consumers (or other stakeholders) with actions that benefit the final user as tax rebates or tax exemptions. These aim to reduce the total investment of the WHS acquisition.

Policy Instrument 9: Fines and tax increases

As a negative form of incentive, fines and tax increase, are used to penalize the non-compliance or abuse of certain actions [48].

Policy Instrument 10: Low-interest loans/third-party financing

Another financial aid suitable for the acquisition of solar water heater (SWH) is to facilitate the access to credit. This option can be complemented with other policy instruments.

Policy Instrument 11: Prizes and Awards

This corresponds to the identification and recognition of best practices or any desired parameters through a prize or award. The prize could be economic or not [49].

All these types of instruments can be used alone or combined. The combinations of instruments is worth mentioning, whereas more than one instrument is applied, and it has been identified in different documents of this review.

Table 4. Policy Instruments list.

Reference	Policy	Country	Year	Detailed Policy Instrument	Policy Instruments Type
(Saskenergy, 2019)	Energy Star Loan Program (SaskEnergy/SaskPower)	Canada	2015	Fiscal incentives	Tax credits/tax incentives
(FortisBC, 2019)	British Columbia Energy Star Water Heater Program + Furnace and Boiler Replacement Program	Canada	2014		
(Arctic Energy Alliance, 2019)	Northwest Territories Energy Efficiency Incentive Program	Canada	2014		
(Ministerio de Industria, 2014)	Tax Rebate	Chile	2016	Tax relief	
(Energywise, 2019)	Warm Up New Zealand: Heat Smart	New Zealand	2009	Grants and subsidies	Grants and Subsidies
(Energywise, 2019)	Insulation Programmes 2009–2018	New Zealand	2009		
(California Solar Initiative, 2019)	California Solar Initiative (CSI)—Thermal Program	California	2017	Rebate	
(Regulator for Energy and Water, 2019)	Solar water heaters	Malta	2010		
(ANME, 2019)	Prosol II	Tunisia	2005	Grant or reimbursable credit	
(Agencia Andaluza de la Energía, 2019)	Andalusia: Sustainable Construction Incentive Programme	Andalusia	2017	Economic Subsidy	
(Enova, 2016)	"Enovatilskuddet" (Enova grants)	Norway	2015		
(BAFA, 2019)	Market Rebate Programme for Renewable Energy	Germany	2012		
(Bydlení, 2019)	Nova Zelena Usporam (New Green Savings/New Greenlight to Savings)	Czech Republic	2013		
(Economy Ministery, 2019)	Programme for partial subsidising of purchased and installed solar water heaters for households	Macedonia	2017	Subsidy for investment	
(Domácnostiam, 2019)	Slovakian incentive programme Green Homes	Slovakia	2019		
(Klima-und Energiefonds, 2019)	Demoprojekte Solarhaus (Demo Projects Solar House)	Austria	2017		
(Residential Energy Efficiency Credit Line, 2019)	Residential Energy Efficiency Credit Line 3 (REECL 3)	Bulgaria	2016	Loan and investment grants	
(LODA/LODA—Lviv Regional State Administration, 2019) [50]	Energy Saving programme for residents of the Lviv region for the year 2013–2016	Ukraine	2013	Low-interest loans	
(Csobod et al. 2009) [51]	Overview and Analysis of Public Awareness Raising Strategies and action on energy savings	Central and Eastern Europe (CEEC)	2009	Informative campaigns	Information, education and awareness campaigns
(IMT), 2009)	A roadmap for creating building energy rating systems in Central Asia	Central Asia	2009	Rating systems	Rating systems
(D'Agostino and Mazzarella, 2019)	What is a Nearly zero energy building? Overview, implementation and comparison of definitions	Europe	2018	Building Index	Building Index
(Dungen, 2011)	Energy Performance Standards How does New Zealand compare with other countries?	New Zealand	2011	Minimum Energy Performance Standard (MEPS)	Standards
(Johnson et al., 2013)	An International Survey of Electric Storage Tank Water Heater Efficiency and Standards	New Zealand	2013	Minimum Energy Performance Standard (MEPS)	
(Turiel, 2000)	Present status of residential appliance energy efficiency standards—an international review	International	2000	Other Standards	
(Pusok and Morris, 2018) [52].	Voluntary Energy Standards: ISO 50001 and the superior energy standards	International	2018	Other Standards	

Table 4. *Cont.*

Reference	Policy	Country	Year	Detailed Policy Instrument	Policy Instruments Type
(Reed, 2002)	National Energy Performance Rating System	USA	2002	Rating System	Rating System
(UNECE, 2018) [53]	Mapping of Existing Energy Efficiency Standards and Technologies in Buildings in the UNECE Region	UNECE Region	2018	Building energy codes; Energy Performance Certification; Labelling	Combined (more than one instrument)
(WEC, 2016)	Energy Efficiency Policies around the World: Review and Evaluation	International	2008	Labelling, Codes & Standards	
(Government of Australia, 2019)	Water Heating Energy Rating	International	2019	Labelling, Codes & Standards	
(Norden, 2015)	Nordsyn—energy labelling requirements for packages of water heaters and solar devices	International	2015	Labelling, Codes & Standards	
(National Housing Bank, 2015) [54]	Capital subsidy scheme for installation of solar thermal systems	India	2014	Soft Loans + Subsidy	
(Ministerio de Industria, 2014)	Solar Plan	Uruguay	2012		
(Conservation, 2019)	The National Financing Mechanism for Solar Water Heaters	Lebanon	2010		
(Facility, 2014)	Western Balkan Sustainable Energy Financing Facility II (WeBSEFF II)	The Western Balkan countries	2013	Loans and Grants	
(Sklad, 2019)	Eco Fund, Slovenian Environmental Public Fund	Slovenia	2009		

The instruments have been organized in Table 4 by the details of the policy or document where they were identified, their detailed instrument and type. Most of the instruments have been gathered from the same policies where the technical measures were identified.

4.2.2. Characterization of Policy Instruments

The identified instruments have been characterized regarding five major aspects and presented in Table 5:

(i) Required competences refers to the regulatory competences required to policy makers for the instrument implementation (buildings, technology development, market development and taxation definition).

(ii) Target group refers to the actors that are most commonly targeted by the different policy instruments (e.g., installers, construction sector and final consumer, technology developers or industry, etc.).

(iii) Most common levels of implementation correspond to the governance level(s) the different policy instruments are commonly implemented, including local, national and international approach.

(iv) Major or key challenges and barriers related to each policy instrument.

(v) The policy instruments retrieved have also been classified in three categories: (a) regulating; (b) informative; and (c) incentivizing. By "regulating" it is meant those instruments that are normally used to formally implement an action, normally through a binding status, compared with informative, whose aim is to provide information to different stakeholders. The instruments enlisted as incentivizing relate to the provision of stimulus (commonly economically) for the measure to be implemented.

Table 5. Policy instruments characterization/relation between technical measures and policy instruments.

Policy Instruments	Required Competences	Target Group	Most Common Levels of Implementation	Challenges/Barriers to Implementation	Categorization of Instrument
1. Minimum Energy Performance Standard (MEPS)	Communication with customers	Technology developers (industry), final consumers	Regional/National	Industry (manufacturers) capability (Johnson et al., 2013 [38]); clients or other stakeholders' disapproval	Regulating
2. Technology Phase-out	Market organization and regulation	Final Consumers	Local, National	Industry and other stakeholders' disapproval	Regulating
3. Rating System/Building Index	Expert qualification & planning	Construction sector	National	Info availability Public's general environmental ignorance and low awareness of social and economic benefits of energy efficiency ((IMT), 2009) Financial costs (D'Agostino and Mazzarella, 2019 [45])	Informative
4. Labelling	Information provision	Final Consumers	Regional/National	Lack of understanding by targeted audience	Informative
5. Certification	Planning	Technology developers (Industry), Final Consumers	National	High demand volume	Informative
6. Information, Education and awareness	Communication with customers	Installers, Construction Sector and Final Consumers	National, Regional, Local	Lack of interest from targeted audience	Informative
7. Grants and Subsidies	Communication with customers	Installers and Final Consumers	National, Regional, Local	Attraction of wrong audience Fiscal Costs (IMF, 2015) private-sector under-investment,	Incentivizing
8. Tax credits/tax incentives	Regulation; Proximity and communication with customers	Installers and Final Consumers	National/Regional	Attraction of wrong audience Market failure and negative effects private-sector under-investment,	Incentivizing
9. Fines and tax increase.	Regulation and planning	Installers and Final Consumers	National, Regional, Local	Not achievement of the policy targets	Incentivizing
10. Low-interest loans/third-party financing	Regulation; Proximity and communication with customers	Installers and Final Consumers	National/Regional	Attraction of wrong audience Supporting the wrong industry development (e.g., foreign vs local)	Incentivizing
11. Prize and Awards	Regulation, planning and economic resources	Installers and Final Consumers	National, Regional, Local	Expensiveness	Incentivizing

4.3. Relation Between Technical Measures and Policy Instruments

The relation between technical measures and policy instruments was assessed in this review by identifying the policy instruments that were adopted to foster the implementation of the different technical measures. This analysis is systematized in Table 6. The detailed technical measures have been enlisted with a corresponding policy instrument. This identification allows for the analysis of which instruments are more often used to promote the implementation of certain measures.

Table 6. Combination of technical measures and policy instruments identified in the literature review.

	Policy Instruments										
Technical Measures	1. MEPS and other standards	2. Technology phase-out	3. Rating system	4. Labelling	5. Certification	6. Information, Education and awareness	7. Grants and subsidies	8. Tax credits/tax incentives	9. Fines and tax increase.	10. Low-interest loans/Third-party financing	11. Prizes and awards
1. Individual/collect-ive optimization of the WHS	X			X							
2. Replacement of WH technologies for more efficient or less-carbon intensive ones							X	X			
3. Installation of solar thermal technologies (to complement WHS)	X		X			X	X			X	
4. Insulation of boilers			X								
5. Heating pipes insulation											
6. Installation of (or replacement for) efficient water appliances			X	X	X						
7. Regular maintenance of Water Heating Systems											
8. Installation of Smart systems	X			X							
9. Reduction in the use of hot water							X				

5. Discussion

The review resulted in the identification of a wide set of technical measures, that go from the installation of devices that improve the system's performance to the replacement of the whole existing system by a new one (more efficient, with a technology shift). The great majority of the technical measures found within the policies and programs reviewed refers to the replacement of the WHS technology or to the installation of devices or equipment that improves the system's performance.

When analyzing the identified measures, it was noted substantial differences and a wide spectrum of possibilities. When assessing the scope of the measures by their applicability in a single vs. multi-apartment, it was found that most measures are not constrained by the type of dwelling, being applicable to both, single and multi-apartment buildings. The installation of solar thermal and photovoltaic systems could be an exception, depending on the roof area available, together with the change from an individual to a collective WHS (Individual/collective optimization of the system). Moreover, when characterizing the measures according to the need for technical or specialized intervention, most of the identified measures require technical expertise to be implemented. There are still a few measures that can be directly implemented by the final user, as the reduction in the use of hot water.

By identifying the policy instruments used to support the implementation of the reviewed WHS policies, it was possible to observe that the instruments mostly used are either economic instruments, education and awareness campaigns or labelling and standards. For instance, regulatory instruments are not commonly adopted in the implementation of policies associated with DHW.

When characterizing the policy instruments under the selected variables, it was found that the competences required for the implementation of each instrument are considerably different. Even so, the competences required for the implementation of instruments under the incentivizing category are relatively similar. This coincidence can be explained by the similarity in their main activities, even if different instruments may require different specific actions for their implementation.

The two most common levels of implementation found are national and regional; not many instruments have been found to explicitly apply a local or international geographical scope.

Regarding the possible challenges or barriers faced by policy makers throughout the instruments' implementation, market-related problems are the most common, whereas the state of the market may be determinant for the instrument choice, e.g., the existing market (or institutional) barriers that are currently hampering the distribution and commercialization of certain technologies.

The large and diverse set of stakeholders that interact with the WHS-related markets may imply a difficulty in choosing the most effective policy instrument(s), as the impact on different stakeholders will vary substantially. Nevertheless, the selection of some instruments may lead to positive externalities, as the economic effect on certain technologies (price decrease), as a consequence of the market development and spread of those technologies (after the use of incentivizing instruments); or the innovation boost derived from the technological research encouragement behind the economic support of certain technologies.

Another important finding of this review refers to the features that need to be taken into account when choosing the technical measures to the promoted and the respective policy instruments. By classifying the instruments and relating them with the technical measures, it was possible to identify how the informative instruments are widely used, both directly and indirectly. Providing information, educating the consumer and other stakeholders and raising the awareness of different actors may be used to demonstrate the benefits and relevance of energy efficiency, help and assist a consumer with the installation of a more efficient device, and even incentivize regular inspections and maintenance work. Informative instruments could also lead to measures that require behavioral changes from final users, as using less hot water.

Moreover, it was also noted that economic instruments are often used to promote the implementation of measures that require engaging consumers, to commit to the policy objectives (partially or totally). Economic instruments are considered to be effective for consumer engagement, probably due to the positive response of consumers to financial rewards, and their willingness to change their energy use in exchange for a monetary compensation [55].

Even so, there are situations where the financial reward may not be enough. For instance, for gauging the attention and commitment of the targeted group, economic stimulus could be insufficient (as when a lack of interest from targeted audience can be experienced). These cases may be overcome by the combination of several policy instruments. For instance, when experiencing a low participation level to initiatives that are based on economic instruments, education and awareness initiatives may be used to foster participation and adhesion to the economic incentives. Therefore, the combination of the instruments should be seen as a possibility, which may have advantages depending on the policy objectives and characteristics of the target group as well as on the available resources.

6. Conclusions

This work led to the identification and classification of nine types of WHS technical measures. These were characterized in terms of scope, level of transformation, level of energy savings and the requirements for training for their implementation. By classifying the measures, it was also possible to understand and systematize the differences among them. After this, decision makers can evaluate their alternatives, taking into consideration the different features.

Moreover, eleven types of policy instruments were also identified and characterized according to the required competences for implementation, the most common levels in which they are implemented, and the challenges that may hamper their implementation. The characterization of policy instruments according to their main features has been performed, enabling policy makers to compare them against the available resources and context characteristics, and to select the most suitable instruments.

Lastly, an identification and assessment of the most common combinations between technical measures and policy instruments was also included in this work. Results show that there has been a preference of using economic instruments to support DHW technologies. A notable relevance and applicability of the "education and awareness campaigns instruments was identified". Complementarity among instruments has been observed, leading to the policy instruments combination with other policy instruments, providing a wider set of alternatives.

- Considering the policy characterization performed in this work, some of the context specificities that should be assessed in order to choose the most adequate measure/instruments include:
- Policy objective (e.g., reduce GHG emissions, energy savings, reduced costs, etc.);
- Main type of dwellings covered by the policy;
- Availability of skilled/trained technical staff;
- Regulatory competences of the entity responsible for the policy implementation;
- Budget/resources availability;
- Policy target group (e.g., end-user, technology provider, technology developer, etc.).

Future work in this topic could include analyzing in detail the cost structure of the measures and instruments (average costs and who pays for it), and identifying/recommending regional differences regarding the choice of instruments to be adopted, based on the local specificities found. In this field there is ongoing work regarding the correlation of policy results with the contextual variables. Starting at the identification of suitable, criteria and indicators to characterize the results [56]. The development of this work should eventually lead to the possibility of adapting policies to the local geography.

Author Contributions: Conceptualization, D.O. and V.L.; formal analysis, D.O., V.L. and I.A.; funding acquisition, D.O.; investigation, D.O.; methodology, D.O. and V.L.; supervision, V.L.; validation, D.O., V.L. and I.A.; visualization, D.O.; writing—original draft, D.O. and I.A.; writing—review & editing, D.O., V.L. and I.A. All authors have read and agreed to the published version of the manuscript.

Funding: This research was funded by the Mexican Consejo Nacional para la Ciencia y la Tecnología grant number 695470, within MIT SES Portugal Doctoral Program.

Conflicts of Interest: The authors declare no conflict of interest.

Abbreviations

DHW	Domestic Hot Water
GHG	Greenhouse Effect Gases
RES	Renewable Energy Sources
SWH	Solar Water Heater
WHS	Water Heating Systems

Appendix A

The identified technical measures have been enlisted with the policy instrument used to implement the policies.

Table A1. Technical measure with policy instrument.

Detailed Technical Measure	Technical Measure Type	Detailed Policy Instrument	Policy Instruments Type	Policy or Document
Device change (from fuel to gas)	Installation/Replacement of heating technologies with more efficient ones	Fiscal/financial incentives	Tax credits/tax incentives	Energy Star Loan Program (SaskEnergy/SaskPower)
				British Columbia Energy Star Water Heater Program + Furnace and Boiler Replacement Program
Device change (from oil boiler; Gas or propane furnace/boiler to more efficient ones)				Warm Up New Zealand: Heat Smart
		Rebate		Insulation Programmes 2009–2018
Installation of renewable-based systems (SWH, photovoltaics, heat pumps, small windmills)				Slovakian incentive programme Green Homes
Installation of solar water heaters, pellet boilers and heat pumps				"Enovatilskuddet" (Enova grants)
Installation of SWH systems, heat pumps, pellets tanks, woodchips equipment		Tax relief		Market Rebate Programme for Renewable Energy
Installation of solar water heaters and heat pumps		Tax credits		Solar water heaters
Installation of Solar water heating systems and other renewable technologies				Eco Fund, Slovenian Environmental Public Fund
Addition Solar thermal systems, biomass boilers and heat pumps		Grant or reimbursable credit		Nova Zelena Usporam (New Green Savings/New Greenlight to Savings)
Solar thermal systems addition	Addition/installation of solar thermal technologies (to complement WHS)	Economic Subsidy		Demoprojekte Solarhaus(Demo Projects Solar House)
				Industrial and Institutional solar thermal collector systems for water heating needs
				Western Balkan Sustainable Energy Financing Facility II (WeBSEFF II)
				Montesol
				Energy Saving programme for residents of the Lviv region for the year 2013-2016
				Prosol II
				Capital subsidy scheme for installation of solar thermal systems

Table A1. *Cont.*

Detailed Technical Measure	Technical Measure Type	Detailed Policy Instrument	Policy Instruments Type	Policy or Document
		Interest-free loans	Low-interest loans/third-party financing	The National Financing Mechanism for Solar Water Heaters
				Mechanism of Renewable Heating Systems and Energy Efficient Measures "Conto termico 2.0"
		Loan and investment grants		Solar Plan
		Low-interest loans		Programme for partial subsidising of purchased and installed solar water heaters for households
		Informative campaigns	Information, education and awareness campaigns	Residential Energy Efficiency Credit Line 3 (REECL 3)
		Rating systems	Rating systems	Energy Efficient Homes Package and Solar Hot Water Rebate
Installation of solar WHS (for new construction buildings)		Other Standards		Andalusia: Sustainable Construction Incentive Programme
Electric hot water cylinder wraps installation Insullation	Insulation of boilers	Rating System	Rating System	Northwest Territories Energy Efficiency Incentive Program
Installation of efficient taps and showerheads	Installation of (or replacement for) efficient water appliances	Building energy codes; Energy Performance Certification; Labelling	Combined (more than one instrument)	Analysis of Energy Efficiency Measures and Retrofitting Solutions for Social Housing Buildings in Spain as a Way to Mitigate Energy Poverty Aranda et al., 2017 [34]
Independent heat substation for heat and hot water preparation; Hot water system balancing; Hot water metering for each final customer of the building; Smart intelligent wireless monitoring and data collection system	Optimization of scale of the system; smart systems installation	Labelling, Codes & Standards		Technical measures to decrease heat energy consumption of final customer in multi-apartment buildings according to Energy Efficiency (Savickas and Bielskus Directive, 2015 [35])
Rinsing dishes with cold water; shorter showers; Insulation of heating pipes	Reduction in the use of hot water; Insulation of heating pipes	Soft Loans + Subsidy		Viable behavioural and technological energy-saving measures (Poortinga and Steg, 2002 [36]

References

1. Novikova, A. *Methodologies for Assessment of Building's Energy Efficiency and Conservation: A Policy-Maker View*; German Institute for Economic Research: Berlin, Germany, 2010.
2. Ürge-vorsatz, D.; Harvey, D.; Mirasgedi, S.; Levine, M.D. Mitigating CO2 Emissions from Energy Use in the World's Buildings. *Build. Res. Inf.* **2007**. [CrossRef]
3. DNV. *Renewables, Power and Energy Use Forecast to 2050*; Det Norske Veritas: Oslo, Norway, 2017.
4. Government of Australia. Water Heating | Energy Rating. 2019. Available online: http://energyrating.gov.au/products/water-heaters (accessed on 22 July 2018).
5. Kegel, M.; Tamasauskas, J.; Sunye, R.; Giguere, D. *Heat Pump Water Heaters in the Canadian Residential Market*; Natural Resources Canada: Varennes, QC, Canada, 2017; pp. 5–15.
6. Research, Fraunhofer Institute for Systems and Innovation; Fraunhofer Institute for Solar Energy Systems; Institute for Resource Efficiency and Energy Strategies GmbH; Observ'ER; Technical University Vienna—Energy Economics Group; TEP Energy GmbH. Quality Review; *Mapping and Analyses of the Current and Future (2020–2030) Heating/Cooling Fuel Deployment (fossil/Renewables)*; European Commission: Brussels, Belgium, 2016.
7. IEA. *Water Heating. Consumer Resources, (June)*; International Energy Agency: Paris, France, 2012; pp. 78–86. [CrossRef]
8. IEA. *Perspectives for the Clean Energy Transition, The Critical Role of Buildings*; International Energy Agency: Paris, France, 2019.
9. Griffiths-sattenspiel, B.; Wilson, W. *The Carbon Footprint of Water*; River Network: Portland, OR, USA, 2009.
10. Eugene Water and Electric Board. *Annual Water Heating Costs & Carbon Footprint*; Eugene Water and Electric Board: Eugene, OR, USA, 2020; p. 80.
11. Denholm, P. *The Technical Potential of Solar Water Heating to Reduce Fossil Fuel Use and Greenhouse Gas Emissions in the United States*; Midwest Research Institute: Golden, CO, USA, 2007.
12. Howlett, M. *What Is a Policy Instrument? Tools, Mixes, and Implementation Styles*; McGill-Queen's University Press: London, UK, 2005.
13. IEA Energy Policy Database. Available online: https://www.iea.org/policies (accessed on 10 February 2018).
14. Solar Thermal World. Available online: https://www.solarthermalworld.org/incentive (accessed on 18 February 2018).
15. Saskenergy. Appliance Financing. Available online: https://www.saskenergy.com/residential/appliancefinancing.asp (accessed on 9 September 2018).
16. FortisBC. Natural Gas Water Heater Rebate. Available online: https://www.fortisbc.com/rebates/home/natural-gas-water-heater-rebate (accessed on 9 September 2018).
17. Energywise. Funding for Heaters and Insulation. Available online: https://www.energywise.govt.nz/funding-and-support/funding-for-heaters-and-insulation/ (accessed on 9 September 2018).
18. Domácnostiam, Z. Podpora využívania obnoviteľných zdrojov energie v domácnostiach. Available online: https://zelenadomacnostiam.sk/sk/ (accessed on 10 September 2018).
19. Enova. Solfanger | Enova. Available online: https://www.enova.no/privat/alle-energitiltak/solenergi/solfanger-/ (accessed on 9 September 2018).
20. BAFA. Homepage. Available online: https://www.bafa.de/DE/Energie/Heizen_mit_Erneuerbaren_Energien/Solarthermie/solarthermie_node.html (accessed on 11 September 2018).
21. Services, R. for E. and W. SWH-national-2011-2017. 2019. Available online: https://www.rews.org.mt/#/en/a/26-swh-national-2011-2017 (accessed on 12 September 2018).
22. Sklad, E. Subvencije in ugodni krediti za okolju prijazne naložbe | Eko sklad. Available online: https://ekosklad.si/ (accessed on 13 September 2018).
23. Bydlení, D. pro úsporné. Nová zelená úsporám. Available online: https://www.novazelenausporam.cz/ (accessed on 19 September 2018).
24. Klima-und Energiefonds. Ausschreibungen. 2019. Available online: https://www.klimafonds.gv.at/ausschreibungen/ (accessed on 13 September 2018).
25. Western Balkans Sustainable Energy Financing Facility. Available online: http://www.webseff.com/ (accessed on 11 September 2018).

26. Ministerio de Industria, E. y M. DNE—Programa de Energía Solar. Available online: http://www.energiasolar.gub.uy/ (accessed on 16 September 2018).
27. ANME. Available online: http://www.anme.tn/ (accessed on 8 September 2018).
28. Conservation, L.C. for E. LCEC—Welcome. Available online: http://lcec.org.lb/ (accessed on 10 September 2018).
29. Economy Ministery. Economy Ministery. Available online: http://www.economy.gov.mk/ (accessed on 12 September 2018).
30. Residential Energy Efficiency Credit Line. Home—Residential Energy Efficiency Credit Line. Available online: http://reecl.org/ (accessed on 14 September 2018).
31. California Solar Initiative. California Solar Initiative—Thermal Program. Available online: https://www.csithermal.com/ (accessed on 17 September 2018).
32. Agencia Andaluza de la Energía. Construcción sostenible | Agencia Andaluza de la Energía. Available online: https://www.agenciaandaluzadelaenergia.es/es/financiacion/incentivos-2017-2020/programa-para-el-desarrollo-energetico-sostenible-de-andalucia/construccion-sostenible (accessed on 19 September 2018).
33. Arctic Energy Alliance Rebates on Energy-Efficient Products. Available online: http://aea.nt.ca/programs/energy-efficiency-incentive-program (accessed on 20 September 2018).
34. Aranda, J.; Zabalza, I.; Conserva, A.; Millán, G. Analysis of Energy Efficiency Measures and Retrofitting Solutions for Social Housing Buildings in Spain as a Way to Mitigate Energy Poverty. *Sustainability* **2017**, *9*, 1869. [CrossRef]
35. Savickas, R.; Bielskus, J. Technical measures to decrease heat energy consumption of final customer in multi-apartment buildings according to Energy Efficiency Directive. *Moksl.—Liet. Ateitis* **2015**, *7*, 461–467. [CrossRef]
36. Poortinga, W.; Steg, L. Viable behavioural and technological energy-saving measures. *Environ. J. Environ. Sci.* **2002**, *17*, 123–135.
37. Itard, L.; Visscher, H. Actual heating energy savings in thermally renovated Dutch dwellings. *Energy Policy* **2016**. [CrossRef]
38. Johnson, A.; Lutz, J.; Mcneil, M.A. *An International Survey of Electric Storage Tank Water Heater Efficiency and Standards*; Southern African Energy Efficiency Convention: Johanesburg, South Africa, 2013.
39. Turiel, I. Present status of residential appliance energy efficiency standards—An international review. *Energy Build.* **2000**, *26*, 5–15. [CrossRef]
40. Van Den Dungen, S. Minimum Energy Performance Standards How Does New Zealand Compare with Other Countries? CSAFE, University of Otago: Dunedin, New Zealand, 2011.
41. UNFSS. *Voluntary Sustainability Standards*; UNFSS: Geneva, Switzerland, 2013; pp. 1–56.
42. Hohne, P.; Kusakana, K.; Numbi, B. A review of water heating technologies: An application to the South African context. *Energy Rep.* **2019**, *5*, 1–19. [CrossRef]
43. Reed, C.A. *National Energy Performance Rating System*; ASHE Magazine: Chicago, IL, USA, 2002; Volume 10.
44. IMT. *A Roadmap for Creating Building Energy Rating Systems in Central Asia I*; IMT: Washington, DC, USA, 2009; Volume 41, pp. 272–278.
45. D'Agostino, D.; Mazzarella, L. What Is a Nearly Zero Energy Building? Overview, Implementation and Comparison of definitions. *J. Build. Eng.* **2019**. [CrossRef]
46. Santiago, H. Promoting behavioral change towards lower energy consumption in the building sector. *Innov. Eur. J. Soc. Sci. Res.* **2011**, *24*, 7–26. [CrossRef]
47. Schaefer, C.; Weber, C.; Voss-Uhlenbrock, H.; Schuler, A.; Oosterhuis, F.; Nieuwlaar, E.; Angioletti, R.; Kjellsson, E.; Leth-Petersen, S.; Togeby, M.; et al. *Effective Policy Instruments for Energy Efficiency in Residential Space Heating? An International Empirical Analysis (EPISODE)*; University of Stuttgart: Stuttgart, Germany, 2000.
48. European Commission. Commission Awards 2003 Prizes to the Best European Projects in Renewable Energy. Available online: https://ec.europa.eu/commission/presscorner/detail/en/IP_04_76/ (accessed on 9 September 2020).
49. WEC. *World Energy Perspectives Energy Efficiency Policies*; World Energy Council: London, UK, 2016; pp. 1–152.
50. LODA/LODA—Lviv Regional State Administration. Available online: https://loda.gov.ua/ (accessed on 19 November 2019).

51. Csobod, É.; Grätz, M.; Szuppinger, P. Overview and analysis of public awareness raising strategies Éva Csobod, 1–33. *J. Build. Eng.* **2008**, *21*, 200–212. [CrossRef]
52. Pusok, K.A.; Morris, J. *Voluntary Energy Standards: ISO 50001 and the Superior Energy Standard*; The Reason Foundation: Los Angeles, CA, USA; Washington, DC, USA, 2018.
53. UNECE. *Mapping of Existing Energy Efficiency Standards and Technologies in Buildings in the UNECE Region*; UNECE: Geneva, Switzerland, 2018.
54. National Housing Bank. *Capital Subsidy Cum Loan Scheme for Installation of Solar Water Heating and Solar Lighting Equipments in Homes*; National Housing Bank: New Delhi, India, 2015.
55. Pothitou, M.; Kolios, A.; Varga, L.; Gu, S. A framework for targeting household energy savings through habitual behavioural change. *Int. J. Sustain. Energy* **2014**, *35*, 686–700. [CrossRef]
56. Leal, V.; Ortiz, D. Energy policy concerns, objectives and indicators: A review towards a framework for effectiveness assessment. *Energies* undereview.

Publisher's Note: MDPI stays neutral with regard to jurisdictional claims in published maps and institutional affiliations.

energies

MDPI

Article

Decomposition Analysis of the Evolution of the Local Energy System as a Tool to Assess the Effect of Local Actions: Methodology and Example of Malmö, Sweden

Isabel Azevedo * and Vítor Leal

Institute of Science and Innovation in Mechanical and Industrial Engineering (INEGI), Faculty of Engineering of University of Porto (FEUP), Rua Dr. Roberto Frias, 400, 4200-465 Porto, Portugal; vleal@fe.up.pt
* Correspondence: iazevedo@inegi.up.pt; Tel.: +351-229-578-710

Abstract: This paper proposes the use of decomposition analysis to assess the effect of local energy-related actions towards climate change mitigation, and thus improve policy evaluation and planning at the local level. The assessment of the impact of local actions has been a challenge, even from a strictly technical perspective. This happens because the total change observed is the result of multiple factors influencing local energy-related greenhouse gas (GHG) emissions, many of them not even influenced by local authorities. A methodology was developed, based on a recently developed decomposition model, that disaggregates the total observed changes in the local energy system into multiple causes/effects (including local socio-economic evolution, technology evolution, higher-level governance frame and local actions). The proposed methodology, including the quantification of the specific effect associated with local actions, is demonstrated with the case study of the municipality of Malmö (Sweden) in the timeframe between 1990 and 2015.

Keywords: energy and climate policy; local energy planning; policy evaluation; local authorities

check for updates

Citation: Azevedo, I.; Leal, V. Decomposition Analysis of the Evolution of the Local Energy System as a Tool to Assess the Effect of Local Actions: Methodology and Example of Malmö, Sweden. *Energies* **2021**, *14*, 461. https://doi.org/10.3390/en14020461

Received: 19 September 2020
Accepted: 13 January 2021
Published: 16 January 2021

1. Introduction

The role of local actions towards the achievement of global targets on climate change mitigation is widely recognized by academics and policy makers. The physical characteristics of local energy systems added to the regulatory competences of local authorities, prompt the local level as an appropriate level of action [1]. The call for a paradigm shift towards more decentralized energy management and use (production/consumption), mentioned in several EU publications (including the Clean Energy Package [2]) and in worldwide discussions, reinforces this relevance of local actions. Local authorities, as the regulatory bodies closest to the citizens, are in a strategic position to promote this transition towards decentralization, including lifestyle changes and local sustainable energy investments.

The recognition of local actions' contribution towards climate change mitigation has been accompanied by an increasingly active role of local actors—including local authorities as well as other local stakeholders, such as local businesses, municipal utilities, citizens associations and individual citizens [3]. For instance, more than 10,000 European municipalities have committed to reduce their GHG emissions level by at least 20%, becoming signatories of the Covenant of Mayors initiative [4,5]. There are also several initiatives promoted at national and EU level with the objective of levering sustainable energy planning and climate change mitigation at the local level, such as the European Energy Award, CIVITAS, Managenergy and others.

While local actions become more and more common, it is necessary to properly assess their impact on greenhouse gas (GHG) emissions, i.e., their contribution towards climate change mitigation. Nonetheless, a thorough assessment of local actions and respective effects is still challenging, given the technical difficulty in properly quantifying the effects of climate related policies. The evaluation of climate policies, in general, is already a challenge given the large interaction with other policies (including economic and demographic associated policies). For instance, ref. [6] illustrates the difficulty in assessing policy effectiveness, focusing on national energy efficiency policies. This difficulty is amplified when dealing with policies implemented at the local level by the fact that they are implemented as a part of a complex multilevel governance framework, which also includes policies implemented at regional, national and international levels. The local energy system is part of a much larger system, influenced by several external factors, and the total change observed is the consequence of all of them. In [7], the authors demonstrate the importance of local characteristics in the definition and implementation of climate change mitigation policies

Current practice regarding the assessment of local actions for climate change mitigation consists mostly of the comparison of the local energy system before and after the policies' implementation, without taking into account external factors, or taking into account, at most, the changes in the national or regional electricity systems [8]. A more complex and rigorous analysis is still not common practice within local policy making, which is a limitation for the evaluation of local actions.

This paper consists of a demonstration of how a more detailed assessment of local actions, taking into account the effects associated external factors by decomposition analysis, can improve current practice on the evaluation of local actions. Herein, local actions refer to energy-related actions implemented in urban environments, i.e., aimed at the reduction of energy-related GHG emissions at the local level (i.e., scope 1 and 2 emissions). The work presented here includes a short review on the state of the art for the evaluation of local actions. The review is then followed by an explanation of how the decomposition of observed changes into the individual effects of the different factors of change, combined with the analysis of the actions taken by local actors, can be used to improve the assessment of local actions. Due to the nature of the urban nature of the most populated areas, and because the paper is energy-related, the emissions model accounts for CO_2 but not for methane. The evaluation methodology is applied to one case study, Malmö, in order to demonstrate its applicability.

The paper is composed by this introduction and four additional sections. The following section is dedicated to a review of the current practice for evaluation of local actions towards climate change mitigation. Then, Section 3 presents an analysis of how the disaggregation of observed changes can improve current evaluation practice, including a summarized description of the disaggregation model that is used and the discussion on how the model can be used as a tool for the assessment of the evolution of local energy systems. Section 4 presents the case study of Malmö (Sweden). Finally, Section 5 summarizes the main findings resulting from this work.

2. Review on the Evaluation of Local Actions for Climate Change Mitigation

The importance of assessing effectiveness, efficiency and relevance of different policies is globally recognized. Policy evaluation is key for the identification and transfer of good practices, as well as for the adjustment of ongoing and development of new policies. There have been several studies focusing on the assessment of energy, climate and environmental policies at different levels, which resulted on the development of a vast set of methodologies to quantify the effects of different policy actions.

Nonetheless, the specific methods for evaluation have not been a recurrent subject for study, at least not as much as would be expected. This might be due to the difficulty in properly assessing their effects. Given the unavoidable interaction with other policies, their contribution to climate change mitigation depends on factors that are not fully controlled by local authorities, and whose disaggregation is neither obvious nor immediate.

Current practice on the evaluation of local actions also reflects this difficulty. The most common assessment of the effects of the Sustainable Energy Action Plans (SEAPs) in the context of the Covenant of Mayors initiative corresponds to the comparison of the energy and emissions inventories before and after the actions' implementation [9,10].

This comparison is relevant, as it provides an insight on the overall evolution of the local energy system, both in terms of energy use and in terms of GHG emissions [11]. However, this is not sufficient to assess the specific impacts or merits of local actions for climate change mitigation.

There are already some studies that go beyond the analysis of observed changes, which provide more detailed analyses of specific actions [12,13]. Such studies use bottom-up models, where the effects are estimated according to the changes caused to the system by the specific policy actions that were implemented at the local level. This alternative approach requires a deep knowledge of the energy system and of the implemented actions, including the changes incurred in the system in physical units (e.g., number of photovoltaic (PV) panels installed, number of persons per kilometer shifted from diesel to electric car, number of appliances/lamps replaced)). Here, the major shortcoming is the difficulty in having the necessary data to establish a reliable and detailed baseline, and to perform a robust bottom-up analysis, especially when assessing the effect of a set of actions covering different sectors.

Thus, in practice, there is still a gap in the assessment of local policy actions, as state-of-the-art practices do not provide methodology for the evaluation of local actions and/or plans which: (1) can be used in a systematic manner with commonly available data and (2) provide an estimate of the real impact of local actions (disaggregated from the effects associated with external factors).

Therefore, taking into account existing limitations from current evaluation practices and considering recent advances in the understanding and representation of the local energy system and respective evolution over time, this paper proposes a new methodology for the assessment of local climate change related actions built upon the top-down decomposition model recently developed and published in [14].

3. Assessing Local Actions Using a Decomposition Analysis of the Evolution of the Energy System

This section describes how the disaggregation of observed changes of a specific local energy system into the individual effects of the different factors of change can contribute to an improved assessment of local actions. First, there is a short description of the model published in [14] and used in the quantification of the individual effects (including the methodology and the main inputs and outputs). Then, an explanation on how this model may be used as an assessment tool for local actions is presented, highlighting innovation compared to current practice.

3.1. Model for the Quantification of Local Actions Effects—Brief Description

The factors of change that are considered include the natural evolution of the local energy system, the higher-level governance framework and local climate change mitigation actions. The natural evolution of the system corresponds to the changes in local population and local economy which are not directly associated with energy and climate policies. Factors associated with higher-level governance framework refer to the energy-related policies implemented at international, national and regional level within the studied timeframe. Finally, local actions comprehend all local energy-related policy actions implemented with the aim of reducing local GHG emissions, including both actions taken under a formal plan and isolated actions. In total, the model disaggregates the observed changes into the effects of 18 individual factors, as identified in Table 1.

Table 1. List of factors of change considered by the disaggregation model for the decomposition of the observed changes in local greenhouse gas (GHG) emissions.

	Explanatory Factor
A	Climate
B	Overall population
C	Population activeness
D	Urbanization
E	Household conditions
F	Transportation habits
G	Local economy (scale and structure)
H	Private consumption
I	Energy prices
Ja	Autonomous savings private consumers
Jb	Energy intensity of economic sectors
K	Energy supply sector
L	Energy taxes
M	Building regulations
N	Transports regulation (Biofuel incorporation)
O	Energy efficiency requirements
P	Other policies
RY	Local Actions

The model itself corresponds to a decomposition analysis that, starting from the energy matrices from the base year, sequentially constructs new energy matrices that result from introducing the changes caused by the different factors of change into the local energy system (Equation (1)). Local actions effects are assumed to be the remaining gap between observed changes and the sum of all the effects from the external factors of change (Equation (2)).

$$\Delta GHG_{\alpha}^{t_0 \to t_1} = \sum_i E_{i,\alpha}^{t_0 \to t_1}, \text{ for } \alpha = \text{municipality and i} = \text{factors of change} \quad (1)$$

$$E_{LocalCCAct,\alpha}^{t_0 \to t_1} = \Delta GHG_{\alpha}^{t_0 \to t_1} - \sum_{i'} E_{i',\alpha}^{t_0 \to t_1}, \text{ for i}' = \text{external factors of change} \quad (2)$$

where ΔGHG is the variation in local GHG emissions between the base year (t_0) and the reporting year (t_1) for the municipality α, and E_i is the effect of factor of change i in local GHG emissions between the base year (t_0) and the reporting year (t_1) for the municipality α.

The individual effects for the different factors of change are estimated by using already known relations (theoretical or empirically proven), taken from the literature, between those factors and the local energy system. These known relations are translated into mathematical formulas that reflect the changes caused by each individual factor of change to the local energy system. Assumptions have been made to adjust the formulas to the existing data and to allow for their applicability in different contexts. Figure 1 presents a schematic representation of the decomposition process and the quantification of the different effects.

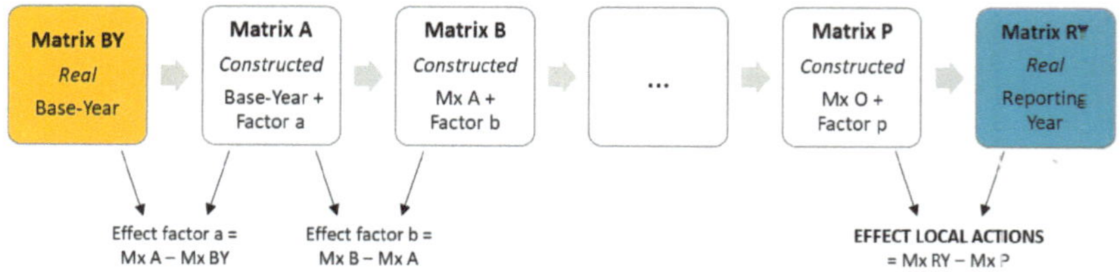

Figure 1. Schematic representation of the decomposition analysis (based on [14]).

The model output consists of the quantification of the individual effects associated with each of the 17 external factors and the overall effect of local climate change mitigation actions in terms of final energy and GHG emissions.

Acknowledging the impossibility of guaranteeing the consideration of all the relevant factors (given the diversity of local contexts) and the risks of assuming the remaining effect as an estimation of the effect of local actions, the methodology includes a set of procedures that improve the robustness of the model results. These correspond to (1) the implementation of the model to municipalities that are within the same (or in a similar) context than the municipality being studied and (2) the comparison between obtained results and reported actions.

A more detailed description of the model, the governing equations and accompanying validation procedures can be found in [14].

3.2. *How to Use the Model as a Tool for the Assessment of Local Actions for Climate Change Mitigation*

The disaggregation model described in the previous section can be a useful tool for the assessment of the local path towards a sustainable energy system. It provides an overview of the observed changes in the local energy system for a specific timeframe, and also a robust estimation of the effect of local actions per sector.

In what concerns the **overall path of the local energy system**, the model outputs can be useful in two distinct ways: first, on the assessment of the evolution of the local energy system over time and second, by assisting on the identification of external synergies and/or barriers regarding the achievement of the established sustainability goals. Concerning the first, the comparison between the energy matrices from the base and the reporting years may provide important insights regarding the evolution of the local energy system, regardless of the factors that have led to the observed changes. This analysis, done per sector and per energy carrier, gives a first overview of the changes that occurred in the local energy system and may result in a set of indicators regarding the unitary energy use and respective GHG emissions and their evolution over time. Concerning the second, as a specific feature of this model, the disaggregation of the observed changes into the individual effects of each factor of change can be used to have a better understanding what contributed more significantly to the observed changes. It can provide relevant insights on the influence of each factor upon the evolution of local GHG emissions, including the identification of which factors are hampering and which are contributing the local path towards a sustainable energy system.

In what concerns the **quantification of the effects associated with the local actions** towards climate change mitigation, their assessment can be used to evaluate the actual contribution of local actions (overall and per sector) for the achievement of a sustainable energy system. The comparison between the results of the disaggregation model and the targets established by local authorities can be seen as a tool to compare what was achieved with what was intended. This analysis may inform decision makers on the need to review, accelerate or simply maintain their actions towards climate change mitigation. On the other hand, a comparison between the ex-ante projections of the implemented actions with the estimated effects provides an assessment of the effectiveness of those actions.

4. Application to the Case Study of Malmö (Sweden)

This section is dedicated to the implementation of the evaluation methodology to the municipality of Malmö (Sweden). This choice was based on the following cumulative criteria: (1) to be a signatory of the Covenant of Mayors, with an approved SEAP (submitted in 2011) and with the first monitoring report already published in 2015; (2) being in the forefront on the availability of data from the local energy system and its evolution over time and (3) having a long history regarding climate change mitigation.

4.1. Characterization of the Case-Study

Malmö is a municipality located in the southern region of Sweden, in Skåne County. In 2015, it had a population of about 323 thousand inhabitants distributed in an area of 157 km^2. The municipality has grown over the last years due to its strategic location—15 min away from Copenhagen (Danish capital) and from Lund (with one of the largest Swedish universities). This growth has been accompanied by a continuous increase in local population, which increased almost 50% over the last 30 years. Malmö is mostly an urban municipality, and the provision of services corresponds to its main economic activity. A brief characterization of the municipality and its evolution between 1990 and 2015 is presented in Table 2.

Table 2. Main socioeconomic indicators of the municipality of Malmö for 1990 and 2015 and their variation over time.

	Base Year 1990		Reporting Year 2015		Variation 1990–2015
Population	233,887		322,574		+38%
Men	111,089	(47%)	158,955	(49%)	
Women	122,798	(53%)	163,619	(51%)	
Less than 15 years old	34,849	(15%)	57,771	(18%)	
Between 15 and 65 years old	149,069	(64%)	215,127	(67%)	
Over 65 years old	49,942	(21%)	49,676	(15%)	
Population density (inhab/km^2)	1491		2056		+38%
Purchase Power per capita *	123.9		184.4		+49%
GVA ** ('000 SEK)	57,612		149,907		+160%

* as a comparison with the national value (i.e., 100 = National average), ** Gross Value-Added, Source: Statistics Central Bureau website (www.scb.se).

In what concerns energy and climate actions, Malmö has been a pioneer, with concerted actions since the 1990s. Its first interventions were done in the context of the Local Investments Program (LIP), a national investment program that promoted the development of local environmental plans and initiatives. Currently, Malmö local administration is committed to become climate-neutral by 2020. Moreover, in the Sustainable Energy Action Plan submitted to the Covenant of Mayors, the city set as a target the supply of local energy use to be exclusively from renewable sources by no later than 2030 [15].

The timeframe considered for this case study is between 1990 and 2015. The choice of the base year is associated with the fact that 1990 was the year considered for the first inventory of local GHG emissions for the municipality of Malmö, being also the base year of the municipality's SEAP. The reporting year corresponded to the most recent year with available statistical data for local energy use. The studied timeframe covers the period when concerted actions taken by the local authority towards climate change mitigation have been more significant.

4.2. Main Input Data

An important stage of the application of the model refers to the collection of input data.

Regarding the construction of the energy matrices, the main sources of input data were the official statistics on energy and end-use consumption published yearly by the Statistics Central Bureau (SCB) [16,17]. Moreover, as there are no empirical data regarding the end-use efficiency for the different energy uses for the municipality of Malmö, the estimations provided by the energy matrices for the Metropolitan Area of Porto [18] (complete and readily available) were used as reference and adapted to the local context. The level of detail from the information available for the Metropolitan Area of Porto, along with the limited impact that the assumptions on end-use efficiency could have on the model results, justify this choice. The values for end-use efficiency were adapted to local context, taking into account existing data on technology options and average efficiency of different appliances in Portugal and Sweden. Table 3 lists the main input data required to build both base and reporting years' energy matrices with the respective data sources.

Table 3. Input data to build the energy and emissions matrices for the municipality of Malmö for 1990 and 2015, with the identification of main sources and data geographical level.

Input Data	Main Source	Geographical Level
Final energy use per sector	SCB (statistics) [16]	Municipality
Distribution of sectorial energy use per end-use	Unander et al. (2004) [19] SEA (Energy in SE 2015) [20] AdEPorto (Matrix 2009) [18]	n.a. Country Country
Average end-use efficiency	AdEPorto (Matrix 2009) [18]	n.a.
Efficiency of energy supply sector	SCB (energy balances) [21] SCB (hydro and wind corrections)	Country/Municipality Country
Carbon content of electricity, heat and oil products	SCB (energy balances) [21] SCB (hydro and wind corrections)	Country/Municipality Country
Carbon content of other energy carriers	IPCC [22]	n.a.

In what concerns the disaggregation of the individual effects per factor of change, an effort was made to collect data as accurate as possible (i.e., specific to the municipality and to the studied timeframe) in order to reduce uncertainty. When data was not available for the specific geographical context, the collected data corresponded to the closest territorial level to which the data was available and, if necessary, the values were adapted to the municipality level using downscaling indicators. Moreover, when specific values for base and/or reporting years were not found, these were estimated from the existing values using trend analysis. Table 4 lists the collected input data along with the respective sources and geographical level.

Table 4. Input data for the decomposition analysis of the municipality of Malmö with the identification of main sources and data geographical level.

Factor of Change	Indicators	Sources	Data Geographical Level
Climate	Heating Degree-Days Cooling Degree-Days	Eurostat [23] Eurostat [23]	Region Region
Population	Local inhabitants	SCB [24]	Municipality
Population activeness	Share of active population Ratio of energy use between households with active and inactive inhabitants per sector	SCB [25] Eurostat [26]	Municipality Country
Urbanization	Share of households in rural area Ratio between energy usage of rural and urban households per sector	- Eurostat [26]	- Country
Household conditions	Number of persons per household Appliances ownership per capita Average household size	SCB [27] Unander et al. (2004) [19] Eurostat [28]	Municipality Country Country
Transportation habits	Average commuting time (min) Share of travels and duration per motive Car ownership ratio Elasticity of leisure travel with car ownership	National Travel Survey [29] National Travel Survey [30] SCB [31] ESRC * [32]	Type of city Country Country Multi-Country
Local economy (scale and structure)	GVA per sector	SCB [33]	Region

Table 4. *Cont.*

Factor of Change	Indicators	Sources	Data Geographical Level
Private consumption	Private purchase power per capita	SCB [34]	Municipality
	Elasticity of residential energy use with private income	Krishnamurthy and Kriström (2015) [35]	Country
	Elasticity of passenger transportation energy use with private income	ESRC [32]	Multi-Country
	Elasticity of freight transportation energy use with private income	ESRC [32]	Multi-Country
Energy prices	Average annual price per energy vector	SEA [20]	Country
	Elasticity of residential energy use with energy prices	Krishnamurthy and Kriström (2015) [35]	Country
	Elasticity of passenger transportation energy use with energy prices	ESRC [32]	Multi-Country
Autonomous savings private consumers	Energy intensity of residential sector for EU countries	Eurostat [36]	EU27
	Energy intensity of passenger transportation for EU countries	Eurostat [36]	EU27
Energy intensity of economic sectors	Energy intensity of services, industry and agriculture and fisheries sectors for EU countries	Eurostat [36]	EU27
Energy supply sector	Annual energy balance	SCB [21,37]	Country/Municipality
	Hydro production normalization index	SCB	Country
	Eolic production normalization index	SCB	Country
Energy taxes	Energy-policy related taxes per vector	SEA [20]	Country
	Elasticity of residential energy use with energy prices	Krishnamurthy and Kriström (2015) [35]	Country
	Elasticity of passenger transportation energy use with energy prices	ESRC [32]	Multi-Country
Building regulations	Average Nic of residential dwellings per building period—before regulation	Hiller (2003) [38] IEA	Country
	Average Nic of residential dwellings per building period—after regulation	IEA ** (Swedish Building Codes)	Climatic Zone
	Average IEE *** of services buildings per building period—before regulation	Hiller (2003) [38] IEA	Country
	Average IEE of services buildings per building period—before regulation	IEA (Swedish Building Codes)	Climatic Zone
	Average dwellings area per building type and building period	SCB [39]	Municipality
	Number of dwellings built and rebuilt per building period	SCB [39]	Municipality
Transports regulations	Obligation to introduce biofuels—shares per year	Legislation [40]	Country
	Substitution of traditional vehicles for other running with biofuels	Progress report under Kyoto [41]	Country
Energy efficiency requirements	Average annual consumption of existing stock per appliance	EU (impact assessment reports) [42–51]	EU level
	Minimum requirements of annual consumption per appliance	EU (impact assessment reports) [52–61]	EU level
Other policies		Progress reports under Kyoto [41,62] Stenqvist and Nilsson (2012) [63] SEA (Energy in Sweden 2015) [20]	Country Country Country

* ESRC—Economic and Social Research Council, Transport Studies Unit; ** IEA—International Energy Agency; *** IEE—Energy Efficiency Indicator.

Recognizing the relevance that assumptions and input data may have on the model results, the methodology includes a fine-tuning process that, as much as possible, ensures the adequacy of the assumptions and the reliability of the input data, in order to guarantee the robustness of the results. The decomposition model was applied to six surrounding municipalities (Helsingborg, Kristianstad, Landskrona, Lund, Trelleborg and Ängelholm) and the obtained results were assessed individually, and in comparison, with each other. This exercise provided additional validation of the results obtained for the municipality of Malmö, by assessing whether plausible results can be obtained for the other municipalities when using the same inputs' data sources and assumptions.

4.3. Model Results for Malmö

4.3.1. Overall Evolution between 1990 and 2015

Figure 2 shows the change in annual final energy per sector (GWh) for the municipality of Malmö, between 1990 and 2015. In what concerns final energy, the overall energy use remained fairly constant over the considered timeframe. However, changes were observed at the sectorial level, with the exception of the transport sector (both passenger and freight) where the use of energy was kept almost constant. The services sector is the only sector with a significant increase in energy use (30%). The remaining sectors show a decrease in energy use over the studied timeframe. The sector with the highest absolute decrease is the residential sector, reducing about 290 GWh/year, what corresponds to 13% of the sector's energy use in the base year. Nonetheless, the agriculture and fisheries sector is the sector with the highest relative decrease, reducing to 55% of the base year's energy use by 2015.

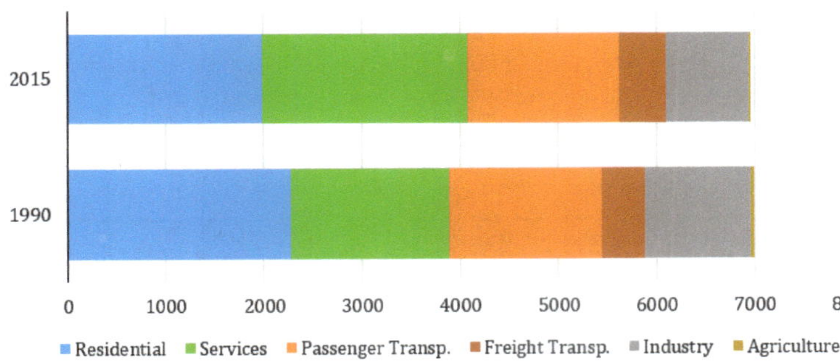

Figure 2. Change in annual final energy per sector (GWh) for the municipality of Malmö, between 1990 and 2015.

The evolution of local GHG emissions between 1990 and 2015 shows a downward trend over all the sectors, as shown in Figure 3. GHG emissions refer to CO_2, CH_4 and N_2O; and the level of emissions in $tCO_{2\,eq.}$ is estimated based on the carbon content of the different energy carriers and the respective global warming potential, as reported in the IPCC Guidelines for National Greenhouse Gas Inventories [22]. Overall, the emissions in the reporting year correspond to 78% of the emissions in the base year. This reduction seems to be a combination of the shift to less carbon intensive energy carriers with the reduction of the carbon content of fossil heat distributed through district heating. Indeed, it is noticeable that, while final energy use in the services sector has increased, the equivalent GHG emissions decreased (almost 30%), due to the significant weight of district heating on the sectorial energy use. In addition, the decrease in GHG emissions in the residential and industry sectors (39% and 30%) is more significant than the decrease in the equivalent final energy use (13% and 22%).

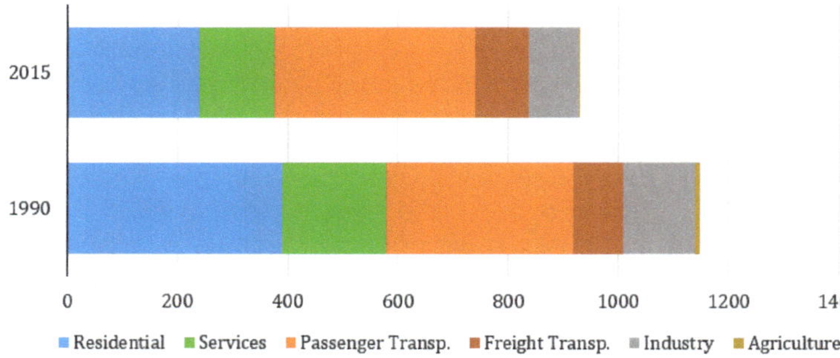

Figure 3. Change in annual GHG emissions per sector (kton$CO_{2\ eq.}$) for the municipality of Malmö, between 1990 and 2015.

Summarizing, the observed changes in the local energy system between 1990 and 2015 correspond to an almost negligible change in final energy use and a considerable decrease in the respective GHG emissions. The reduction of GHG emissions (19%) may be justified by the decarbonization of the heat production sector and by slight shifts of energy carriers. Despite the significant increase in the number of local inhabitants within the studied timeframe, the level of final energy use was kept constant and the equivalent GHG emissions decreased, which implies a decrease in the level of final energy use and GHG emissions per capita. The average use of final energy per capita decreased almost 30% between 1990 and 2015, while the observed reduction in GHG emissions per capita (41%) was considerably higher than the overall reduction (19%). The main energy indicators for the municipality of Malmö are presented in Table 5, for both base year and reporting year, in order to provide an overview of the observed evolution of the local energy system.

Table 5. Energy indicators for the base year (1990) and the reporting year (2015) for the municipality of Malmö.

Main Indicators of the Local Energy System	1990	2015	Variation
Total final energy use (GWh)	6992	6962	−0.4%
Final energy use per capita (MWh/cap)	29.9	21.6	−28%
Total GHG emissions ($tCO_{2\ eq.}$)	1,147,659	929,051	−19%
GHG emissions per capita ($tCO_{2\ eq.}$/cap)	4.9	2.9	−41%
Final energy for residential buildings (GWh)	2280	1988	−13%
Final energy for services buildings (GWh)	1606	2084	+30%
Final energy for transports (GWh)	1995	2030	+1.8%
Final energy for industry (GWh)	1074	839	−22%
Final energy for agriculture and fisheries (GWh)	36	20	−45%
Sector responsible for the highest share of energy use	Residential Buildings 33%	Services Buildings 30%	
Sector responsible for the highest share of GHG emissions	Residential Buildings 34%	Passenger Transport. 39%	
Most used energy carrier	Oil Prod. 35%	Electricity 35%	

4.3.2. Disaggregated Effects Associated with the Several Factors of Change

When applying the decomposition analysis to the case study of Malmö, the changes that occurred in the local energy system between 1990 and 2015 were disaggregated into 17 external factors plus the effects associated with local climate change mitigation actions [14]. Figure 4 presents the estimated effects, in terms of GHG emissions, per factor of change.

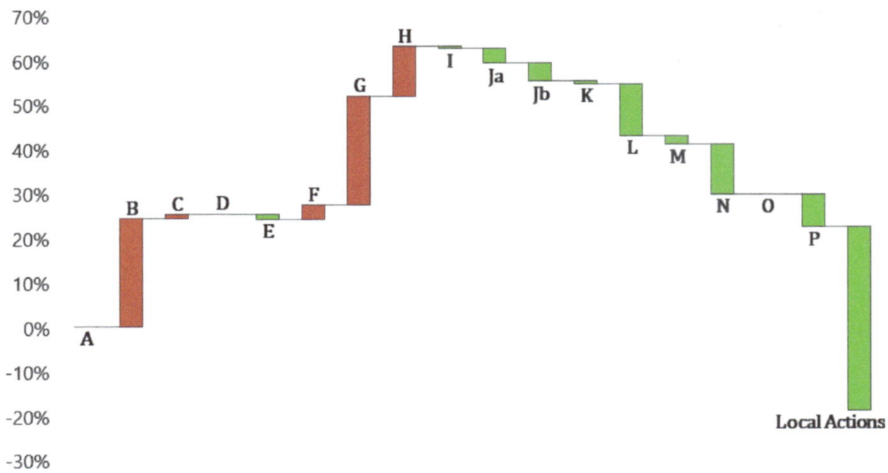

Figure 4. Disaggregation of the observed changes in local GHG emissions, between 1990 and 2015, per factor of change for the municipality of Malmö, as a percentage of the base year emissions level. (A—Climate; B—Population; C—Activeness; D—Urbanization; E—Household conditions; F—Transportation habits; G—Local economy; H—Private consumption; I—Energy Prices; Ja—Autonomous savings private consumers; Jb—Energy intensity economic sectors; K—Energy supply sector; L—Energy taxes; M—Building regulations; N—Transport regulations; O—Energy efficiency requirements; P—Other policies; Local Actions).

From the analysis of the graphic, it is clear that the largest contributions towards the reduction of local GHG emissions are:

(1) Energy taxes, which are all the taxes applicable to energy products, which are associated with energy and climate policies. In Malmö, these include the energy tax (with the goal of improving energy efficiency), the CO_2 tax (implemented to promote the shift to less carbon intensive energy carriers) and a special tax (associated with GHG emissions, other than CO_2)).

(2) Transport regulations, which in the specific case of Malmö include the incorporation of biofuels in the traditional fuel mixtures and the incentives implemented to promote the substitution of traditional fuels light-weight vehicles with vehicles running with biofuels).

(3) Local climate change mitigation actions. Nonetheless, the contribution of local actions is significantly higher than the ones from the remaining factors. This could be explained by the long history of Swedish municipalities, in general, and Malmö, specifically, in having an active role towards climate change mitigation, a role that has been highly supported by national initiatives.

It was also estimated by the model that a few factors of change had a counteracting effect towards the reduction of GHG emissions. Here, the demographic evolution and the changes in activity level and structure of the local economy were the ones with a more significant effect, corresponding to an increase of 24.1% and 24.5%, respectively, of local GHG emissions compared to the base year's level.

Moreover, when looking at Figure 4, it is worth noticing that the first insights on the overall evolution of the local energy system did not reveal the changes that occurred in the demographic and socio-economic dynamics of the municipality. Even if the energy matrices from the base and reporting years did not reveal abrupt changes in the local energy system, the disaggregation into individual effects showed that there were, in fact, significant changes in the activity level and structure of the local economy as well as in the population size, and that these were counteracted by the effects of other factors of change.

4.4. Assessment of the Local Path towards Climate Change Mitigation

4.4.1. SEAP Malmö—Proposed Actions and Implementation Status

The municipality of Malmö was a pioneer in recognizing the importance of local actions towards climate change mitigation and environmental sustainability. Since the 1990s, the local authority has had an active role on the promotion of climate change mitigation by acting on their own infrastructure and by prompting others to act. Since then, there different plans and strategies were developed towards a sustainable future in terms of energy (but not only).

On the vision and goals established for Malmö's energy system, the most recent publication refers to the Covenant of Mayors adhesion, in 2008, where the municipality established the target to reduce local energy use per capita by 20% (compared to average annual use during the period 2001 to 2005) and GHG emissions by 40% (compared to the 1990 level) by 2020. In the Sustainable Energy Action Plan submitted to the Covenant (herein, SEAP Malmö), Malmö's city authority also committed to at least 50% of the local authority's energy use to be from renewable energy sources, with a large proportion being produced locally [15]. The proposed actions covered the different sectors and focused on achieving a more effective energy use and switching to renewable energy sources. Table 6 presents a summary of the technical measures proposed by the local authority.

From 1997 up to today, the city authority publishes an annual report on the local actions towards local environmental sustainability [64–80]. These documents allowed for an appropriate understanding of the efforts being done to achieve the vision described above. Appendix A (Figure A1) presents a summarized timeline of the actions taken by the local authority towards a more sustainable energy system. Between 2013 and 2015, a survey on the status of implementation of the actions proposed by the city authority was performed. The respective activity reports present an overview of the survey outcomes, concluding that there is still a large number of actions ongoing or not yet implemented that should have been completed by the end of 2015.

Table 6. Technical measures proposed in the Sustainable Energy Action Plan (SEAP) Malmö.

Technical Measures (as Proposed by the Local Authority)
Decarbonize local electricity production
Decarbonize local heat production
Buildings Reduce energy needs Use of energy from renewable energy sources for heat Microgeneration from renewable energy sources
Municipal buildings Reduce energy needs Use of energy from renewable energy sources Reduce electricity use Microgeneration from renewable energy sources
Transports Reduce transportation needs Use of biofuels in public transports Shift to soft modes Shift to public transports Use of biofuels in private cars Optimization of transportation flow
Industry Decrease energy intensity Increase energy efficiency

4.4.2. Overall Pathway

Table 7 presents a comparison between the observed changes and the goals established for 2020 by the local authority in the SEAP Malmö, for the local energy system and specifically for the local public sector. This comparison shows that the municipality is indeed moving towards the established goals, especially in what concerns the reduction of final energy use per capita. The evolution of the final energy use associated with the infrastructure owned by the municipality is the exception, given that an increase of 13% in the respective energy use was observed while the established goal was a reduction of 30%. However, this evolution can be justified by the increase in local energy services, due to the municipality's population growth. The comparison between observed changes (as assessed by the model) and the projected evolution of local GHG emissions provides an insight on the alignment of the actual evolution of the local energy system with the goals established for the municipality of Malmö (also considering national and international initiatives). This corresponds to the assessment of the overall distance to the projected pathway, regardless of the factors of change that have caused such evolution.

Table 7. Comparison of the overall GHG emissions reduction observed (1990–2015) and the goals established by the municipality of Malmö in the SEAP Malmö (1990–2020).

	SEAP Malmö By 2020	**Observed Changes By 2015**
Local energy system		
Reduction of final energy per capita (2001–2005 level)	20%	17%
Share of energy from renewable sources	50%	24%
Reduction of GHG emissions (1990 level)	40%	19%
Energy use in municipality owned infrastructure		
Reduction of final energy (2001–2005 level)	30%	−13%
Share of energy from renewable sources	100%	61%

Regarding the reduction of local GHG emissions, the observed reduction by 2015 corresponds to about half of the reduction target established for 2020. Moreover, as visible in the plot presented in Figure 5, if a linear evolution over time is considered (dotted line), it is possible to say that the observed reductions in local GHG emissions are behind what was projected by the SEAP. However, considering a trajectory where the decrease in GHG emissions would be smaller up to 2008 (when Malmö adhered to the Covenant of Mayors), the observed reductions would be in line with the established goal for 2020.

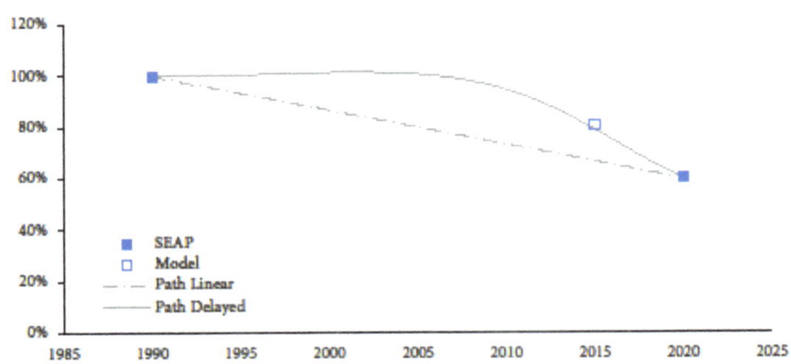

Figure 5. Projected reduction of local GHG emissions by 2020 as presented in the SEAP Malmö and the observed reduction in 2015 (compared to 1990 level), and two possible paths to achieve the desired reduction by 2020.

4.4.3. Local Actions' Contribution

The expected contribution of each specific measure (or group of measures) towards the goals established by the local authority is not specified in the Environmental Program of Malmö. Thus, as referred previously, it is not possible to compare the projected outcomes with the estimated effects of local actions. Being so, a comparative analysis between the estimated effects associated with local actions and the overall goals established by the municipality is performed instead. This analysis is also useful to understand the level of contribution of local actions towards the overall targets.

On one hand, it was estimated that the effects associated with local actions between 1990 and 2015 corresponded to a reduction in GHG emissions of over 470 thousand tons of $CO_{2\ eq.}$, which is equivalent to 41% of the base year's level of emissions. On the other hand, the municipality established the goal of reducing GHG emissions by 40% compared to 1990 level, by 2020. Thus, it could be said that the contribution of local actions already exceeds the reduction targets established by the local authority, even if the overall reduction of GHG emissions is still behind.

Indeed, as visible in Figure 6, in the case of Malmö, the sum of the effects associated with all the external factors (including natural evolution of the local energy system and higher-level policies) corresponds to an increase in the municipality's GHG emissions. This may be justified by the high growth in terms of local demography and economy, which neutralizes the positive effects of national and international policy actions. Moreover, the overall effect of external factors partially counteracts the contribution of local actions towards climate change mitigation. Thus, the mere assessment of the overall change in GHG emissions would lead to an underestimation of the contribution of local actions towards climate change mitigation.

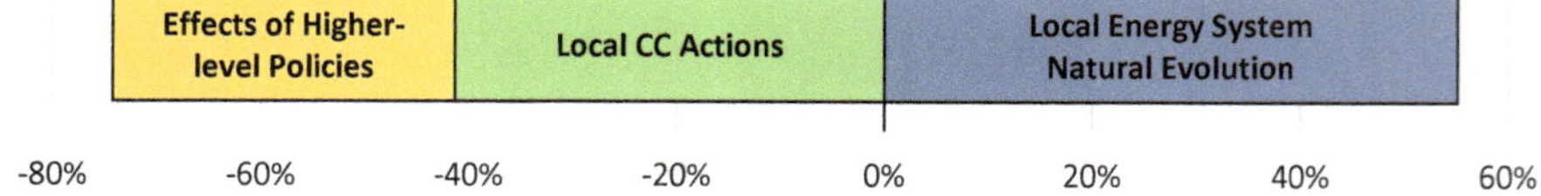

Figure 6. Change in GHG emissions for the municipality of Malmö ($ktCO_{2\ eq.}$/year) due to local actions and to external factors and the overall change as estimated by the disaggregation model (1990–2015) and as projected by the SEAP (1990–2020).

This analysis illustrates the importance of disaggregating the observed changes at the local level in order to properly understand and assess the effect of local actions and the respective contribution on the overall reduction of local GHG emissions. In this specific case, the assessment of the overall reduction of GHG emissions could be misleading, as it would hint that the actions promoted by the municipality of Malmö did not contribute towards climate change mitigation, while the actual contribution of local actions, even if behind what was planned and proposed in the SEAP, was still significant.

4.5. Main Findings

From the comparison between the estimated results and the goals defined by local authority, it was observed that the level of local GHG emissions is still far from what is targeted by the municipality. Nonetheless, the values per capita seem to be more satisfactory. The total level of emissions is not justified by ineffectiveness on action but rather by the actual population growth, which is higher than what was considered when developing the SEAP Malmö.

Moreover, when disaggregating the observed changes into the effects of local actions and the ones of external factors of change, the analysis shows that the local contribution to the reduction of GHG emissions nearly reached the proposed reduction goal. The gap between the accomplished and desired GHG emissions reduction is due to the effects associated with external factors, which counteracted the effect associated with local climate change mitigation actions.

Thus, this comparative analysis showed the importance of properly quantifying the actual effects of local actions. Indeed, the difference between the two comparative analyses revealed how the mere assessment of observed total changes can be misleading.

5. Conclusions

This work proposes a methodology for the assessment of the local path towards climate change mitigation, and of the contribution of local actions towards this path, based on the disaggregation model presented by Azevedo [14]. The model provides a decomposition of the observed total changes of a local energy system into the individual effects of the several factors of change, including external factors and local actions for climate change mitigation. This may be an important asset for the evaluation of local energy systems and their evolution over time. The deeper understanding of what is behind the observed total change in local GHG emissions can help local authorities, as well as higher-level policy makers, to assess current policies and develop future ones.

The model was applied to the municipality of Malmö, for the timeframe 1990–2015.

On the overall change, an almost negligible change in final energy use and a reduction of 22% in the level of GHG emissions was observed. Given the large increase in the number of local inhabitants in the studied timeframe, the changes in energy use and associated GHG emissions per capita were considerably different, with a reduction of 28% and 41% respectively.

When disaggregating the observed change into the effects of the different factors of change, it was estimated that demographic evolution and changes in the local economy were the main drivers for increase of the GHG emissions. On the opposite trend, local climate change mitigation actions were the main driver for decrease in GHG emissions between 1990 and 2015 (41%). Initiatives promoted at the national level, as new energy taxes and transport regulations, also contributed to the reduction in the local GHG emissions, by 12% and 11% respectively.

The fact that the estimated impact of local actions on GHG emissions was a decrease of 41%, while the total observed decrease was only 22%, clearly shows the importance of disaggregating the contribution of local actions from the effects associated with external factors. The assessment of the overall change alone can be misleading, not being representative of the real contribution of local actions.

Moreover, the disaggregation by sector and energy carrier estimated by the model can also be used to assess the contribution of specific local actions towards the reduction of local GHG emissions. The comparison between projected outcomes and the estimated effects may be seen as a tool to assess whether the effects of already implemented actions are in line with the respective projections, identifying which mechanisms are effective and which need to be improved or even replaced.

Overall, this paper, with the implementation of the model to the municipality of Malmö, shows the importance of going beyond the quantification of the changes in the local level of GHG emissions over time. The assessment of the contribution of local policy actions towards climate change mitigation could be significantly improved with the disaggregation of the overall change in GHG emissions into the individual effects of the different factors of change.

Finally, considering that the proposed methodology only provides an estimate of the overall effect of local actions, further research on the individual contribution of specific policy instruments is still needed. Indeed, it would be beneficial to estimate the actual contribution of the different policy mechanisms implemented at a more disaggregated level to better assess their effectiveness and efficiency. In order to do so, it would be interesting to develop a bottom-up methodology that would use the overall effects of local actions as well as detailed information on the actions' implementation to estimate these individual effects and to assess their contribution in terms of effectiveness and efficiency. Another area where more work may be useful is the assessment of the precision of input data, and of the typical impacts of data lack of precision upon the model results.

Author Contributions: Conceptualization, I.A. and V.L.; formal analysis, I.A. and V.L.; funding acquisition, I.A.; investigation, I.A.; methodology, I.A.; supervision, V.L.; validation, I.A. and V.L.; visualization, I.A.; writing—original draft, I.A.; writing—review & editing, I.A. and V.L. All authors have read and agreed to the published version of the manuscript.

Funding: The research was funded by "Fundação para a Ciência e Tecnologia" (FCT—Portugal), through the PhD Scholarship PD/BD/105862/2014, granted in the frame of the MIT Portugal Program.

Data Availability Statement: The data presented in this study are available in Azevedo (2019).

Conflicts of Interest: The authors declare no conflict of interest.

Appendix A. Local Actions of the Municipality of Malmö

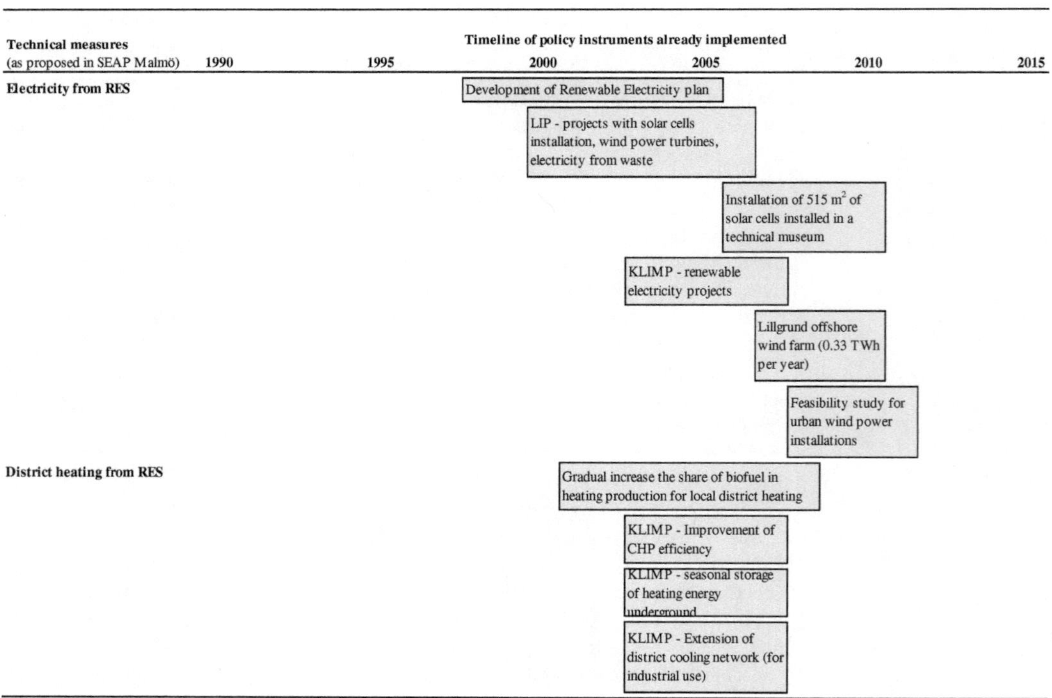

Figure A1. *Cont.*

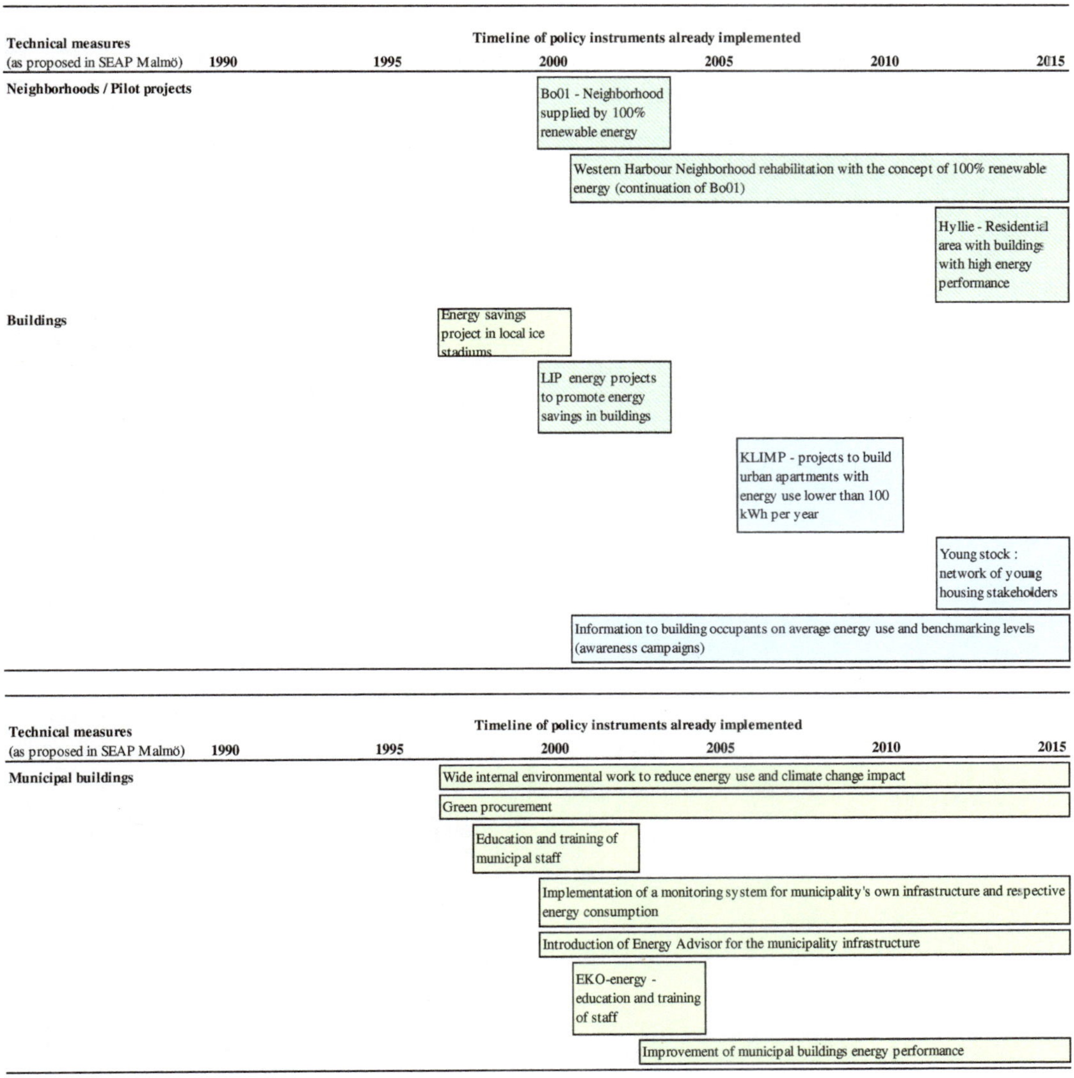

Figure A1. *Cont.*

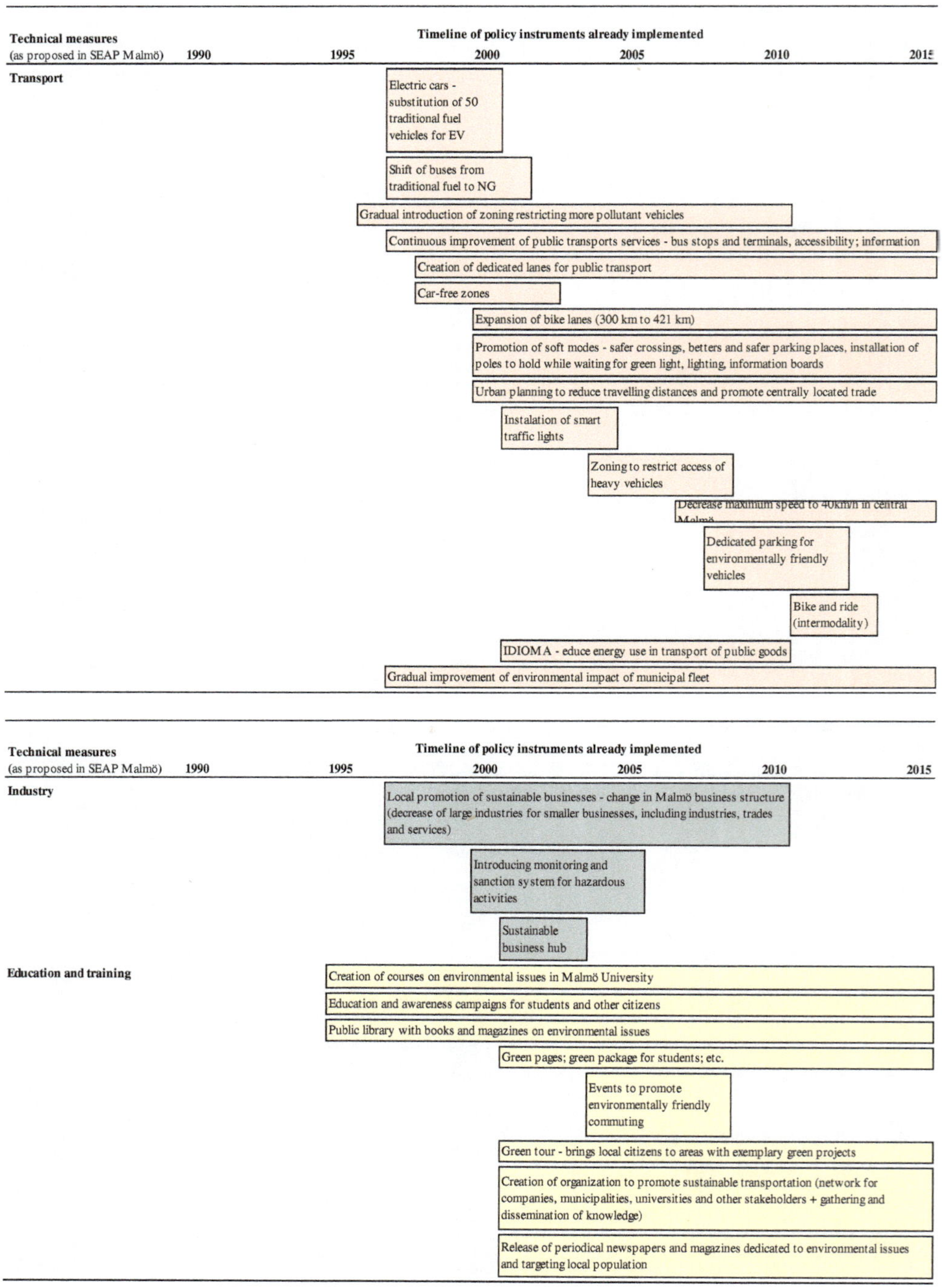

Figure A1. Timeline of policy instruments implemented by the municipality of Malmö per technical measure, between 1990 and 2015.

References

1. Meeus, L. Smart Cities initiative: How to foster a quick transition towards local sustainable energy systems. In *THINK Project, EU 7th Framework Programme*; European University Institute: Florence, Italy, 2011.
2. EC, European Commission. *Clean Energy for all Europeans*; Publications Office of the EU: Luxembourg, 2019.
3. Kern, K.; Alber, G. Governing climate change in cities: Modes of urban climate governance in multi-level systems, in Competitive cities and climate change. In Proceedings of the 2008 OECD Conference Proceedings, Milan, Italy, 9–10 October 2008; pp. 171–196.
4. CoM, Covenant of Mayors, Covenant of Mayors—Official Website. 2020. Available online: https://www.eumayors.eu/en/ (accessed on 14 September 2020).
5. Domorenok, E. Voluntary instruments for ambitious objectives? The experience of the EU Covenant of Mayors. *Environ. Politics* **2019**, *28*, 293–314.
6. Bertoldi, P.; Mosconi, R. Do energy efficiency policies save energy? A new approach based on energy policy indicators (in the EU Member States). *Energy Policy* **2020**, *139*, 111320.
7. Palermo, V.; Bertoldi, M.P.; Apostolou, A.K.; Rivas, S. Assessment of climate change mitigation policies in 315 cities in the Covenant of Mayors initiative. *Sustain. Cities Soc.* **2020**, *60*, 102258. [CrossRef]
8. AdEPorto and CMPorto. *Plano de Acção para a Energia Sustentável da Cidade do Porto*; AdEPorto: Porto, Portugal, 2010.
9. Millard-Ball, A. Do city climate plans reduce emissions? *J. Urban Econom.* **2012**, *71*, 289–311. [CrossRef]
10. DG Energy. *European Commission, Energy Solutions for Smart Cities and Communities—Lessons Learnt from the 58 Pilot Cities of the CONCERTO Initiative*; Steinbeis-Europa-Zentrum der Steinbeis Innovation gGmbH, Ed.; European Commission: Brussels, Belgium, 2014.
11. Hsu, A.; Tan, J.; Ng, Y.M.; Toh, W.; Vanda, R.; Goyal, N. Performance determinants show European cities are delivering on climate mitication. *Nat. Clim. Chang.* **2020**, *10*, 1015–1022. [CrossRef]
12. de Melo, C.A.; Jannuzzi, G.d.M.; Ferreira Tripodi, A. Evaluating public policy mechanisms for climate change mitigation in Brazilian buildings sector. *Energy Policy* **2013**, *61*, 1200–1211. [CrossRef]
13. Lin, J. Evaluating the effectiveness of urban energy conservation and GHG mitigation measures: The case of Xiamen city. *China. Energy Policy* **2010**, *38*, 5123–5132. [CrossRef]
14. Azevedo, I. A methodology for ex-post evaluation of local climate change mitigation actions under a multi-level governance framework. In *Faculty of Engineering*; Universidade do Porto: Porto, Portugal, 2019.
15. Ruben, A.; Pitkä-Kangas, L. *Environmental Programme for the City of Malmö 2009-2020. Malmö Stad Miljoprogram*; Malmö Stad: Malmö, Sweden, 2012.
16. SCB, Statistiska Centralbyrån. Energidata (MWh) Efter Region, Kategori, Energityp och år (1990). 2018. Available online: http://www.statistikdatabasen.scb.se/pxweb/sv/ssd/START__EN__EN0203/EnergiKommKat/?rxid=a88e0C4e-ac34-496c-a2a6-43498221fd7a (accessed on 8 February 2018).
17. SCB, Statistiska centralbyrån. Slutanvändning (MWh) Efter Region, Förbrukarkategori, Bränsletyp och år (2015). 2018. Available online: http://www.statistikdatabasen.scb.se/pxweb/sv/ssd/START__EN__EN0203/SlutAnvSektor/?rxid=a88e004e-ac34-496c-a2a6-43498221fd7a (accessed on 8 February 2018).
18. AdEPorto, Agência de Energia do Porto. *Matriz de energia da Área Metropolitana do Porto—Norte do Douro*; AdEPorto: Porto, Portugal, 2014.
19. Unander, F. Residential energy use: An international perspective on long-term trends in Denmark, Norway and Sweden. *Energy Policy* **2004**, *32*, 1395–1404. [CrossRef]
20. SEA, Swedish Energy Agency. *Energy in Sweden*; SEA: Stockholm, Sweden, 2015.
21. SCB, Statistiska centralbyrån. Elproduktion och bränsleanvändning (MWh), Efter Län Och Kommun, Produktionssätt Samt Bränsletyp. 2018. Available online: http://www.statistikdatabasen.scb.se/pxweb/sv/ssd/START__EN__EN0203/ProdbrEl/?rxid=092d056e-8b3a-4223-aa6e-7a0a76625bd6 (accessed on 30 January 2018).
22. IPCC. *2006 IPCC Guidelines for National Greenhouse Gas Inventories*; IPCC: Geneva, Switzerland, 2006.
23. Eurostat. *Cooling and Heating Degree days by NUTS 2 (Annual Data)*; Eurostat Database: Luxembourg, 2018.
24. SCB, Statistiska Centralbyrån. Population by Region, Marital Status, age and Sex. Year 1968–2017. 2018. Available online: http://www.statistikdatabasen.scb.se/pxweb/en/ssd/START__BE__BE0101__BE0101A/BefolkningNy/?rxid=a88e004e-ac34-496c-a2a6-43498221fd7a (accessed on 23 January 2018).
25. SCB, Statistiska centralbyrån. Inactive Population as a Percentage of the Total Population. 2018. Available online: http://www.statistikdatabasen.scb.se/ (accessed on 15 March 2018).
26. Eurostat. *Mean Consumption Expenditure per Adult Equivalent*; Eurostat Database: Luxembourg, 2017.
27. SCB, Statistiska Centralbyrån. Number of Dwellings by Region and Type of Building (Including Special Housing). 2018. Available online: http://www.statistikdatabasen.scb.se/pxweb/en/ssd/START__BO__BO0104/BO0104T01/?rxid=a88e004e-ac34-496c-a2a6-43498221fd7a (accessed on 24 January 2018).
28. Eurostat. *Average Household Size—EU-SILC Survey*; Eurostat: Luxembourg, 2017.
29. Trafikanalys. RVU Sverige—Den Nationella Resvaneundersökningen. 2018. Available online: https://www.trafa.se/kommunikationsvanor/RVU-Sverige/ (accessed on 1 February 2018).
30. *Trafikanalys, RVU Sverige—den Nationella Resvaneundersökningen 2014–2015*; Trafikanalys & Sveriges Officiella Statistic: Stockholm, Sweden, 2016.

31. SCB, Statistiska Centralbyrån. Passenger Cars in Use by Region and Type of Ownership. Year 2002–2017. Available online: http://www.statistikdatabasen.scb.se/pxweb/en/ssd/START__TK__TK1001__TK1001A/PersBilarA/?rxid=a88e004e-ac34-496c-a2a6-43498221fd7a (accessed on 24 January 2018).
32. Hanly, M.; Dargay, J.; Goodwin, P. *Review of Income and Price Elasticities in the Demand for Road Traffic*; University College London: London, UK, 2002.
33. SCB, Statistiska Centralbyrån. Bruttoregionprodukt (BRP, ENS2010), Löpande Priser, Mnkr Efter Region och år. 2018. Available online: http://www.statistikdatabasen.scb.se/pxweb/sv/ssd/START__NR__NR0105__NR0105A/NR0105ENS2010T01A/?rxid=28ea630e-d9ef-490a-b634-f13b39de2187 (accessed on 30 January 2018).
34. SCB, Statistiska Centralbyrån. Disposable Income of Households (ESA2010) by Region (NUTS1-3) and Transaction Item. 2018. Available online: http://www.statistikdatabasen.scb.se/pxweb/en/ssd/START__NR__NR0105__NR0105A/NR0105ENS2010T02A/?rxid=9d75735b-e013-4215-8337-8cb45f99ce42 (accessed on 14 February 2018).
35. Krishnamurthy, C.K.B.; Kriström, B. A cross-country analysis of residential electricity demand in 11 OECD-countries. *Resour. Energy Econom.* **2015**, *39*, 68–88. [CrossRef]
36. Eurostat. *Simplified Energy Balances—Anual Data EU27*; Eurostat: Luxembourg, 2017.
37. SCB, Statistiska Centralbyrån. Fjärrvärmeproduktion och Bränsleanvändning (MWh), efter län och Kommun, Produktionssätt Samt Bränsletyp. 2018. Available online: http://www.statistikdatabasen.scb.se/pxweb/sv/ssd/START__EN__EN0203/ProdbrFj/?rxid=8145d8f1-79df-479a-b96f-82e03dcd62a5 (accessed on 8 February 2018).
38. Hiller, C. Sustainable energy use in houses: Will the energy use increase with time? In *Study of Literature and Computer Estimations*; Department of Building Physics, Lund Institute of Technology: Lund, Sweden, 2003.
39. SCB, Statistiska Centralbyrån. Number of Dwellings by Region, Type of Building and Period of Construction. 2018. Available online: http://www.statistikdatabasen.scb.se/pxweb/en/ssd/START__BO__BO0104/BO0104T02/?rxid=352ec506-e10c-488c-8a79-cf1ae3e9176a (accessed on 24 January 2018).
40. EU, European Parliament and Council of the European Union. *Directive 2003/30/EC of the European Parliament and of the Council of 8 May 2003 on the Promotion of the Use of Biofuels or Other Renewable Fuels for Transport*; Official Journal of the European Union: Brussels, Belgium, 2003.
41. MSD. *The Swedish Report on Demonstrable Progress under the Kyoto Protocol*; MSD: Stockholm, Sweden, 2015.
42. EC, European Commission. *Commission Staff Working Document—Full Impact Assessment Accompanying Document to the Commission Regulation Implementing Directive 2005/32/EC of the European Parliament and of the Council with Regard to Ecodesign Requirements for Fluorescent Lamps without Integrated Ballast, for High Intensity Discharge Lamps, and for Ballasts and Luminaires Able to Operate Such Lamps, and Repealing Directive 2000/55/EC of the European Parliament and of the Council*; European Commission: Brussels, Belgium, 2009.
43. EC, European Commission. *Commission staff Working Document—Full Impact Assessment Accompanying Document to the Commission Regulation Implementing Directive 2005/32/EC with Regard to Ecodesign Requirements for Televisions*; European Commission: Brussels, Belgium, 2009.
44. European Commission. *E., Commission Staff Working Document—Full impact Assessment Accompanying Document to the Commission Regulation Implementing the Directive 2005/32/EC of the European Parliament and of the Council with Regard to Ecodesign Requirements for Non-Directional Household Lamps*; European Commission: Brussels, Belgium, 2009.
45. EC, European Commission. *Commission Staff Working Document—Full Impact Assessment Accompanying Document to the Proposal for a Commission Regulation Implementing Directive 2005/32/EC with Regard to Ecodesign Requirements for Household Refrigerating Appliances*; European Commission: Brussels, Belgium, 2009.
46. EC, European Commission. *Commission Staff Working Document—Impact Assessment Accompanying Document to the Draft Commission Regulation Implementing the Directive 2009/125/EC of the European Parliament and of the Council with Regard to Ecodesign Requirements for Household Dishwashers*; European Commission: Brussels, Belgium, 2010.
47. EC, European Commission. *Commission Staff Working Document—Impact Assessment Accompanying Document to the Draft Commission Regulation Implementing the Directive 2009/125/EC of the European Parliament and of the Council with Regard to Ecodesign Requirements for Household Washing Machines*; European Commission: Brussels, Belgium, 2010.
48. EC, European Commission. *Commission Staff Working Document—Full Impact Assessment Accompanying the Document Proposal for a Commission Regulation Implementing the Directive 2009/125/EC of the European Parliament and of the Council with Regard to Ecodesign Requirements for Air Conditioners and Comfort fans*; European Commission: Brussels, Belgium, 2012.
49. EC, European Commission. *Commission Staff Working Document—Impact Assessment Accompanying the Document Commission Regulation Implementing Directive 2009/125/EC of the European Parliament and of the Council with regard to ecodesign requirements for Directional Lamps, Light Emitting Diode Lamps and Related Equipment*; European Commission: Brussels, Belgium, 2012.
50. EC, European Commission. *Commission staff working document—Impact Assessment Accompanying the Document Commission Regulation Implementing the Directive 2009/125/EC of the European Parliament and of the Council with Regard to Ecodesign Requirements for Household Tumble Driers*; European Commission: Brussels, Belgium, 2012.
51. EC, European Commission. *Commission Staff Working Document—Full Impact Assessment Accompanying the Document Proposal for a Commission Regulation Implementing the Directive 2009/125/EC of the European Parliament and of the Council with Regard to Ecodesign Requirements for Computers, Servers and Displays*; European Commission: Brussels, Belgium, 2013.

52. EC, European Commission. *Commission Regulation (EC) No 244/2009 of 18 March 2009 Implementing Directive 2005/32/EC of the European Parliament and of the Council with Regard to Ecodesign Requirements for Non-Directional Household Lamps (Text with EEA Relevance)*; Official Journal of the European Union: Brussels, Belgium, 2009.
53. EC, European Commission. *Commission Regulation (EC) No 245/2009 of 18 March 2009 Implementing Directive 2005/32/EC of the European Parliament and of the Council with Regard to Ecodesign Requirements for Fluorescent Lamps without Integrated Ballast, for High Intensity Discharge Lamps, and for Ballasts and Luminaires Able to Operate Such Lamps, and Repealing Directive 2000/55/EC of the European Parliament and of the Council (Text with EEA Relevance)*; Official Journal of the European Union: Brussels, Belgium, 2009.
54. EC, European Commission. *Commission Regulation (EC) No 642/2009 of 22 July 2009 Implementing Directive 2005/32/EC of the European Parliament and of the Council with Regard to Ecodesign Requirements for Televisions (Text with EEA Relevance)*; Official Journal of the European Union: Brussels, Belgium, 2009.
55. EC, European Commission. *Commission Regulation (EC) No 643/2009 of 22 July 2009 Implementing Directive 2005/32/EC of the European Parliament and of the Council with Regard to Ecodesign Requirements for Household Refrigerating Appliances (Text with EEA Relevance)*; Official Journal of the European Union: Brussels, Belgium, 2009.
56. EC, European Commission. E., *Commission Regulation (EU) No 1015/2010 of 10 November 2010 Implementing Directive 2009/125/EC of the European Parliament and of the Council with Regard to Ecodesign Requirements for Household Washing Machines (Text with EEA Relevance)*; Official Journal of the European Union: Brussels, Belgium, 2010.
57. EC, European Commission. *Commission Regulation (EU) No 1016/2010 of 10 November 2010 Implementing Directive 2009/125/EC of the European Parliament and of the Council with Regard to Ecodesign Requirements for Household Dishwashers (Text with EEA Relevance)*; Official Journal of the European Union: Brussels, Belgium, 2010.
58. EC, European Commission. *Commission Regulation (EU) No 1194/2012 of 12 December 2012 Implementing Directive 2009/125/EC of the European Parliament and of the Council with Regard to Ecodesign Requirements for Directional Lamps, Light Emitting Diode Lamps and Related Equipment (Text with EEA Relevance)*; Official Journal of the European Union: Brussels, Belgium, 2012.
59. EC, European Commission. *Commission Regulation (EU) No 206/2012 of 6 March 2012 Implementing Directive 2009/125/EC of the European Parliament and of the Council with Regard to Ecodesign Requirements for Air Conditioners and Comfort Fans (Text with EEA Relevance)*; Official Journal of the European Union: Brussels, Belgium, 2012.
60. EC, European Commission. *Commission Regulation (EU) No 932/2012 of 3 October 2012 Implementing Directive 2009/125/EC of the European Parliament and of the Council with Regard to Ecodesign Requirements for Household Tumble Driers (Text with EEA Relevance)*; Official Journal of the European Union: Brussels, Belgium, 2012.
61. EC, European Commission. *Commission Regulation (EU) No 617/2013 of 26 June 2013 Implementing Directive 2009/125/EC of the European Parliament and of the Council with Regard to Ecodesign Requirements for Computers and Computer Servers (Text with EEA Relevance)*; Official Journal of the European Union: Brussels, Belgium, 2013.
62. Ministry of the Environment Sweden. *Sweden's Sixth National Communication on Climate Change—Under the United Nations Framework Convention on Climate Change*; ME: Stockholm, Sweden, 2014.
63. Stenqvist, C.; Nilsson, L.J. Energy efficiency in energy-intensive industries—An evaluation of the Swedish voluntary agreement PFE. *Energy Eff.* **2012**, *5*, 225–241. [CrossRef]
64. *Malmö Stad, Miljöbokslut för Malmö Stad 1997*; Malmö Stad: Malmö, Sweden, 1998.
65. *Malmö Stad, Miljöbokslut för Malmö Stad 1998*; Malmö Stad: Malmö, Sweden, 1999.
66. *Malmö Stad, Miljöbokslut för Malmö Stad 1999*; Malmö Stad: Malmö, Sweden, 2000.
67. *Malmö Stad, Miljöbokslut för Malmö Stad 2000*; Malmö Stad: Malmö, Sweden, 2001.
68. *Malmö Stad, Miljöbokslut för Malmö Stad 2001*; Malmö Stad: Malmö, Sweden, 2002.
69. *Malmö Stad, Miljöredovisning för Malmö Stad 2002*; Malmö Stad: Malmö, Sweden, 2003.
70. *Malmö Stad, Miljöredovisning för Malmö Stad 2003*; Malmö Stad: Malmö, Sweden, 2004.
71. *Malmö Stad, Miljöredovisning för Malmö Stad 2004*; Malmö Stad: Malmö, Sweden, 2005.
72. *Malmö Stad, Miljöredovisning för Malmö Stad 2005*; Malmö Stad: Malmö, Sweden, 2006.
73. *Malmö Stad, Miljöredovisning för Malmö Stad 2006*; Malmö Stad: Malmö, Sweden, 2007.
74. *Malmö Stad, Miljöredovisning för Malmö Stad 2007*; Malmö Stad: Malmö, Sweden, 2008.
75. *Malmö Stad, Miljöredovisning för Malmö Stad 2008*; Malmö Stad: Malmö, Sweden, 2009.
76. *Malmö Stad, Malmö Stads Miljöredovisning 2012*; Malmö Stad: Malmö, Sweden, 2013.
77. *Malmö Stad, Malmö Stads Miljöredovisning 2013*; Malmö Stad: Malmö, Sweden, 2014.
78. *Malmö Stad, Miljöredovisning 2014*; Malmö Stad: Malmö, Sweden, 2015.
79. *Malmö Stad, Miljöredovisning 2015*; Malmö Stad: Malmö, Sweden, 2016.
80. *Malmö Stad, Miljöredovisning 2016—Uppföljning av Miljöprogram för Malmö Stad 2009–2020*; Malmö Stad: Malmö, Sweden, 2017.

energies

MDPI

Article

Integrated Performance Optimization of Higher Education Buildings Using Low-Energy Renovation Process and User Engagement

Abdullahi Ahmed [1,*], Monica Mateo-Garcia [2], Andrew Arewa [3] and Kassim Caratella [1]

[1] Institute for Future Transport and Cities, Coventry University, Coventry CV1 5FB, UK; ab7095@coventry.ac.uk
[2] Faculty of Computing, Engineering and the Built Environment, Birmingham City University, Birmingham B5 5JU, UK; Monica.MateoGarcia@bcu.ac.uk
[3] School of Energy, Construction and Environment, Coventry University, Coventry CV1 5FB, UK; arewaandrew@yahoo.com
* Correspondence: Abdullahi.Ahmed@coventry.ac.uk; Tel.: +44-247-765-0884

Abstract: Building performance improvement through low-energy renovation traditionally involves building performance diagnostics of the existing building, technology evaluation, selection and implementation. Effective building performance diagnostics, post-retrofit assessment and user engagement are essential to deliver performance as well as achieving socio-economic and environmental benefits at every stage of the renovation project life cycle. User's views are often ignored when renovating a building, causing sub-optimal energy performance, user comfort and wellbeing. This paper seeks to critically evaluate the low-energy renovation process and the role of user and stakeholder engagement in the strategic implementation of low-energy retrofit technologies for performance improvement of higher education buildings. The research focuses on renovation methodology, innovative materials/systems and end-user engagement throughout the renovation project phases (pre-renovation, the renovation process and post renovation). A mixed research method was adopted, which includes building performance modelling, monitoring and user evaluation questionnaires pre and post-renovation. The research is part of European Union (EU)-funded project, targeting 50% reduction in energy consumption using innovative materials and technologies in existing public buildings. The surveys allow comparative analysis of comfort levels and user satisfaction as an indicator of the efficacy of renovation measures. A new renovation process and user engagement framework was developed. The findings suggest that there is a direct relationship between retrofit intervention, improving energy performance of low-carbon buildings and the comfort of occupants. The technologies and strategies also appear to have different impacts on user satisfaction.

Keywords: low-energy renovation; indoor environmental quality (IEQ); energy efficiency; user satisfaction; stakeholder engagement

check for updates

Citation: Ahmed, A.; Mateo-Garcia, M.; Arewa, A.; Caratella, K. Integrated Performance Optimization of Higher Education Buildings Using Low-Energy Renovation Process and User Engagement. *Energies* **2021**, *14*, 1475. https://doi.org/10.3390/en14051475

Academic Editor: Vítor Leal

Received: 31 January 2021
Accepted: 17 February 2021
Published: 8 March 2021

1. Introduction

The function of buildings is to provide a comfortable internal environment that meets occupant satisfaction and wellbeing with optimum use of energy. Existing building stock across the European Union is ageing and failing to meet expected energy efficiency and environmental performance standards. Buildings account for 40% of total energy consumption in Europe [1]; therefore, improving energy efficiency and occupant comfort in existing buildings is essential to reduce energy dependency and greenhouse gas emissions. Indeed, the replacement rate for existing building stock is very low (1–2% per year). Arguably, one of the most cost-effective measures to meet energy reduction targets is to address the performance of existing buildings [2]. However, there are several challenges in the process. Recent building performance evaluation studies highlight the existing gap between predicted and actual energy consumption. They also reveal failure of buildings to meet essential energy, environmental and occupant comfort performance standards [3,4]. One of the reasons for the performance gap is human behaviour during construction and building use, highlighting the significance of user and stakeholder engagement at every stage of the project life cycle.

Policy debate seems to be more focused on energy savings and emission reductions, rather than empirical research regarding end-user perspective and wellbeing [5] To contribute to energy reduction in the building sector, various green rating systems have been established globally to evaluate sustainability of construction projects [6]. Buildings certified by these rating systems are considered to consume less energy, providing a better living environment and contributing to the overall reputation of the properties [7]. However, research shows that green building designs do not automatically guarantee occupant wellbeing and satisfaction [8]. Current sustainability tools tend to focus on technical aspects such as energy consumption, water use or materials, prioritising quantitative over qualitative factors. Yet it is the qualitative factors by which most people judge buildings and their environments, hence the measure of human satisfaction needs to play a bigger role when evaluating the performance of buildings.

There is limited research focusing on the process of end-user engagement relative to indoor comfort levels and general management of buildings before and after building renovation. Technology Strategy Board retrofit for the future report [9] argues that "engaging residents from the start can increase their understanding and acceptance of the works, and this can be a defining factor for success", highlighting the importance of continuous refining of the process of user and stakeholder engagement in building retrofit. This study seeks to determine effectiveness of low-energy renovation technologies and user engagement methodology in achieving energy performance improvement, indoor environmental quality and general management of existing higher education buildings. The study is part of European funded project on retrofitting solutions and services for the enhancement of energy efficiency in public edifications (RESSEEPE). The project developed and tested a variety of passive and active technologies to improve the energy and environmental performance of public buildings in three European cities representative of the breadth of EU climate conditions: Coventry (United Kingdom), Barcelona (Spain) and Skelleftea (Sweden). This paper focuses on the research carried out at two demo-sites buildings located at Coventry University in the United Kingdom.

2. Literature Review

2.1. Low-Energy Renovation in Higher Education Buildings

Further and higher education (FHE) is a fast growing sector, with student numbers increasing by 44% over in the past ten years; with annual energy costs for the FHE sector estimated at around GBP 400M, resulting in CO_2 emissions of around 3.1 million tonnes per year [10]. Many universities own a considerable number of 1960s and 1970s buildings; as a result, they are facing problems of out-of-date building stock that are not fit for purpose [11]. Sustainable renovation can be a more viable, practical and potentially affordable solution compared to complete demolition and reconstruction [12].

A number of strategies have been identified as key to deliver performance in deep renovation of existing buildings. According to Ma et al. [13] the low-energy renovation of existing buildings has many stages including: "project setup and pre-renovation survey, energy auditing and performance assessment, identification of renovation options, site implementation, commissioning, validation and verification". The success of each of these stages has significant impact on the efficacy of the renovation intervention measures. Other researchers have proposed similar approaches, such as Piaia et al. [14], who have developed a procedure for deep renovation consisting on four stages (4M): mapping, modelling, making and monitoring. However, each of these stages will only succeed with the support and engagement of key stakeholders, relative to their feedback and insights, that will ensure successful evaluation and implementation of the renovation measures.

This is arguably one of the most important elements for reducing performance and gap and unintended consequences of low-energy renovation.

2.2. Users Perception of Indoor Environments

Knowledge about effects, latest advances in low-carbon design, impact of building technologies on users' health and wellbeing is still limited [15]. Energy consumption of buildings depends significantly on the criteria used for maintaining indoor environmental quality (IEQ) (temperature, ventilation and lighting), building design and operation. Kang and Mak [16] found that thermal comfort, indoor air quality, acoustics and lighting levels have proven to be important factors that significantly affect building users' comfort, wellbeing and work performance. Giddings et al. [17] found other factors such as the provision of artwork, personal control of temperature and ventilation to be significant in increasing stimulation and user satisfaction. They also found that the most significant factors to consider during design stage are user choice of layout, design and décor and break areas, suggesting that these factors should be incorporated in pre-renovation diagnostics.

Traditionally, users' perception about indoor conditions is often measured through feedback questionnaires. Occupant feedback questionnaires have been used as part of post-occupancy evaluation protocols (POE), as they provide an understanding of user satisfaction [18]. Meir et al. [19] categorised the benefits of POE into short, medium and long term. Short-term benefits include obtaining users' feedback on problems in buildings and the identification of solutions; medium-term benefits include feed-forward of the positive and negative lessons learned into next building life cycle; long-term benefits aim at the creation of databases, update, upgrade and generation of planning and design protocols paradigms. Hay et al. [20] argued that the role of POE in systematic learning from previous projects is critical "to improving building performance, resulting in a built environment that better fits the needs of clients, end users, wider society and the environment".

Baird [21] measured user perception in sustainable buildings and found that "refurbished buildings can rate very highly and, in some cases, surpass new buildings, and there are indications that a design process that includes the users can result in better perception scores". In renovation projects, there is often change in internal layout mainly for space optimisation and utilisation purposes. These changes and their impact on the current users of the building should be of paramount importance. Malkoc and Ozkan [22] argue that assessing reaction of existing users using post-occupancy evaluation is the most efficient way to enhance space quality and important for developing future design. This demonstrates the need to engage the users at different stages of design and construction to achieve maximum satisfaction and energy performance.

2.3. Occupants' Satisfaction in Educational Buildings

There is a growing concern about indoor environmental quality in educational premises. Many research studies show that indoor environment conditions can affect productivity and learning performance of individuals in non-domestic buildings [23]. Students in buildings with good environmental conditions earn test scores 5–17 percent higher than scores for students in substandard buildings [24] and have up to 14 percent lower student suspension rates [25]. Considering this, internal conditions and user comfort play a critical role in teaching and learning. Kim et al. [26] used student post-occupancy evaluation to establish space choice and rejection model. They established three categories of relationships: space-oriented relationship with space environmental performance and spatial form; user-oriented relationships with capacity and locational accessibility; equipment-oriented relationships with equipment adequacy and equipment conditions. In every major renovation project, these relationships should be clear to support informed performance improvement and maximise occupant's satisfaction; otherwise, uncomfortable occupants may take adaptive actions to improve their comfort, which often leads to sub-optimal performance.

Despite the evidence on the effects of poor indoor environments in productivity and learning performance, energy refurbishment projects in higher education buildings tend to focus on energy savings, giving less attention to occupants' comfort and engagement. Existing research in energy refurbishment in higher education buildings present optimized methodologies for reducing energy consumption [27,28], performance gap [29] and preserving the historical value and building usability while improving energy savings [30,31]. The lack of studies incorporating a holistic approach where the views of stakeholders are incorporated demonstrates the need of research in the area of user engagement, as a key element to minimise the energy gap and ensure user comfort and satisfaction. Thus, renovation of higher education buildings can be used for academic exercise to understand occupants' satisfaction, behaviour and patterns of use before and after renovation processes.

3. Materials and Methods

Multi-methods research design involving use of case studies and QUAN-QUAL concept [32] was adopted; meaning that quantitative method is the lead data collection instrument, while qualitative data are used to support and validate the quantitative findings. The study adopted sequential explanatory type of mixed methods design strategies, because the research inquiry is designed to explain relationship between end-user engagement, indoor comfort levels and general management of the buildings before and after renovation. Research ethics approval was sought from Coventry University Research Ethics Committee, and it was granted.

The research monitors and analyses the implementation of low-energy renovation technologies, comfort levels and user satisfaction in Coventry University demo sites. Building performance modelling, monitoring and user evaluation questionnaires have been used before and after renovation. The surveys are designed to investigate user perception, develop a strategy for responding to negative changes, improve user comfort, measure the efficacy of the renovation actions, learning from the process and use the variables studied for improving future renovations. Integrated Environmental Solution Virtual Environment (IES-VE) has been used to model and simulate the performance of the building and several renovation technologies. The renovation technologies have been implemented in two real case study buildings located at Coventry University.

3.1. Case Study Location and Description

The study location was in Coventry, United Kingdom. Two buildings, namely, Richard Crossman (RC) and Sir John Laing Building (JL) owned by Coventry University have been used for the experiments. Coventry University and Coventry City Council own about 70–80% of the built assets within the city centre. Figure 1 shows the location of the case study buildings and the city centre.

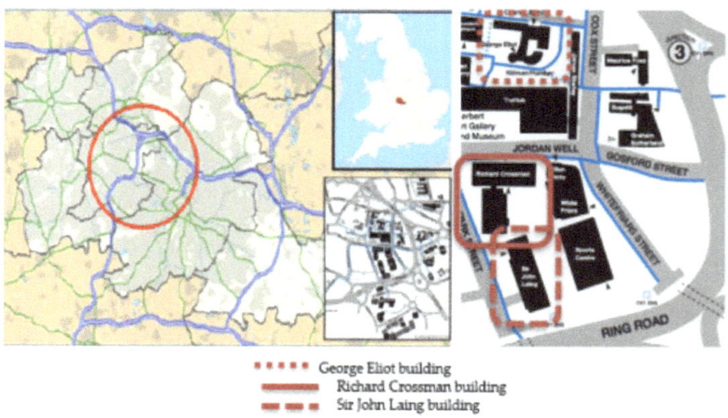

Figure 1. Location of Coventry and Coventry University.

Building Physical Properties

The case study buildings are typical 1970s buildings constructed of brick and single glazed metal frame windows. JL is a two-storey building with net floor area of 3660 m^2, and RC is a five-storey building with a net floor area of 9306 m^2. The façade in both buildings consists of brick masonry cavity wall with 6mm single glazing metal frame windows and concrete structure. Both buildings have a Display Energy Certificate Rating of "C", which is an average energy efficiency rating. Table 1 shows full description of the case study buildings, pictorial view, year of completion, net area, electricity, gas and water consumption and carbon footprint. Recent planning approach for Coventry City Council moving forward is the recognition of a city living lab status, "establishing Coventry as a test-bed, incubation hub and international showcase for low carbon innovations" [33]. The living lab status is key element to this project. Thus, building selected encompasses a living lab ethos, acting as a live experimental facility for several innovative technologies. The heating energy data are based on the net heated floor area using the Display Energy Certificate calculation methodology. Carbon emission factor provided by UK Government to support Carbon Reduction Commitment (CRC) reporting has been used to calculate the total carbon emission (electricity—0.34885 kg CO_2/kWh, natural Gas—0.2042 kg CO_2/kWh and water—0.344 kg CO_2/m^3.) [34].

Table 1. Building characteristics.

Building	John Laing Building	Richard Crossman Building

Year of completion	1970	1971
Net area (m^2)	3660	9306
Electricity (kWh/m^2/year) *	94	116
Gas (kWh/m^2/year) * (heated floor area)	129	129
Water (m^3/annum) *	957	2462
Carbon Footprint (tonnes/year) *	282	841

* Consumption and emissions per year.

3.2. Decision-Making Process

Due to the complexity of building renovation, many factors influence the selection of technologies. Figure 2 shows the decision-making criteria that informed the selection of the technologies and strategies for the demonstration buildings, most of these criteria involve user and stakeholder engagement. The decision-making criteria includes buildings related structural and Operation and Maintenance (O&M) requirements, organisational financial constraints, policy and regulatory limitations. Using this methodology, a decision-making matrix was developed, as shown in Table 2, to compare different materials and technology solutions to provide an initial basis for selecting the most efficient, cost effective and most beneficial solutions for each demo site to achieve the best possible renovation interventions based on the energy and indoor environmental quality requirement of the client.

Figure 2. Decision-making criteria.

Table 2. Decision-making matrix.

-	-	-	**T3.2 Isolation Strategies for Energy Conservation**		**T3.3 Solar Strategies for Energy and Heat Recovery**			**T3.4 Strategies for Thermal Energy Storage**				**T3.6 Efficient Lighting**	
	Criteria	**Demo Site Owner Weight**	**Aerogel-Super Insulat-ing Mortars**	**VIP Pan-els**	**EC/ PV Win-dow**	**Venti lated Façade BIPV**	**Electr ical Stor-age/ Solu-tions**	**Solar AC**	**PCM Ther-mal Stor-age**	**Pass ive Cool-ing**	**Heat Re-cov-ery**	**Ind oor LED Light-ing**	**Urb an LED Light-ing**
-	**Technical Feasibility**	10	8	2	8	2	5	2	5	3	2	8	10
-	**Certificated?**	5	2	2	5	5	0	5	5	5	5	5	5
Key Attributes	Initial cost	4.00	4	1	1	1	2	2	1	3	2	2	2
	Operating Cost and Maintenance	4.00	5	5	4	2	2	2	4	4	2	4	4
	Environmental Impact (Potential CO2 reduction)	5.00	2	2	3	5	3	5	3	3	4	2	4
	Service Lifetime (Durability)	3.00	5	5	4	3	2	3	4	4	3	4	4
Construction	Construction Works (5=Small amount of work, 1=A lot of work)	3.00	3	2	2	1	3	2	3	3	1	2	4
	Specialist contractors require ment for installation (Y = 1, N = 5)	1.00	4	4	4	1	5	4	4	5	5	5	5
Other specifications	Aesthetics	4.00	4	4	4	1	3	1	5	4	5	3	4
	Space requirements	3.00	4	3	5	3	2	1	1	2	1	5	5
	Air Quality	2.00	3	3	3	3	3	3	3	3	4	3	3
	Noise	3.00	2	5	4	3	5	3	5	3	2	3	3
	Control Capability	3.00	5	5	5	3	5	3	2	2	5	5	5
	Environment Friendly (Embodied Carbon)	3.00	3	4	4	5	2	5	3	3	4	2	4
	Compatibility with the existing systems	3.00	4	2	4	1	1	1	2	3	1	1	3
	Total	56	240	168	250	150	164	154	200	185	165	228	280
	Weighted Vote		4.29	3.00	4.46	2.68	2.93	2.75	3.57	3.30	2.95	4.07	5.00
	Decision		Y	N	Y	N	N	N	Y	Y	N	Y	Y

3.2.1. Decision-Making Matrix

From the decision-making process, a decision-making matrix was derived for comparing different solutions to provide an initial ranking of the selection of the most efficient, cost effective and most beneficial solutions for each demo site to achieve the best possible retrofitting interventions. The decision-making matrix includes the different intervention materials considered for the retrofit, whilst marking it against the decision making. The matrix provides an overall cumulative score for each technology for each case study site (taking into consideration location, climatic conditions and use), which contributes towards selection of the optimum technology selection for the most beneficial renovation. The decision-making matrix uses both technical and financial feasibility, client and stakeholder perspectives and energy and environmental performance potentials.

3.2.2. Strategies for Technology Selection

A feasibility study of potential technologies was conducted with a target for reducing building energy demand. The construction materials and technologies proposed relate to envelop insulation, building services, passive solutions and renewable energy systems, ranging from absolute state of the art to market ready materials and technologies. Table 3 shows the list of technologies considered for the two demonstration buildings.

Table 3. Summary of materials and technologies.

Technology	Description	Image
EPS-G Panel	This technology is an improved thermal insulation panel, which is made from a self-extinguishing expanded polystyrene, which is used in the TRADITERM® external thermal insulation system. This is a great insulator, with characteristics being lightweight, workable and a very low conductivity [35].	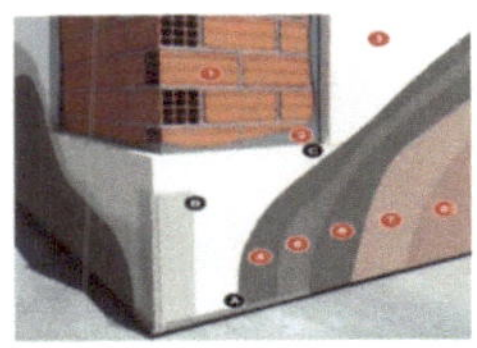
Aerogel-Based Insulating Mortar	Combines its properties with cementitious materials to provide low thermal conductivity at <0.020 W/mK. RESSEEPE aims to admix aerogels into cement to design a super-insulating mortar. The insulation performance of the aerogel was preserved [36].	
Vacuum-Insulated Panels	Vacuum Insulated Panels (VIP)panels consist of a mesoporous core typically fumed silica, which is wrapped in a multilayer laminate foil barrier. They have a low value of thermal conductivity (λ = 0.003–0.004 W/mK) [37].	
Solar PV	Solar photovoltaics (PV) allow the production of electricity from sunlight. The conversion happens in the PV cell, where a semiconductor generates a direct current (DC electricity). This happens when it is exposed to light and in turn due to the photovoltaic effect. The electricity produced can be used on the spot (off-grid systems also called stand-alone PV systems) or into the grid (grid connected systems) or both [38].	

Table 3. *Cont.*

Technology	Description	Image
PCM Seasonal Thermal Energy Storage	This technology stores and releases thermal energy during the process of melting and freezing. When they freeze, they release large amounts of energy. When they melt, energy is absorbed from the environment when changing from solid to liquid [39].	
EC Windows	Changes the light transmission properties in a controlled and reversible manner through a small electric current which flows through the device. This technology can reduce energy expenses by 19 and 48% in cooling and lighting demand. They are considered smart windows [40].	
Ventilated Façade	This is a construction system consisting of the attachment of an outer skin of ventilated cladding to a new or existing building which avoids thermal bridges and improves thermal and acoustic performance of the envelope. The ventilated facade generates electric power through the vertical PV [41].	
LED Lighting	Light-emitting diodes (LEDs) are semiconductor diodes, which emit light when a voltage is applied. LEDs are more efficient, durable, versatile and longer lasting than incandescent lighting and compact fluorescent lighting (CFL) [42].	
BIPV	Trina Solar modules were selected as the PV panels for a vertical installation. The selected PV modules TSM-PDG5 by Trina Solar have dimensions of 1685 × 997 mm^2 with a thickness of 6 mm. This technology was implemented in combination with the ventilated façade [43].	
Roof Insulation	Kingspan TR27 insulation bonded in Sarnacol adhesive with a U value of 0.18 W/m^2K was selected as roof insulation for RC. An adhered system is to be installed using Sarnafil G410-18ELF Lead Grey with integral 300 g/m^2 polyester fleece as the main roof sheet and Sarnafil G410-15EL Lead Grey or S327-15EL Lead Grey for all detail work flashings [44].	

After the feasibility analysis, a twin strategy for implementation and testing of these technologies was developed. The first strategy was based on a whole-building renovation in RC building. In this strategy, advanced market-ready technologies for low-carbon renovation were implemented at a large scale. The second strategy was based on the design and implementation of innovative technologies for low-carbon renovation in selected areas of the JL building to test their efficacy in a living lab demo site.

Table 3 summarises all the materials and technologies considered for the renovation project. The innovative retrofit technologies include those that have been technically advanced, adapted, within the project; this includes aerogel-based insulating mortar, vacuum-insulated panels, ventilated façade and PCM seasonal thermal energy storage. These new technologies have been combined with other market ready technologies, such as Solar photovoltaics (PV), Electrochromic (EC) windows, light-emitting diode (LED) lights and Expanded Polystyrene with Graphite (EPS-G) insulation panels.

3.2.3. Strategic Intervention during Building Renovation

The low-carbon technologies implemented in RC and JL buildings have been summarised in Table 4, which includes LED lighting, solar photovoltaic panels, a building management system (BMS), double-glazed windows and thermal insulation for RC building; whereas for JL building, a range of fabric state-of-the-art technologies are included, such as aerogel-based mortar, vacuum-insulated panels, ventilated façade, EPA-G panels and passive PCM tube. They have been selected based on the decision-making criteria and decision-making matrix presented in Figure 2 and Table 2 in response to the key energy and IEQ challenges identified in the case study buildings.

Table 4. Summary of retrofit technologies.

Technology (m^2)	Demo-Site	
	Richard Crossman Building	**John Laing Building**
EPS-G Panels	-	57
Aerogel-Based Insulating Mortar	-	57
Vacuum-Insulated Panels	-	56
Solar PV	9395	-
Seasonal Thermal Energy Storage (Water and PCM)	-	-
EC Windows	-	56
Ventilated Façade	-	28
LED Lighting	2600	-
High-Efficiency Windows	9395	28
BIPV	-	57
Solar Thermal Collectors—UPC	-	-
Solar Thermal Collectors	-	-
Roof Insulation	934	-
Total Area of Site Affected	**9395 (m^2)**	**3660 (m^2)**

3.3. *Quantitate and Qualitative Data Collection*

The research design entails user comfort evaluation exercise and energy evaluation using both quantitative and qualitative research methods. The data collection instrument includes questionnaires distributed to occupants before, during and after the renovation to collect data on perceived comfort, level of user control and level of user engagement in the renovation process shown in Table 5. The questionnaires have been distributed using hardcopies and BOS online platform. Population sample includes both students and staff of Coventry University. All data have been collected from October 2015 to February 2017. The questionnaires were designed following the standards EN 15251 and ISO 7730 [45,46]. The indoor environmental parameters included in the survey followed the recommendations of EN 15251, which identifies parameters for monitoring the indoor environment as recommended in the Energy Performance of Buildings Directive. These parameters refer to thermal environment, indoor air quality, humidity, lighting and noise. ISO 7730 was followed for assessing the general thermal sensation and degree of discomfort (thermal dissatisfaction) of occupants, through the analysis of the 7-point thermal sensation scale, based on the heat balance of the human body.

Table 5. Research method for user comfort evaluation.

Process Followed		Parameters Analysed
Before renovation	User satisfaction survey	User characteristics: role, age, gender, preference, interests User experience: Indoor Environmental Quality (IEQ) Level of control General maintenance
During renovation	User satisfaction survey of the renovation process *	User characteristics: role, age, gender, preference, interests Evaluation of the renovation process: Level of engagement Level of disruption
After renovation	User satisfaction survey	User characteristics: role, age, gender, preference, interests User experience: IEQ Level of control General maintenance Assessment of the IEQ improvement

* Only in Richard Crossman building.

The questionnaire consists of four parts (Table 6). Part 1 targets user information, including age, gender, seat position, occupancy hours and the main work activity of respondents. Parts 2, 3 and 4 include a quantitative set of questions based on a 7-point scale and open-ended questions to capture qualitative views of users such as their experiences pre, during and post building renovation. Part 2 involves the perception of respondents' relating to four key IEQ conditions (air quality, thermal environment, lighting environment and acoustic environment), rating 1 (very dissatisfied) to 7 (very satisfied). Temperature and air quality are rated 1 (very hot) to 7 (very cold) and from 1 (fresh) to 7 (stale), respectively. Part 3 investigates the level of user control of environmental conditions in the work area. Participants were asked to rate the level of heating, cooling, ventilation and lighting control from 1 (no control) to 7 (full control). Part 4 assesses the overall building performance from 1 (unsatisfactory) to 7 (satisfactory). An additional set of questions (part 5) have been included for RC to evaluate the level of satisfaction with the renovation process regarding indoor environmental quality improvement, disruption and level of user engagement during the design and construction stages. The stratified sampling method was used to target staff participants in RC building. The questionnaire has been distributed using staff mailing list, after completion of renovation works to assess IEQ improvement, level of engagement and project disruption (Table 2). The questionnaire was designed to make participants reflect on the changes between pre and post renovation environmental conditions.

Table 6. User satisfaction survey structure.

Part 1	Part 2	Part 3	Part 4	Part 5 *
Individual Factors	**IEQ Aspects**	**Personal Control**	**Overall Building Conditions**	**Richard Crossman Renovation**
Age Gender Seat position Hours in the building	Temperature Air quality Lighting Noise	Heating Cooling Ventilation Lighting	Comfort Facilities Health	Indoor environment improvement Level of engagement Disruption

3.4. Energy Modelling

Building performance modelling and simulation has been used for building performance diagnosis, selection and optimisation of the different technologies. The building performance diagnosis methodology includes predicting potential impact of renovation technologies on the building energy and environmental performance has been assessed using building energy performance analysis simulations software IES Virtual Environment (IESVE), whole building modelling and simulation software (Software by Integrated Environmental Solutions Limited, Glasgow, UK). There are a number of building and system modelling software tools used within industry and academia for the predictive analysis of building systems and their impact on energy and environmental performance. Each program has unique features in terms of modelling resolution, solution algorithms, intended target audience, modelling options and ease of use vs. flexibility [47]. IES Virtual Environment (IESVE) is an in-depth suite of integrated analysis tools for the design and retrofit of buildings, which is widely used for research and industrial applications within the building services industry. IES-VE software 2018 version [48] has been used for the analysis. Crawley et al. [49] categorized the IESVE as one of the software that has undergone most rigorous validation studies in addition to other software such as EnergyPlus (Software by Department of Energy's (DOE) and National Renewable Energy Laboratory (NREL) USA)., ESP-r (Software by University of Strathclyde, UK), ICE ((Software by Ice Edge Business Solutions, Ltd., USA) and TRNSY(Software by University of Wisconsin, USA) after robust critical comparison of their features and capabilities for building energy simulation programs. The first stage was to establish building information for RC and JL using legacy data as well as different retrospective surveying methods. The building data have been used to create an intelligent BIM models using Autodesk REVIT and IESVE dynamic energy simulation model with the building geometry, materials, buildings systems and building occupancy and use profile (Figure 3).

Figure 3. Building simulation IESVE models of John Laing and Richard Crossman.

To maximise accuracy of the building performance prediction, three different sets of simulations were performed. The first set of simulations for energy savings were carried out on the pre retrofitting stage where no technologies had been installed to establish energy performance baseline scenario. The next set of simulations were carried out with the technologies proposed for each demo site, taking into consideration the exact location and installation of the technologies. The final set of simulations provided an extrapolation where the performance of the technologies were extrapolated to the whole building scale. The simulation results had been collected and analysed and used at different stages of the renovation project. The results also form the basis recommendations and prioritisation of technology selection and for the future verification and correct use of these technologies on the final installation places (demo sites).

4. Result Analysis

The estimated population in the buildings are 310 and 1378 for JL and RC, respectively, with staff population considered as stable. Indeed, students' population are transient and, therefore, difficult to estimate with certainty, and it was estimated based on the average student population of the various modules using spaces in the buildings. Table 7 shows the distribution and characteristics of building users that responded to the questionnaire both before and after the various renovation actions were implemented, in JL, and after renovation was completed in RC.

Table 7. Characteristics of respondents.

Building		John Laing				Richard Crossman
		Architecture Studio		Offices		
		Before	After	Before	After	
Questionnaires distributed		35	35	20	20	-
Number of questionnaires fully completed		32	30	18	13	48
Response %		91.43%	85.71%	90%	65%	48%
Age	Under 30	33	28	0	0	2
	Over 30	3	2	19	13	45
Sex	Male	21	17	18	12	7
	Female	14	13	1	1	41
Days per week in the building		4.93	4.27	4.71	4.46	3.81
Hours per day		7.53	6.45	7.97	7.92	7.21
Hours per day at desk		5.96	5.42	5.39	5.31	5.04

As shown in Table 7, the response rate was higher in the JL Building (between 65 and 91.43%) than in the RC Building (48%). Perhaps, differences in response rate are because hard copy questionnaires were distributed and collected at JL, whereas online questionnaires were used in RC Building. The questionnaires in JL were distributed in person during class time, in the case of the Architecture Studio, or office by office. This action led to a considerable increase in the response rate.

4.1. *Quantitative Data Analysis*

SPSS version 25 and MS Excel have been used to analyse questionnaire data. The data analysis revealed Cronbach's alpha sigma value of 0.71; meaning that the internal reliability of the quantitative data is very good. For emphasis and better understanding of the study, data collected were classified into five categories namely: individual factors, IEQ aspects, personal control, overall building conditions and renovation process. Subsequently, the data were analysed based on each individual building.

4.1.1. John Laing Building

Table 8 summarises the descriptive statistics of JL results pre- and post-renovation. Questions A2.1 to A5.3 are described in Table A1 in the Appendix A. Table 8 shows the statistical analysis of the data based on 7-point scale from 1 (uncomfortable) to 7 (comfortable). The results infer that classrooms participants (students) noticed a slight improvement in the overall thermal conditions in winter (A2.3), with a comfort mean value of 4.17 after renovation compared to 3.47 before renovation; the overall conditions in summer (A2.6), with a mean value of 4.17 after renovation compared to 3.81 before renovation. Temperature conditions (A2.1 and A2.4) improved, with the indoor environment being warmer in winter and cooler in summer. The result reveals improvement in user satisfaction with overall comfort condition in the building (A5.1) with a mean value of 4.00 after renovation compared to 3.51 before renovation and a healthier indoor environment (A5.3) with a mean value of 3.60 after renovation compared to 3.40 before renovation. In terms of spread, IEQ responses after renovation have a lower variance, and therefore, standard deviation, than before renovation. This means that there is no significant change in IEQ conditions post renovation, but there is a consensus among students that indoor conditions have improved. Findings from the study also infer that installation of a passive low-carbon technology such as PCM tubes made a positive impact on the indoor environment, because students identified indoor air temperature to be more stable both in winter and summer (A2.2 and A2.5). Note: PCM technology helps to reduce peak temperatures to a more stable indoor environment.

Table 8. Descriptive statistics for IEQ responses in John Laing Building.

	Classrooms						Offices					
	Before			After			Before			After		
	Mean	Variance	Standard Deviation	Mean	Variance	Standard Deviation	Mean	Variance	Standard Deviation	Mean	Variance	Standard Deviation
A2.1	4.83	1.66	1.29	4.13	1.22	1.11	4.86	0.81	0.90	4.60	0.80	0.89
A2.2	4.30	3.46	1.86	3.73	1.93	1.39	5.29	4.24	2.06	4.00	4.00	2.00
A2.3	3.47	2.05	1.43	4.17	1.87	1.37	4.43	2.29	1.51	4.40	1.80	1.34
A2.4	3.41	1.25	1.48	3.77	0.46	0.68	2.71	0.90	0.95	2.40	0.80	0.89
A2.5	4.26	2.66	2.02	4.00	1.59	1.26	5.14	3.81	1.95	3.60	5.30	2.30
A2.6	3.81	2.31	1.52	4.17	0.76	0.87	3.29	1.24	1.11	3.40	2.30	1.52
A5.1	3.51	2.08	1.44	4.00	1.38	1.17	3.29	2.24	1.50	3.60	1.80	1.34
A5.2	4.03	3.26	1.81	3.83	1.94	1.53	3.29	3.57	1.89	3.80	2.70	1.64
A5.3	3.40	1.60	1.26	3.60	0.94	0.97	3.57	0.62	0.79	3.60	0.80	0.89

Similarly, staff identified significant improvement in the overall comfort of the offices (A5.1), with a mean value of 3.60 after renovation compared to 3.29 before renovation works as illustrated in Table 8. The perception of users regarding indoor temperature and air quality in winter show an improvement (A2.1 and A2.2) with a less cold environment and more stable air temperature. However, temperature conditions in summer (A2.4) did not show significant improvement with a mean value of 2.40 after renovation compared to 2.71 before renovation (rated on a 7-point scale from 1 (too hot) to 7 (too cold)). Additionally, variance is higher after renovation for summer conditions, which indicates a considerable performance difference between renovated offices. This might be related to the fact that PCM tubes were installed only in two offices, where staff benefitted from the reduction of up to 4 degrees Kelvin in the hottest days of summer. Despite the temperature reduction, the general user perception for the PCM renovated offices did not show any significant improvement in the summer temperature satisfaction (A2.4). However, an improvement in temperature stability was observed (A2.5) and overall conditions in summer (A2.6), as shown in Table 9. This might be linked to lack of night-time ventilation, because it was observed that rooms with PCM were warmer during early morning periods compared to rooms without PCM. Possibly, the issue could be avoided using effective night ventilation to remove the heat absorbed in the PCM during hot summer afternoons.

4.1.2. Richard Crossman Building

As mentioned in the research design, additional questions related to key problems before renovation, improvement of indoor environmental parameters and levels of stakeholder engagement with the renovation work were added to allow for comparison of the thermal comfort before and after the renovation. Figure 4 shows the indoor environmental parameters identified by study participants as major problems before low-carbon retrofit and level of improvement of the same factors after renovation. Respondents rated thermal comfort as the most important as the main indoor environmental problem before renovation 34.18%, followed by lighting/visual comfort at 21.52%, noise 13.92% and air quality by 8.86%. After renovation, visual comfort was identified as most significant improvement rated 41.67%; followed by lighting 25%, thermal comfort 16.67%, noise 12.50% and air quality 4.17%. The results obtained show that, although thermal comfort and noise have experienced some improvement, visual/lighting comfort experienced more significant improvement. This assessment is positive, considering the fact that visual comfort and lighting were identified as key environmental problems before renovation works.

Table 9. IEQ descriptive statistics for offices with PCM and control rooms in the JL Building.

	Offices Not Retrofitted						Offices Retrofitted with PCM					
	Before			After			Before			After		
	Mean	Variance	Standard Deviation	Mean	Variance	Standard Deviation	Mean	Variance	Standard Deviation	Mean	Variance	Standard Deviation
A2.1	4.36	2.05	1.43	4.13	2.41	1.55	5.5	0.33	0.58	4.5	1.00	1.00
A2.2	4.18	2.76	1.66	4.88	0.98	0.99	6.25	0.92	0.96	4	5.33	2.31
A2.3	4.36	2.85	1.69	3.75	2.21	1.49	4.25	2.92	1.71	4.25	2.25	1.50
A2.4	2.58	1.36	1.16	2.75	2.50	1.58	2.75	1.58	1.26	2.5	1.00	1.00
A2.5	5.17	2.52	1.59	4.88	2.41	1.55	6	0.67	0.82	4.25	4.25	2.06
A2.6	3.17	2.33	1.53	2.25	1.07	1.04	2.75	0.92	0.96	3.5	3.00	1.73
A5.1	4.33	1.33	1.15	3.63	1.41	1.19	2.75	2.92	1.71	3.25	1.58	1.26
A5.2	3.83	1.06	1.03	3.25	1.64	1.28	3	4.00	2.00	3.25	1.58	1.26
A5.3	3.92	1.17	1.08	2.75	0.79	0.89	3.25	0.92	0.96	3.5	1.00	1.00

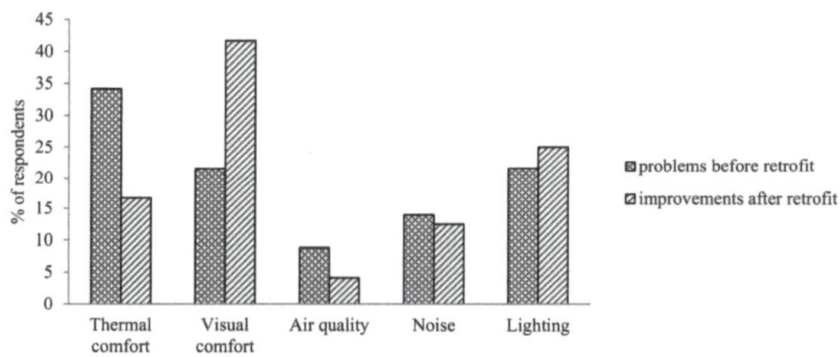

Figure 4. Improvement of indoor environmental parameters in Richard Crossman.

Table 10 presents findings relating to IEQ questions. The assessment of overall thermal conditions in winter (A2.3) and summer (A2.6) is average, with mean values of 3.27 and 3.39, respectively. A5.3 has a low variance in relation to the other variables. All winter conditions (A2.1, A2.2 and A2.3) have a variance higher than 2.0, which highlight the difficulty of achieving uniformity and consistency of thermal comfort with new open-plan office layout design. Overall, the values obtained show that users are more satisfied with thermal conditions in summer compared to winter.

Table 10. Descriptive statistics for IEQ responses in the Richard Crossman Building.

	Mean	Variance	Standard Deviation
A2.1	4.62	2.83	1.68
A2.2	5.74	2.24	1.50
A2.3	3.27	2.07	1.44
A2.4	3.02	1.50	1.22
A2.5	5.25	2.10	1.45
A2.6	3.39	1.93	1.39
A5.1	3.46	2.00	1.41
A5.2	3.42	1.99	1.41
A5.3	2.96	1.06	1.03

Significant improvement (mean value of 3.25) was observed for overall assessment of indoor environmental quality improvement (B.8), shown in Figure 5, similar to the overall IEQ of the building (A5.1), which is 3.46. Though variance for most variables is greater than 2.0, IEQ questions were included in the same questionnaire with additional questions about engagement in the renovation process and disruption experienced, which may have led to some bias in user responses.

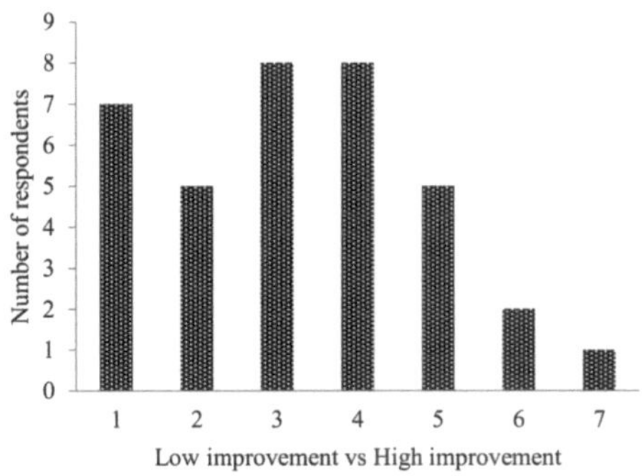

Figure 5. Overall assessment of the indoor environmental quality improvement. Mean rank 3.25, variance 2.58, standard deviation 1.61.

4.2. *Qualitative Data Analysis*

Questionnaires had open-ended questions to providing opportunity to express views in more detail. To ensure validity of qualitative process; three main areas were addressed: production (questionnaire design, circulation and data recording), presentation (replicability, valid inference and data arrangement) and interpretation (data discussion). Content analysis is used for qualitative data.

Textual contents of the open-ended questions were transcribed into manuscript, inputted into NVivo 12 software and coded using key study themes. For example, when participants were asked to express their view concerning whether they observed noticeable improvements since building renovation; a host of issues was raised about the subject matter. Key issues observed by participants are mainly related to aesthetic improvements and new office arrangements due to new office layout and space allowance rather than indoor environmental aspects. For example, some comments highlight an increase in the noise pollution experienced, causing disruption and affecting staff concentration levels. This is due to the new office layout after renovation (from small offices to open-plan offices), which means a higher density of occupancy per unit area causing significant changes to the working environment compared to pre-renovation. This further emphasizes the need for robust user engagement during renovation planning to ensure user awareness and input into planned changes and the possible positive and negative impact of these changes so they can adapt to them when reoccupying the building post renovation.

In terms of lighting conditions, study participants claim to miss natural light because windows were small and have to rely on artificial light. Additionally, the new deeper open-plan offices have significantly less daylight penetration compared to perimeter offices before renovation. Though, users generally felt that lighting levels improved significantly during post renovation, due to the installation of new strip lighting, they find these lights to be very bright, causing glare and complaints of headaches and migraines. Overall, participants claim that there was significant improvement in indoor air quality, thermal comfort and room lights after renovation.

4.3. *Energy Performance Result*

Several simulations had been carried out regarding the demo sites, with simulations for a pre-renovation, post-renovation and post-renovation with renovation extrapolation. Through the development of the energy models and the building performance analysis simulations, the results summarised in Table 11 are derived showing the results of building energy performance based on the three scenarios for each of case study building.

Table 11. Energy performance result of both buildings.

Performance Parameter	Richard Crossman Building				John Laing Building			
	Pre	Post	Post Full	Change	Pre	Post	Post Full	Change
Boilers energy (MWh)	2593.34		749.83	0.71	418.76	399.30	371.25	0.11
Total system energy (MWh)	3180.57		1097.08	0.66	448.84	428.90	401.35	0.11
Total lights energy (MWh)	0.00		0.00		0.00	0.00	0.00	
Total equip energy (MWh)	0.00		0.00		0.00	0.00	0.00	
Total nat. gas (MWh)	2593.34		749.83	0.71	418.76	399.30	371.25	0.11
Total electricity (MWh)	1103.26		1168.41	−0.06	30.08	30.10	30.10	0.00
Total Carbon Emissions (Kgco2)	1,132,751.00		632,847.00	0.44	106,064.00	101,614.00	95,810.00	0.10
Total energy (MWh)	3696.60		1885.39	0.49	448.84	428.90	401.35	0.11
Total energy (MWh/m^2)	0.39		0.20	0.49	0.12	0.12	0.11	0.11
Total energy (KWh/m^2)	393.46		200.68	0.49	122.63	117.19	109.66	0.11
Total grid disp. Elec (Mwh)	0.00		−32.84					

Three scenarios have been presented in JL building, which include pre- and post-renovation results, because the renovation did not cover the total building area. A third scenario was created, and the results have been extrapolated based on the assumption that the total floor area of the building will be renovated, this has been referred to as "post full". The results for Richard Crossman Building have only two scenarios simulated, due to the full-scale renovation plans; therefore, the extrapolation and the real intervention are the same. The results show the Richard Crossman Building having energy savings of 49%, which is a significant improvement compared to the base case scenario. The improvement is down to a mix of technologies targeting the building envelope, glazing and lighting. This includes the use of light-emitting diodes (LED) lighting, changing of the glazing and windows to a more efficient glazing system and frames and the improvement of the roofing material.

Furthermore, all three simulations were carried out for the JL Building, with the results showing a 10.58% savings in terms of energy performance. The savings are not as significant compared to the RC Building because the JL Building has been chosen as a living lab testing facility for the innovative renovation solutions. Different materials were selected at small scale for testing their efficacy. Further analysis is required to compare the different innovative materials at full extrapolation to find the most empowering innovative technology for optimum performance.

The results of the modelling of RC show significant reduction in total energy consumption for the entire building in the region of 49%, which meets the initial project objective of 50% post retrofit energy reduction. The modelling shows an increase in electricity consumption in the retrofit scheme due to increase in air-condition in areas that where otherwise naturally ventilated. Even though there is slight increase in electricity consumption, this will be upset by the 75 kWp Solar PV system integrated in the RC building.

4.4. Proposed New Methodology for User Comfort Evaluation

The evaluation of project implementation, decision-making process and the results obtained from the user comfort evaluation questionnaires led to the development of a new renovation process and stakeholder engagement protocol for managing low-energy renovation projects. The framework proposes the stages of continuous stakeholder engagement and information flow for effective building performance improvement and user satisfaction across renovation projects life cycle for higher education buildings. The framework has been structured in four stages: pre-renovation, during renovation and handover and post-occupancy/in-use stages. Figure 6 shows the proposed building renovation project framework with communication process flows between building occupants and the estates and facilities management team. The process has been broken down into four core stages:

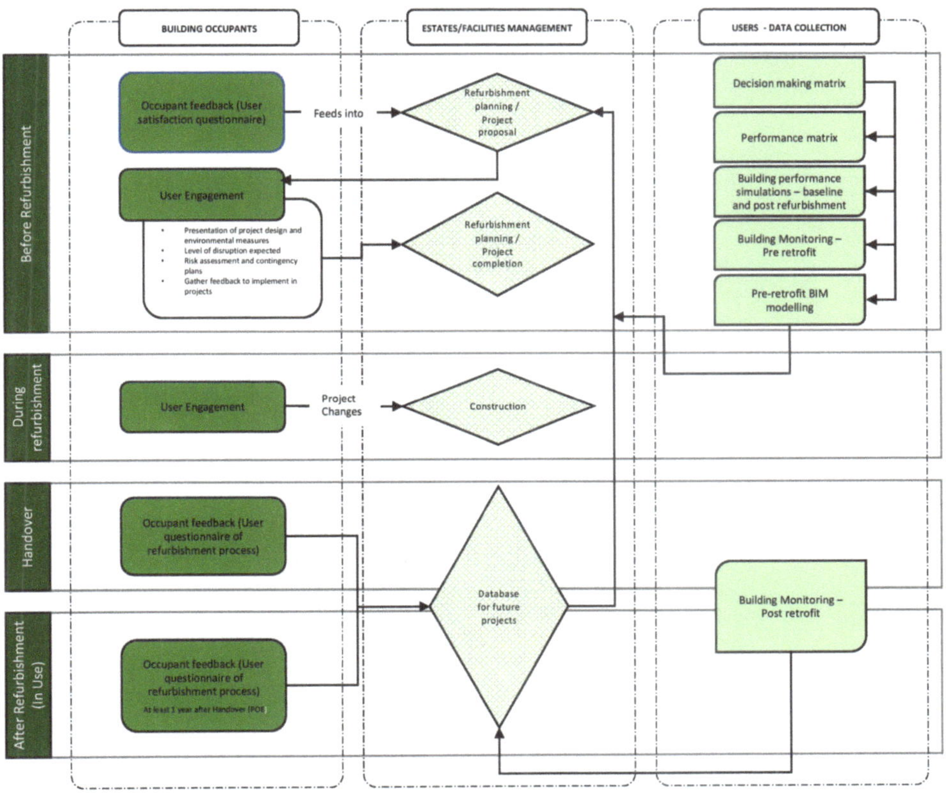

Figure 6. Proposed renovation project user engagement and evaluation framework.

Stage 1: Pre-renovation. This is the building performance diagnostic stage; the estates and facilities management team engage users to assess general building conditions and aspects of building environmental quality. This will feed into the project planning and be the basis for proposing alternative renovation designs. The estates management team further engages building users in relation to renovation plan and timelines, alternative space arrangement and contingency planning during renovation, before finalising the renovation design and planning. During this phase, the development and use of the decision-making and performance matrix should be developed and applied. Energy and indoor environmental data should be monitored and analysed where there is no existing building energy and environmental management system (BEEM). The application of Internet of Things (IoT)-enabled devices will make this process easier in building renovation projects without existing BEEM systems. Pre-renovation retrospective modelling and performance simulation should be planned and carried out at this stage.

Stage 2: Renovation/construction process: The construction team enhances continuous communication and engagement within timeline agreed in stage 1. Any potential construction work that may cause disruption or alter quality of internal environment such as noise, air quality (e.g., particulate matter) and cooling and heating systems should be adequately communicated. Workspace allocation and changes to construction timeline should be discussed and carefully communicated with all stakeholders. A two-way communication platform for reporting any significant deviation from to agreed protocol should be available construction team and building users.

Stage 3: Handover and occupancy stage: Estates management team prepares for handover and occupancy. This should involve an effective and smooth soft-landing process. Before occupants are fully back into the building, there should be effective communication and training relating to the new technologies and systems installed, especially the control systems and aspects of user control for opening and closing windows, lighting, HVAC, etc. Additionally, occupant feedback regarding the entire construction process, alternative accommodation arrangements and the process of moving back into the facility should be evaluated. A clear communication channel should be created for users to send feedback regarding any problems with operating any systems and for general maintenance requests.

Stage 4: Post-occupancy/in-use: Post-occupancy evaluation of the building should be done at least up to a year after handover using a range of objective and subjective building assessments. The subjective assessment focusses on the views of building users relating to the quality of the internal environment and the general building standard of operation. The estate/facilities management team should maintain a two-way user communication and feedback channels. A database should be created to document the findings from handover and post-occupancy evaluation, which will be helpful in optimising future maintenance and building renovation planning, design and implementation.

5. Conclusions

The paper presents a methodology for the implementation of low-energy renovation using innovative materials and technologies. The purpose is to improve the energy/environmental performance and user satisfaction in existing higher education facilities. The paper sets out to evaluate the role of technology, process and people in achieving the socio-economic and environmental benefits of low-energy renovation technologies and processes. By bringing together these components, it is essential to have a systematic approach from project inception to guide the renovation planning, technology integration and evaluation.

Building renovation is essential for improving energy and environmental performance, comfort and wellbeing of users. User evaluation questionnaires have been used to investigate end-user comfort and satisfaction in two existing higher education buildings with varying degrees of low-carbon renovation. Surveys were circulated before and after the refurbishment, providing data on user comfort and engagement at different stages of the project. The study shows the importance of using a holistic approach to meet not just the energy reduction targets, but also to improve the health and wellbeing of building occupants.

Overall, findings from the study suggest that there is a relationship between building performance improvement and an increase in the thermal comfort of occupants. However, user engagement at different stages of a project is essential for maximising the socio-economic and environmental benefits of low-energy renovation. Findings from the study infer that end-user engagement at the early stage is highly recommended, for smooth space configuration and control of indoor environment, which can be translated into a better comfort perception.

Where end-user views have not been implemented for technical and financial reasons, it is essential to engage them in constructive discussions regarding the new scheme and its potential impacts as well as adaptation measures necessary to mitigate effect of changes. Learnings from the renovation project and feedback from stakeholders were used to propose a new renovation and communication framework. Continuous engagement is vital to renovation process, not only for understanding user perception, but to improve user comfort and quicker response to adverse changes.

Ultimately, the study also indicates efficacy of renovation and its potential to achieve up to a 50% reduction in energy consumption through a mix of systematic and robust planning, diagnostics and selection of the most empowering combination of active and passive materials and technologies. Successful low-energy renovation can be achieved if a holistic process that gives strong consideration for both energy and user satisfaction is implemented.

Author Contributions: Conceptualization, A.A. (Abdullahi Ahmed) and M.M.-G.; methodology, A.A. (Abdullahi Ahmed) and A.A. (Andrew Arewa); software, A.A. (Abdullahi Ahmed) and K.C.; validation, M.M.-G. and A.A. (Abdullahi Ahmed); formal analysis, M.M.-G. and A.A. (Andrew Arewa); investigation, A.A. (Abdullahi Ahmed), M.M.-G.; resources, A.A. (Abdullahi Ahmed); data curation, M.M.-G. and A.A. (Andrew Arewa); writing—original draft preparation, A.A. (Abdullahi Ahmed) and M.M.-G.; writing—review and editing, A.A. (Abdullahi Ahmed) and K.C.; visualization, M.M.-G. and K.C.; project administration, A.A. (Abdullahi Ahmed), M.M.-G. and K.C.; funding acquisition, A.A. (Abdullahi Ahmed) and K.C. All authors have read and agreed to the published version of the manuscript.

Funding: This research was funded by European Union Framework 7, grant number 609377.

Institutional Review Board Statement: The study was conducted according to the guidelines of the Declaration of Helsinki, and approved by the Institutional Ethics Committee of Coventry University (Ethics code P38547 and approved 11/2016)

Informed Consent Statement: Informed consent was obtained from all subjects involved in the study.

Data Availability Statement: The data presented in this study are available on request from the corresponding author.

Acknowledgments: We would like to acknowledge the contribution of all project partners for their participation and ideas during the development of the project.

Conflicts of Interest: The authors declare no conflict of interest.

Appendix A

Table A1. Questions analysed.

	Topics	Rating Scale
	Temperature and air quality	
A2.1	Temperature in winter	Too Hot = 1 to Too cold = 7
A2.2	Air in winter	Stable = 1 to 7 = Varies through day
A2.3	Overall conditions in winter	Uncomfortable = 1 to Comfortable = 7
A2.4	Temperature in summer	Too Hot = 1 to Too cold = 7
A2.5	Air in summer	Stable = 1 to 7 = Varies through day
A2.6	Overall conditions in summer	Uncomfortable = 1 to Comfortable = 7
	Overall building performance	
A5.1	How do you rate the overall comfort of the building environment?	Unsatisfactory = 1 to Satisfactory = 7
A5.2	In the building as a whole, do the facilities meet your needs?	Unsatisfactory = 1 to Satisfactory = 7
A5.3	Do you feel less or more healthy when you are in the building?	Uncomfortable = 1 to Comfortable = 7
	Additional questions for Richard Crossman building	
B.1	In your opinion what were the key indoor environmental problems with the building before retrofit?	Thermal comfort/Visual comfort/Air quality/Noise/Lighting
B.2	What level of engagement did you have with the refurbishment project before construction works started?	Just informed/I was engaged in the process/None/Other
B.3	What level of engagement did you have during the construction works?	Just informed/I was engaged in the process/None/Other
B.4	Were you provided with clear information about potential disruptions during retrofit works?	Yes/No/Some
B.5	What was the level of disruption experienced during refurbishment?	Moving from your offices/Changing lecture rooms/Noise/Pollution/Other
B.6	Would you have liked to be more engaged in the refurbishment process?	Yes/No/I don´t know
B.7	Which of the following improvements are noticeable since refurbishment?	Thermal comfort/Visual comfort/Air quality/Noise/Lighting/Aesthetics/Toilets/Other
B.8	What is your overall assessment of the indoor environmental quality improvement?	Low improvement = 1 to High improvement = 7
B.9	What is your overall assessment of the quality of building improvement?	Low improvement = 1 to High improvement = 7

References

1. The Directive 2010/31/EU of the European Parliament and of the Council of 19 May 2010 on the energy performance of buildings, Official Journal of the European Union, L153. 2010. Available online: http://eur-lex.europa.eu/LexUriServ/LexUriServ.do?uri=OJ:L:2010:153:0013:0035:EN:PDF (accessed on 24 February 2021).
2. Baker, N. *The Handbook of Sustainable Renovation: Non-Domestic Buildings*; Earthscan: London, UK, 2009; pp. 3–4.
3. Gupta, R.; Gregg, M. Empirical evaluation of the energy and environmental performance of a sustainably designed but under-utilised institutional building in the UK. *Energy Build.* **2016**, *128*, 68–80. [CrossRef]
4. Johnston, D.; Farmer, D.; Brooke-Peat, M.; Miles-Shenton, D. Bridging the domestic building fabric performance gap. *Build. Res. Inf.* **2016**, *44*, 147–159. [CrossRef]
5. Storey, J.B.; Pedersen Zari, M. Factor X-Wellbeing as A Key Component of Next Generation Green Buildings. In Proceedings of the 12th Rinker International Conference Rethinking Sustainable Construction 2006, Sarasota, FL, USA, 19–22 September 2006.
6. Doan, D.T.; Ghaffarianhoseini, A.; Naismith, N.; Zhang, T.; Ghaffarianhoseini, A.; Tookey, J. A critical comparison of green building rating systems. *Build. Environ.* **2017**, *123*, 243–260. [CrossRef]
7. Yu, S.M.; Tu, Y. Are Green Buildings Worth More Because They Cost More? In *NUS Institute of Real Estate Studies Working Paper (IRES2011-023)*; National University of Singapore: Singapore, 2011.
8. Altomonte, S.; Schiavon, S. Occupant satisfaction in LEED and non-LEED certified. *Build. Environ.* **2013**, *68*, 66–76. [CrossRef]
9. Technology Strategy Board (TSB). Retrofit for The Future: Reducing Energy Consumption in Existing Homes—A Guide to Making Retrofit Work. Available online: https://assets.publishing.service.gov.uk/government/uploads/system/uploads/attachment_data/file/669113/Retrofit_for_the_future_-_A_guide_to_making_retrofit_work_-_2014.pdf (accessed on 10 July 2019).
10. Carbon Trust, Further and higher education Training colleges and universities to be energy efficient. Sector Overview Report. 2014. Available online: https://se-ed.co.uk/edu/wp-content/uploads/2014/03/Carbon-Trust-advice-FE-HE.pdf (accessed on 21 April 2019).
11. Association of University Directors of Estate (AUDE). The Legacy of 1960′s University Buildings. 2008. Available online: https://www.sustainabilityexchange.ac.uk/legacy-of-1960s-buildings-aude-research-project (accessed on 1 July 2016).
12. Painting, N.J.; Piroozfar PA, E.; Farr, E.R.P. Refurbishment of higher education premises: Stakeholder engagement in the process and product. In Proceedings of the 30th Annual ARCOM Conference, Portsmouth, UK, 1–3 September 2014; pp. 955–964. Available online: https://www.arcom.ac.uk/-docs/proceedings/ar2014-0955-0964_Painting_Piroozfar_Farr.pdf (accessed on 28 October 2018).
13. Ma, Z.; Cooper, P.; Daly, D.; Ledo, L. Existing building retrofits: Methodology and state-of-the-art. *Energy Build.* **2012**, *55*, 889–902. [CrossRef]
14. Piaia, E.; Turillazzi, B.; Longo, D.; Boeri, A.; di Giulio, R. Plug-and-Play and innovative process technologies (Mapping/Modelling/Making/Monitoring) in deep renovation interventions. *TECHNE J. Technol. Archit. Environ.* **2019**, *18*, 215–225. [CrossRef]
15. Ucci, M.; Yu, C.W.F. Low-carbon buildings, health, and wellbeing: Current perspectives and critical challenges. *Indoor Built Environ.* **2014**, *23*, 335–339. [CrossRef]
16. Kang, S.; Ou, D.; Mak, C.M. The impact of indoor environmental quality on work productivity in university open-plan research offices. *Build. Environ.* **2017**, *124*, 78–89. [CrossRef]
17. Giddings, B.; Thomas, J.; Little, L. Evaluation of the Workplace Environment in the UK, and the Impact on Users′ Levels of Stimulation. *Indoor Built Environ.* **2013**, *22*, 965–976. [CrossRef]
18. Royal Institute of British Architects (RIBA). *Post Occupancy Evaluation and Building Performance Evaluation Primer*; RIBA: London, UK, 2016; Available online: https://www.architecture.com/-/media/gathercontent/post-occupancy-evaluation/additional-documents/ribapoebpeprimerpdf.pdf (accessed on 18 January 2019).
19. Meir, I.A.; Garb, Y.; Jiao, D.; Cicelsky, A. Post-occupancy evaluation: An inevitable step toward sustainability. *Adv. Build. Energy Res.* **2009**, *3*, 189–219. [CrossRef]
20. Hay, R.; Samuel, F.; Watson, K.J.; Bradbury, S. Post-occupancy evaluation in architecture: Experiences and perspectives from UK practice. *Build. Res. Inf.* **2018**, *46*, 698–710. [CrossRef]
21. Baird, G. Users′ perceptions of sustainable buildings e Key findings of recent studies. *Renew. Energy* **2015**, *73*, 77–83. [CrossRef]
22. Malkoc, E.; Ozkan, M.B. Post-occupancy Evaluation of a Built Environment: The Case of Konak Square (Izmir, Turkey). *Indoor Built. Environ.* **2010**, *19*, 422–443. [CrossRef]
23. AlHorr, Y.; Arif, M.; Katafygiotou, M.; Mazroei, A.; Kaushik, A.K.; Elsarrag, E. Impact of indoor environmental quality on occupant wellbeing and comfort: A review of the literature. *Int. J. Sustain. Built Environ.* **2016**, *5*, 1–11. [CrossRef]
24. Earthman, G.I. School Facility Conditions and Student Academic Achievement. In UCLA′s Institute for Democracy, Education, and Access. Williams Watch Series: Investigating the Claims of Williams v. State of California: Document wws-rr008-1002. 2002. Available online: https://escholarship.org/uc/item/5sw56439 (accessed on 1 July 2019).
25. Boese, S.; Shaw, J. New York State School Facilities and Student Health, Achievement, and Attendance: A Data Analysis Report. 2005. Available online: https://files.eric.ed.gov/fulltext/ED510053.pdf (accessed on 1 July 2019).
26. Kim, T.W.; Cha, S.; Kim, Y. Space choice, rejection, and satisfaction in university campus. *Indoor. Built. Environ.* **2018**, *27*, 233–243. [CrossRef]

27. Ascione, F.; Borrelli, M.; de Masi, R.F.; de'Rossi, F.; Vanoli, G.P. Energy refurbishment of a University building in cold Italian backcountry. Part 2: Sensitivity studies and optimization. *Energy Procedia* **2019**, *159*, 10–15. [CrossRef]
28. Bellia, L.; Borrelli, M.; de Masi, R.F.; Ruggiero, S.; Vanoli, G.P. University building: Energy diagnosis and refurbishment design with cost-optimal approach. Discussion about the effect of numerical modelling assumptions. *J. Build. Eng.* **2018**, *18*, 1–18. [CrossRef]
29. Figueiredo, A.; Kämpf, J.; Vicente, R.; Oliveira, R.; Silva, T. Comparison between monitored and simulated data using evolutionary algorithms: Reducing the performance gap in dynamic building simulation. *J. Build. Eng.* **2018**, *17*, 96–106. [CrossRef]
30. Cho, H.M.; Yun, B.Y.; Yang, S.; Wi, S.; Chang, S.J.; Kim, S. Optimal energy retrofit plan for conservation and sustainable use of historic campus building: Case of cultural property building. *Appl. Energy* **2020**, *275*, 115313. [CrossRef]
31. de Santoli, L.; Mancini, F.; Clemente, C.; Lucci, S. Energy and technological refurbishment of the School of Architecture Valle Giulia, Rome. *Energy Procedia* **2017**, *133*, 382–391. [CrossRef]
32. Creswell, J.W.; Creswell, J.D. Mixed Method Procedure. In *Research Design: Qualitative, Quantitative, and Mixed Methods Approaches*, 5th ed.; Sage: London, UK, 2018; pp. 213–246.
33. City Lab Coventry. Available online: www.openlivinglabs.eu/livinglab/city-lab-coventry (accessed on 27 October 2017).
34. GOV.UK. Greenhouse Gas Reporting: Conversion Factors 2017. 2017. Available online: https://www.gov.uk/government/publications/greenhouse-gas-reporting-conversion-factors-2017 (accessed on 7 September 2018).
35. Grupopuma. TRADITERM®EPS-G PANEL. 2018. Available online: https://www.grupopuma.com/en-WW/products/show/traditerm-panel-eps-g-ww-en (accessed on 24 September 2019).
36. Sarawade, P.B.; Kim, J.; Hilonga, A.; Kim, H.T. Production of low-density sodium silicate-based hydrophobic silica aerogel beads by a novel fast gelation process and ambient pressure drying process. *Solid State Sci.* **2010**, *12*, 911–918. [CrossRef]
37. Va-Q-Tec Limited, Vacuum Insulated Panel (VIP). Available online: https://va-q-tec.com/technologie/vakuumisolationspaneele/ (accessed on 1 May 2019).
38. Fthenakis, V.M.; Kim, H.C.; Frischknecht, R.; Raugei, M.; Sinha, P.; Stucki, M. IES Photovoltaic Power System Report Annual Report: IEA-PVPS T12-02:2019. 2019. Available online: https://iea-pvps.org/wp-content/uploads/2020/05/IEA-PVPS-AR-2019-1.pdf (accessed on 2 February 2020).
39. Ure, Z. Thermal Storage: PCM Products. Available online: http://www.pcmproducts.net/files/TES-2008.pdf (accessed on 3 February 2020).
40. Sbar, N.L.; Podbelski, L.; MoYang, H.; Pease, B. Electrochromic dynamic windows for office buildings. *Int. J. Sustain. Built Environ.* **2012**, *1*, 125–139. [CrossRef]
41. Barbosa, S.; Ip, K. Perspectives of double skin façades for naturally ventilated buildings: A review. *Renew. Sustain. Energy Rev.* **2014**, *40*, 1019–1029. [CrossRef]
42. Lenk, R.; Lenk, C. *Practical Lighting Design with LEDs*, 1st ed.; The Institute of Electrical and Electronics Engineers: Piscataway, NJ, USA, 2011.
43. TSM-PDG5. The Most Durable Module—Technical Brochure. Available online: https://www.enfsolar.com/Product/pdf/Crystalline/51650b738832c.pdf (accessed on 20 July 2020).
44. Kingspan. Flat Roof Insulation | Thermaroof TR27 LPC/FM. 2020. Available online: https://www.kingspan.com/gb/en-gb/products/insulation/insulation-boards/therma/thermaroof-tr27-lpc-fm (accessed on 24 July 2020).
45. BS EN 16798. Energy Performance of Buildings. Ventilation for Buildings. Indoor Environmental Input Parameters for Design and Assessment of Energy Performance of Buildings Addressing Indoor Air Quality, Thermal Environment, Lighting, and Acoustics. British Standard Organisation Module M1-6. 2019. Available online: https://shop.bsigroup.com/ProductDetail/?pid=000000000030297474 (accessed on 3 December 2019).
46. ISO 7730, Ergonomics of The Thermal Environment—Analytical Determination and Interpretation of Thermal Comfort Using Calculation of The PMV and PPD Indices and Local Thermal Comfort Criteria International Standard Organisation. 2005. Available online: https://www.iso.org/standard/39155.html (accessed on 24 February 2021).
47. Loonen, R.C.; Favoino, F.; Hensen, J.L.; Overend, M. Review of current status, requirements and opportunities for building performance simulation of adaptive facades. *J. Build. Perform. Simul.* **2017**, *10*, 205–223. [CrossRef]
48. IES-VE. Integrated Environmental Solution Virtual Environment Software®, Version 2018, Glasgow UK. 2018. Available online: https://www.iesve.com/software (accessed on 24 February 2021).
49. Crawley, D.; Hand, J.; Kummert, M.; Griffith, B. Contrasting the Capabilities of Building Energy Performance Simulation Programs. *Build. Environ.* **2008**, *43*, 661–673. [CrossRef]

MDPI
St. Alban-Anlage 66
4052 Basel
Switzerland
Tel. +41 61 683 77 34
Fax +41 61 302 89 18
www.mdpi.com

Energies Editorial Office
E-mail: energies@mdpi.com
www.mdpi.com/journal/energies